CIGRE Green Books

Compact Studies

Series Editor

CIGRE, International Council on Large Electric Systems (Cigré), Paris, France

CIGRE presents their expertise in compact professional books on electrical power networks. These books are of a self-contained concise character, covering the entire knowledge of the subjcct within power engineering. The books are created by CIGRE experts within their study committees and are recognized by the engineering community as the top reference books in their fields.

More information about this subseries at https://link.springer.com/bookseries/15383

Masoud Farzaneh • William A. Chisholm

Techniques for Protecting Overhead Lines in Winter Conditions

Dimensioning, Icephobic Surfaces, De-Icing Strategies

cigre

Springer

Masoud Farzaneh
University of Quebec
Chicoutimi, QC, Canada

William A. Chisholm
Kinectrics Associate
Toronto, ON, Canada

ISSN 2367-2625 ISSN 2367-2633 (electronic)
CIGRE Green Books
ISSN 2509-2812 ISSN 2509-2820 (electronic)
Compact Studies
ISBN 978-3-030-87454-4 ISBN 978-3-030-87455-1 (eBook)
https://doi.org/10.1007/978-3-030-87455-1

This Springer imprint is published by the registered company Springer Nature Switzerland AG
The registered company address is: Gewerbestrasse 11, 6330 Cham, Switzerland

Foreword

Overhead lines are exposed to all kind of weather like wind, ice, snow, and other environmental influences from topography and soil conditions, and, as such, must be designed and protected adequately.

Several overhead line dimensions are influenced by the effect of winter conditions. The most visible are the distances between the phase conductors and their offset to each other to provide sufficient clearances by deflection from wind and in case of accretion of snow and ice—also in combination with wind. Clearances to ground, buildings, and other objects, also to neighbouring ones, have to be kept under such conditions too.

Lines designed for severe winter conditions have to tolerate a wider range of conductor tension as the conductors contract at low temperatures and in addition heavy accretion of ice and snow adds extra loads and tensions. Effects of climate change are predicted to have an influence on the winter conditions now used for overhead line design.

Some older lines may have difficulties maintaining safe clearances at low tension situations in the summer as a result of conductor elongations due to creeping. Much of this creep accumulates from winter conditions under ice and wind loadings. Utilities are using advanced dynamic line rating methods by taking measurement in representative sections of a line, and inferring the measured sags, tensions, and clearances for all other spans. The principally same coupling can be used to monitor ice accretion and "sleet jump" when the ice eventually falls off the conductors and let them jump off. In such situations extra clearance and horizontal phase offset are helpful against falling short of distances between unloaded and loaded phases with respect to groundwires.

This Green Book summarizes the problems and challenges, and links to the historic literature that shows anti-icing and de-icing schemes, using configurations and methods which helped utilities to operate problematic lines of the 1920s into the 1950s. With increased penetration of HVDC equipment, it seems that utilities can now extend the scope of line de-icing initiatives with controlled remote short circuits.

There has also been progress in the area of coatings to prevent or reduce ice accretion on conductors and insulators. Such superhydrophobic materials can bead up and repel water, and even become self-cleaning. This book guides the users through an understanding of the many factors in superhydrophobicity, leading to an appreciation that it may be a necessary but is not a sufficient condition for ice-phobicity, a term defined and developed fully in the book.

The book is also a review of CIGRE Technical Brochure 438 and 631 and many other CIGRE contributions which touch on the problems of operating overhead lines in winter conditions.

I want to thank the authors for their work on this publication which continues the successful series of CIGRE Green Books.

Vienna, Austria

Herbert Lugschitz
CIGRE B2 Chairman

Preface

Electrical utilities operate overhead lines in a wide range of climate conditions. Each utility makes adaptations in design features to local conditions. The most visible manifestation of this approach is in the number of overhead groundwires placed above the phase conductors for lightning protection in summer. Similar levels of reliability are obtained in areas of low lightning flash density—with unshielded designs—and in areas of high lightning flash density—with twin overhead groundwires above and additional grounding conductors below the phases.

The adaptations that utilities make for winter conditions are more subtle. Regions with large changes in ambient temperature from summer to winter may use shorter spans to accommodate sag changes and to avoid extremely high tensions when the ambient temperature drops below –30 °C. Regions with several annual exposures to modest levels of ice accretion usually include electrical clearances for large-amplitude galloping. The double-circuit transmission towers in these regions have a characteristic outboard offset of the middle phase conductors compared to towers in regions with no possibility of ice accretion. Regions with long periods of winter conditions, accompanied by marine, road salt or industrial pollution accumulation, may require insulators with longer dry arc and leakage distance to avoid electrical flashovers or wood-pole fires when the temperature eventually rises above freezing.

CIGRE has considered all aspects of overhead line design in its many Technical Brochures but has not consolidated and linked the findings explicitly for winter conditions. The consolidation in this volume serves as a practical working tool for engineers who are now facing unfamiliar problems with ice or extreme winter weather. The best example of the need for links among disciplines is in the thermal rating aspects, which are generally a limitation to summer operations. The mechanical state of an overhead conductor is greatly affected by the long-term plastic deformation—creep—especially of the aluminium layers. Creep in turn is affected by the number of times the conductor has been chilled to low temperature—raising tension—and by the number of times the conductor has been exposed to combinations of heavy ice accretion, high wind loads and combined wind-on-ice loads over its service life. Creep leads to reduced line tension, reduction in the "knee point temperature" where the steel takes over the tension load, and changes to the angles at which the insulators hang vertical, which affect tension equalization within stringing sections. High temperature operation is also an important consideration in de-icing methods that use high current.

When atmospheric ice accretes on overhead line conductors, different phenomena can be observed which can lead to small perturbations or catastrophic damage depending on the severity of the ice storm and the intensity of the coincident wind speed. These phenomena can be divided in two categories: moderate and severe icing events.

In some moderate icing events affecting lines in open terrain and exposed to transverse winds, ice deposits may not be large enough to pose an immediate risk of damage to the conductor. However, under specific conditions which mainly depend on the type, direction, and speed of the wind, as well as the shape and dimensions of the ice sleeve, an aerodynamic instability can be created. Depending on the magnitude of the forces involved, conductor galloping can occur (CIGRE TF B2.11.06, 2007-06). This is characterized by large oscillatory movements generally in an elliptical trajectory in the conductor transverse plane. Under conductor galloping, phase-to-phase, or ground wire—phase flashovers may occur, leading to widespread power outages and potential physical damage to conductors. The resulting conductor motions may also lead to high stresses on towers, with possible damage to hardware, tower steelwork and insulator tension strings, as well as conductor fatigue.

During severe icing events, glaze ice deposited on conductors increases their weight resulting in heavy loads on the structural components of power lines with possible conductor breakages and tower collapses or phase clashing causing outages and conductor damage. A recent example of the catastrophic consequences of a severe ice storm is the one that struck a relatively narrow swath of land from Eastern Ontario to southern Quebec to Nova Scotia in Canada, and bordering areas from Northern New York to Southeast Maine in the United States, in January 1998. With up to 100 mm of ice on the ground in the Montreal area, this extreme ice storm exceeded the design limits of the Hydro-Quebec electrical line system (CIGRE WG B2.06, 2008). These limits had been raised after 1961 when a storm in the same region deposited 30–40 mm of ice on the OHL system. Nevertheless, many power lines broke and over 600 towers collapsed in chain reactions under the weight of the ice and the effects of conductor failure, leaving more than 1.5 million people without electricity for up to a month (CIGRE WG B2.06, 2008).

In the past, utilities have adopted two different philosophies regarding ice accretion on power lines. Some have accepted the necessity to build strong lines that can withstand the largest ever recorded ice/wind load and have sufficient phase spacing (Motlis 2002). This philosophy required the use of passive methods like vibration dampers, anti-galloping devices, and changes in geometry, dimension, structural and mechanical strength of line components, etc. Other utilities have chosen to develop and use methods which prevent ice-induced damages on overhead lines (Motlis 2002). This latter philosophy is based on the use of active methods to reduce the ice loads on conductors. From their own experience, these electric utilities have developed methods and strategies for reducing ice problems. These strategies can be divided into two main families commonly identified as anti-icing methods and de-icing methods. Generally, anti-icing methods are employed before or early during an ice storm whereas de-icing methods are activated late during or after the ice storm when on-site interventions are required.

However, the severe ice storms in North America, Europe and China in recent years highlighted the fact that de-icing approaches and strategies are difficult to implement when a large zone is affected, and that new strategies based on the combination of active and passive methods had to be developed. This renewed awareness has thus allowed the emergence of new anti-icing and de-icing methods as well as questions concerning the standards used in the construction of overhead electrical lines in critical areas with high severe ice storm recurrence. Another related issue concerns the climate changes that increase the likelihood of ice storms in regions where they were not expected when the power grid was designed.

In general, de-icing techniques belong to one of the following four categories:

- Passive methods based on natural forces
- Active coatings and devices
- Mechanical methods based on breaking down the ice
- Thermal methods based on ice melting.

The feasibility of the most suitable method for a particular line or network must be evaluated. This evaluation is done by considering the following factors:

- the performance requirements of the line in the power grid
- the degree to which the method can be applied to a given line
- the basic energy requirements
- the efficacity/risk and security level
- the cost of the associated infrastructure and operation.

The passive methods do not require any external energy other than from natural sources: wind, gravity, solar radiation, and temperature variations. As such, passive techniques do not hinder ice formation directly, but help to limit its problematic effects. For example, a technique using counterweights to block the rotation of the conductors, and another approach that allows the conductors to slide in their suspension clamps or be dropped for a given overload, do not prevent ice deposition. The former helps only to reduce to some extent the quantity of ice that is deposited, while the latter helps to eliminate or reduce the static and dynamic loads transferred to adjacent supports.

Active coating methods require electrical energy to be effective. For example, some of them use the heat losses from dielectric materials covering a conductor or are based on the use of ferromagnetic coatings to heat the conductor surface. The goal of these methods is to prevent significant ice accretion on conductors.

The mechanical methods are aimed at removing accreted ice by mechanical means. These are ad-hoc methods applied locally in specific spans and terrain conditions. They include methods like rolling, scraping, induced shocks. These methods are automated, or manual where iced spans are accessible, and require the use of helicopters where spans are less accessible. Because of their makeshift character, the safety and efficacity of these techniques are not proven, and therefore none are presently recommended as an effective de-icing method on a large scale. They are used typically in emergency situations and with limited success, to de-ice the lines where accretions are significant and there is a risk of a second icing event occurring before the ice is shed.

The thermal methods are based on the concept of heating line conductors by electrical current to force ice melting and shedding. Some of these methods can also be used for preventing ice accretion by maintaining the external conductor surface at a temperature above 0 °C during the icing event. Observations of the mechanical state of the conductor have moved from a niche technology, such as vertical or horizontal load cells to monitor conductor weight or line tension respectively, to mainstream applications that can process sags and clearances of conductors immediately from a smart-phone photograph. One aspect for protecting overhead lines in winter conditions relates to monitoring, so that suitable countermeasures are taken as ice accumulates, persists, melts, and drops off the conductors, insulators, and towers. To some extent, the permanent monitoring systems used for dynamic line rating in summer and for managing energy from wind turbines can also satisfy some of the important monitoring requirements in winter conditions. As a first step, this book recommends that users with DLR systems monitor the inputs and outputs of the process closely during periods of severe winter weather.

Better understanding of the dynamic thermal behaviour of overhead conductors can give fresh insight to the traditional de-icing methods, based on reconfiguring the power system to increase the current flow into vulnerable or threatened lines. In advance of a predicted ice storm, the current flow has a function of "anti-icing". Once ice has already formed around the conductors, the current flow needs to perform a "de-icing" function. The continuing penetration of HVDC equipment into power systems provides a new set of options for anti-icing and de-icing configurations.

Experience with overheating of conductors and sudden changes in sag, related to the cascade of melted ice from the conductor, is reviewed and Joule methods are complemented by a survey of mechanical methods for de-icing.

The last three decades saw the development of a variety of surface engineering techniques and advanced coatings with properties like hydrophobicity, self-cleaning, icephobicity and anti-corrosion. These technological advances can help the power industry in several ways. Coatings can reduce the risk of electrical flashovers on polluted and ice-covered insulators by encapsulating pollution. Coatings can reduce the level of partial discharge activity and corona noise from conductors, fittings and insulators, activity that acts as an initiating energy source for flashovers in melting conditions. Coatings address problems of long-term corrosion of metal elements. Advanced coatings may alleviate mechanical problems caused by ice and snow accretion on overhead power networks. Some of these technologies, such as icephobic coatings discussed in detail, have the potential to increase the reliability of existing transmission assets and to reduce the capital cost of new power network construction.

Chicoutimi, Canada | Masoud Farzaneh
Toronto, Canada | William A. Chisholm

Acknowledgments

The authors thank the members of the CIGRE Working Groups B2.29, B2.44 and B2.69 for their contributions in developing, respectively, CIGRE Technical Brochures 438, 631, 837 and 838. We especially acknowledge our memories of Yakov Motlis, a major contributor to CIGRE TB 438, and more recently of Dale Douglass, both specialists in Joule heating and overhead line thermal rating.

We also thank Herbert Lugschitz, SC B2 Chairman and Yves Maugain, Green Book Chief Editor, as well as Konstantin Papailiou, Bernard Dalle, and Svein Fikke for their encouragement and support.

Members, Contributors, Reviewers			
CIGRE WG B2.29 (TB 438)	**CIGRE WG B2.44 (TB 631)**	**CIGRE WG B2.69 (TB 838)**	
		Chapter 2 (Anti-icing coatings)	**Other Chapters**
Convenor	*Convenor*	*Convenor*	
Masoud Farzaneh	Masoud Farzaneh	Masoud Farzaneh	Boris Adum
Secretary	*Secretary*	*Secretary*	Criistina C. Amorim
Franc Jakl	Hélène Gauthier	Alberto Pigini	Neelesh Arona
Mohsen P. Arabani	Siavash Asadollahi	*Assistant Secretary*	Samuel Arturo Asto Soto
Arni Jón Elíasson	Gustau Castellana	Boris Adum	Cristina Chemelli
Svein M. Fikke	Árni Jón Elíasson	*Anti-Icing Chapter*	Árni Jón Eliasson
Angel Gallego	Christiaan Engelbrecht	Boris Adum	Jean-Marie George
Asim Haldar	Svein M. Fikke	Cristina Chemelli	Norbert Handinger
Masanori Isozaki	Corinne Greyling	Árni Jón Éliasson	Franc Jackl
Majid Kermani	Igor Gutman	Franc Jakl	Qi Li
Laszlo Kollar	Tomohiro Hayashi	Pietro Marcacci	Francis Liros
Robert Lake	Reza Jafari	Feipeng Wang	Pietro Marcacci
André Leblond	Franc Jakl	Fanghui Yin	Juan Maya
Geoff McDougall	Zhidong Jia	*Reviewers*	Yasushi Okawa
Fethi Menini	Herbert Lugschitz	William A. Chisholm	Giovanni Pirovano
Malcolm Minchin	Nishal Mahatho	Yuko Kuranari	Feipeng Wang
Masataka Mito	Gelareh Momen	Anne Williams	Fanghui Yin
Laure Pellet	Andrew Phillips		Christian Ziegler
Zsolt Péter	Luis Riera		
Dario Ronzio	Vladimir Shkaptsov		
Javier Santana Lopez	Noriyoshi Sugawara		
Charles C. Ryerson	Natalia Vaga		

Members, Contributors, Reviewers			
CIGRE WG B2.29 (TB 438)	**CIGRE WG B2.44 (TB 631)**	**CIGRE WG B2.69 (TB 838)**	
		Chapter 2 (Anti-icing coatings)	**Other Chapters**
Vladimir Shkaptsov Sergey Turbin Christophe Volat J. Brian Wareing Hubert Zangl *Reviewers* Dale Douglass Rafael Garcia Herbert Lugschitz Ghyslaine McClure	J. Brian Wareing Xiaoxing Wei *Reviewers* Henry Hawes David Horseman Zibby Kieloch		

Message from the Secretary General

Dear Readers,

From 1969 to 2021, CIGRE has published more than 840 Technical Brochures.

A Technical Brochure is a document produced and edited by a CIGRE Working Group, following its specific Terms of Reference. It is published by CIGRE Central Office and available from the CIGRE online library, e-cigre, one of the most comprehensive, accessible databases of relevant technical literature on power engineering.

Between 40 and 50 new Technical Brochures are published yearly, and these Brochures are announced in Electra, CIGRE's bimonthly journal.

They are accessible from e-cigre free of charge for CIGRE members or at a certain price for non-members, except for publications older than three years, which are free for everybody.

In 2011, Dr. Konstantin Papailiou proposed the Technical Council, the Green Book concept to valorise the collective work of CIGRE Study Committees accumulated over many decades, by putting together all the Technical Brochures of a given field in a single book.

In 2014, CIGRE published its first Green Book, the one on Overhead Lines, paving the way to a new CIGRE publication collection.

In 2015, CIGRE decided to cooperate with Springer for the edition and publication of its next Green Books, as "Major Reference Works", and to distribute them through the vast network of this well-known international publisher.

In 2016, the collection enriched itself with a new category of Green Books, the CIGRE "Compact Series," to satisfy the needs of the Study Committees when they want to publish shorter, concise volumes.

The first CIGRE Compact Book, prepared by Study Committee D2 under the title "Utility Communication Networks and Services", became the best seller of the collection.

The concept of the CIGRE Green Books series has continued to evolve, with the introduction of a third subcategory of the series, the "CIGRE Green Book Technical Brochures" (GBTB).

Like for the other publications of CIGRE, e-cigre provides the references of all the books of the Green Books series, which anyone can order from the SPRINGER's platform. A special discount applies for CIGRE members.

Each of the co-authors of a Green Book will receive a free copy in recognition of their contribution to this collective work.

This book prepared by professor Masoud Farzaneh (Editor) and Dr William A. Chisholm (Co-editor) fits into CIGRE plans to co-publish new Green Books by the different Study Committees, with a goal to expand the series at a pace of one or two volumes per year.

In a few final words I would like to thank and congratulate all the authors, contributors and reviewers of this specific publication on the 'Techniques for Protecting Overhead Lines in Winter Conditions'.

Phillipe Adam
Secretary General

Contents

About the Authors

Masoud Farzaneh is an IEEE Life Fellow, IET Fellow, CIGRE Fellow, EIC Fellow, Fellow of the Canadian Academy of Engineering (CAE) and Fellow of the Royal Society of Canada. His research has focused on elucidating the complex multi-physics phenomena resulting from the interactions of electric fields, liquid water and ice or snow on ground wires, phase conductors and insulators. Advanced laboratory tests, multidisciplinary mathematical modelling, and novel numerical implementations, as well as field studies of natural icing, have supported this research and led to associated technological innovations. He has been President of IEEE DEIS (2013), chair and contributor to IEEE Standards 1783 and 1820 as well as Editor-in-Chief of IET High Voltage Journal. He has been Member of CIGRE Canada executive committee, Convenor of CIGRE WG B2.29, B2.44, B2.69, as well as contributor to CIGRE TB 179, TB 256, TB 291, TB 322, TB 438, TB 631, TB 645, TB 837, and TB 838. He is a prolific author with close to 300 papers in archival journals, 400 papers in international refereed conference proceedings, four books and 15 book chapters as well as 350 technical articles and reports; for a lifetime total of almost 1200 publications. Six of his publications have been recognized as best articles in various international forums. On 55 occasions, Prof. Farzaneh has also given keynote lectures. As mentor, he has trained about 250 highly qualified personnel, including 48 Ph.Ds., 44 masters and 38 postdoctoral. His contributions and achievements in research and teaching have been recognized by several prestigious prizes and awards at national and international levels. In recognition of his outstanding contribution and impact, the Masoud Farzaneh Award was established in 2010 by University of Quebec in Chicoutimi in his honour.

William A. Chisholm received his B.A.Sc. in Engineering Science from the University of Toronto in 1977, a Master of Engineering in 1979 and a Ph.D. in Electrical Engineering from the University of Waterloo in 1983. His primary research in 30 years at Kinectrics, the former Ontario Hydro Research Division, was in lightning protection, grounding, thermal rating of conductors and icing flashover performance of insulators. He has co-authored three books, seven book chapters, eight CIGRE Technical Brochure chapters, four IEEE Standards and more than 20 refereed journal papers. In 2020, he made the Stanford University "Top 2% Scientists" list, ranking all disciplines. He served as chair of the IEEE Power and Energy Society, Transmission and Distribution Committee in 2013–2014. As a result of his achievements, Dr. Chisholm received an IEEE "Best Standard" award (1999), IEEE Fellow award (2009), the Masoud Farzaneh Award (2014), the Claude de Tourreil Memorial Award (2017) and IEEE Life Fellow award (2021). He convenes CIGRE Study Committee B2/C4 Joint Working Group 76 on transmission line rebuilding and refurbishing projects.

1 Introduction

This chapter provides the vocabulary for characterizing the many forms of ice deposits on overhead power lines and goes on to use these descriptors to illustrate problems from cold-climate regions in many countries.

1.1 Mechanisms of Ice Adhesion

The two main influences on ice adhesion are the electrostatic forces present at the ice-solid interface and the surface roughness of the substrate.

As concerns the electrostatic forces, materials, or coatings with very low dielectric permittivity such as fluoropolymers exhibit low ice adhesion strengths. As for substrate roughness, it plays a crucial and complex role in ice adhesion strength. In the presence of strong mechanical interlocking, ice cracking and even ice debonding can be observed due to excessive pressure build-up resulting from air trapped in closed pores. Therefore, the engineering of an optimum surface roughness, with superhydrophobic properties for instance, can create enough internal stress to debond ice, particularly when the texture is composed of a low dielectric material.

Other parameters should also be considered to fully understand the ice adhesion mechanism: temperature and the type of precipitation (e.g. impact velocities of supercooled water droplets) that influence water penetration within a three-dimensional surface structure, as well as the shape and size of the ice crystal grains affecting ice adhesion strength.

1.2 Processes of Natural Ice and Snow Shedding

Ice shedding from ground wires and phase conductors, that is any phenomenon of ice mass reduction, may occur through the following mechanisms:

M. Farzaneh and W. A. Chisholm, *Techniques for Protecting Overhead Lines in Winter Conditions*, Compact Studies, https://doi.org/10.1007/978-3-030-87455-1_1

Ice melting starts on the external part of the ice at air temperature above 0 °C. This stage is associated with a low shedding rate which is mainly influenced by air temperature, solar radiation, and wind velocity. Then, after a short period, melting occurs at the ground wire/conductor–ice interface resulting in ice chunks dropping off under the effect of wind and gravity.

Ice sublimation consists of the release of vapour molecules from the ice surface to the ambient air. The process depends on the gradient between the water vapour concentration for saturated conditions and that of moist air. The most important atmospheric parameters influencing sublimation are the relative humidity of air, air temperature, and wind velocity.

Ice mechanical breaking is a consequence of adhesive or cohesive failure of the accretion. Sudden ice shedding can cause the conductors to jump sometimes leading to flashovers, especially when they are vertically arranged. It can also cause damage to the line and supporting structure.

Snow shedding from ground wires and phase conductors differs from ice shedding mainly because the process and structure of ice and snow accretion are different. The principal condition of snow growth is that adhesive forces between the snow and ground wire/conductor surface as well as cohesive forces between the snowflakes are high enough to keep the snowflakes together on the ground wire/conductor. Snow sheds from the ground wire/conductor when aerodynamic and gravitational forces exceed these adhesive and cohesive forces.

1.3 Operative and Design Systems for Anti-icing (AI) and De-Icing (DI)

In general, AI and DI techniques belong to one of the following four categories:

(i) passive methods based on natural forces or physical geometry
(ii) active coatings and devices
(iii) mechanical methods based on breaking down the ice
(iv) thermal methods based on ice melting.

A brief overview of these method categories is now presented.

1.3.1 Passive Methods

Passive methods for de-icing are ones which do not require an external source of energy but rather use natural forces such as wind, gravity or solar radiation, or phase/circuit geometry. Consequently, they can function on both energized and non-energized conductors as well as ground wires. Passive methods include most of the anti-icing methods used to prevent or reduce the formation or development of wet snow and ice on conductors or ground wires. To achieve this, different strategies are used: (i) weakening ice adhesion strength, (ii) preventing freezing of supercooled water droplets on impact, (iii) using a combination of specific devices for limiting the impact of ice overload on conductors, and (iv) exploiting natural forces such as wind, gravity, or solar radiation in order to limit the adverse effects of ice loads on overhead lines. Some of these methods are already effective for wet snow but their efficacity for ice needs to be studied.

1.3.2 Active Coating Methods

Active coating methods require some electrical energy to be effective. One of the proposed methods makes use of losses in a specific dielectric coating covering the entire surface of the conductor. By choosing an adequate dielectric coating among ferroelectric materials, it is possible to maintain the surface conductor temperature above the freezing point to melt the ice/substrate interface. However, this would require the use of a higher frequency, 60 kHz, instead of the normal 50 or 60 Hz service voltage frequency. This condition is problematic as the use of such high frequencies can lead to electromagnetic perturbations and other problems and needs additional investigation. However, most of the proposed methods are not available commercially. In the same area, other methods are based on the use of a ferromagnetic coating for the purpose of sustaining a positive temperature of the energized conductor surface. Instead of absorbing energy from the electric field, the ferromagnetic coating absorbs energy from the magnetic field, which is at a maximum at the surface of the conductor.

1.3.3 Mechanical Methods

Mechanical methods refer to any method involving ice breaking to accelerate ice shedding. In most cases, they can be considered as DI methods as they are used to speed the shedding process after snow pack or ice has formed on conductors or ground wires. It has been demonstrated that mechanical methods require around 100,000 times less energy than thermal methods to force ice shedding. Generally, most of the mechanical methods are based on two strategies. One strategy consists in breaking the ice by scraping it and the second in releasing energy from shock waves, vibrations, or ground wire/conductor twisting to pull or fling off the ice. One of the main advantages of mechanical methods is their ease of application compared to thermal methods. In fact, mechanical methods are preferred for timely and fast intervention to de-ice short critical sections of a power network. However, in the absence of any precise instructions, mechanical methods involving significant bending of ground wires/conductors should be avoided for optical ground wires to prevent damage to the optical fibres.

1.3.4 Thermal Methods

Thermal methods include all non-natural methods causing the ice to melt to force ice shedding. They consist in the heating of line conductors or ground wires to prevent ice accretion or for de-icing purposes. It is recognized worldwide as the most efficient engineering approach to minimize the consequences of severe ice storms on overhead lines. Some of these methods can be used for anti-icing purposes to prevent supercooled water droplets from freezing during their impact on the conductor surface. In that case, less energy is required for anti-icing than for de-icing. Thermal methods can be divided in two categories: (i) methods based on pure Joule effect and (ii) methods based on dielectric losses, radiative waves, and external heat sources, which are discussed in detail in this book.

1.3.5 New Materials

Recent advances in ice adhesion physics, materials science and in analytical tools have spurred new interest in developing materials having enhanced specific properties. Icephobic materials are no strangers to such developments. It would be extremely difficult, if ever possible, to formulate a truly icephobic material (ice adhesion strength = 0). However, by drastically reducing its adhesion strength, ice may be subsequently removed using a minimum amount of work or heat or could even detach itself under its own weight. Such strategies, described in detail in this book, have great potential as efficient AI techniques.

1.3.6 New Weather Monitoring Systems

In recent years, the development of meteorological weather forecasting models of the atmosphere has reached a level of accuracy and details that makes it possible to estimate the amount of ice already accreted or to forecast the expected ice build-up on standard objects. The most important aspect with this technique is to insert local scale models into a global weather forecasting model by nesting. Since it is not possible to insert abrupt changes in model scale from several tens of kilometres (global models) to the order of hundreds of metres (local scale models) in one step because of stability in the models, such transitions are made by nesting of intermediate models.

Utility experiences with dynamic line rating systems (DLR) have sparked additional interest in weather-based operating practices. The complex interactions between summer ratings and adverse winter conditions add uncertainty to the DLR process that, to some extent, can be resolved using instrumentation such as tension, clearance, sag, or insulator tilt monitoring. The instrumented overhead transmission lines can function as long-wire anemometers, reporting average conditions along stringing sections as well as indications of trouble from rapid changes of state.

1.4 Types of Atmospheric Ice Accretion and Their Characteristics

To understand anti- and de-icing methods and apply them to overhead line (OHL) power networks, it is important to have a good knowledge of the different types and processes of ice and snow accretion on ground wires and conductors. These types and processes are described in the CIGRE TB 291 "Guidelines for meteorological icing models, statistical methods and topographic effects", issued by CIGRE SCB2 WG16 "Meteorology for overhead lines" (CIGRE TF B2.16.03, 2006). Other references are CIGRE TF 22.06.01 (2001), Farzaneh (2008), IEC Standard 60826 (2003), ISO (2017), Poots (2000). Some excerpts from these sources clarify the physical processes, definitions, and terminology used this report.

The formation of ice and snow on overhead power lines may come from cloud droplets, rain drops, snow or water vapour. Cloud droplets are a constitutive part of fog, while rain drops and snowflakes are associated with freezing rain and snow falls, respectively. Accordingly, icing is classified into two main types:

(i) *in-cloud icing (or rime icing),* which is caused by supercooled, suspended cloud water droplets, and
(ii) *precipitation icing,* which is caused by rain drops or snowflakes that freeze or stick to the icing body.

The accretion types which occur during in-cloud icing are soft rime, hard rime, and glaze due to supercooled cloud droplets. Glaze due to freezing rain, wet snow and dry snow are the accretion types which occur during precipitation icing. These icing types are principally governed by atmospheric parameters such as air temperature, wind speed, precipitation rate, relative humidity of the air, liquid water content (LWC) of the air or snowflakes, and size of water droplets or snow flakes (CIGRE TF B2.16.03, 2006; Makkonen, 2000; Poots, 1996; Sakamoto, 2000).

Rime is formed by the impaction and freezing of supercooled droplets on a substrate. When air temperature is well below 0 °C, supercooled droplets possess small momentum and air pockets are created between the freezing droplets. This type of deposit is known as *soft rime and* has low density and weak adhesion. When the droplets possess greater momentum, or the contact and freezing time is greater, a denser structure with stronger adhesion, known as *hard rime,* is formed. In these cases, the released latent heat of fusion is ventilated away before new droplets impinge on the same area, thus maintaining the surface temperature below the freezing point; the ice growth is therefore classed as *dry*.

Glaze **ice** is formed when the released heat of fusion is not dissipated before the next impingement and is therefore raising the substrate's surface temperature up to 0 °C, so that the droplet contact and freezing time is sufficiently long for a water film to form on the accreting surface, resulting in *wet* ice growth. This type of ice accretion has the greatest density and the strongest adherence to conductors. Figure 1.3 shows the visible difference between dry rime (left) and wet glaze ice (right).

Frost or hoar frost, is a deposit of ice crystals formed when water vapour in the air condenses on a substrate of temperature below 0 °C. Increase in gravity load due to frost is normally small but the effective wind area can be increased significantly and cause adverse effects if the accumulation persists. Frost can also cause high energy losses due to corona effects on HV and UHV lines.

Dry snow accretes at subfreezing temperatures. It has low density and low adhesion on OHL conductors. Therefore, it appears rarely, and when the wind speed is very low, i.e. below 2 m/s, it sheds naturally as the wind speed increases.

Wet snow accretion is observed when the air temperature is slightly above the freezing point, usually between 0.5 and 2 °C, and may occur under any wind speed. The density of wet snow varies in a wide range (see Table 1.1) but is normally significantly higher than that of fresh snow on the ground, mainly due to wind pressure effects and LWC. The adhesion of wet snow to conductors is relatively poor at air temperature above the freezing point; it may become very strong, however, if the temperature drops below 0 °C after the accretion, which is quite common.

Table 1.1 summarizes, in general terms, the main characteristics of accretion types described by the terms above (CIGRE TF B2.16.03, 2006). The ice accretion on an exposed structure can also be constituted by a mixture of two or more atmospheric ice types mentioned below, depending on the meteorological parameters and/or successive transformations (melting, re-solidification, etc.).

An accretion may be a mixture of two or more ice types, for instance soft rime, hard rime, and wet snow, due to variations in the meteorological parameters during the icing event. Various shapes, densities, adhesion strengths, etc., result accordingly. The density measurement for mixed ice types is the mean value of the sample.

Among the listed icing types, glaze, hard rime and wet snow, with a relatively high-water content, have a strong adhesion force when they accumulate on exposed structures. Some examples of severe accidents from icing with strong adhesion force include the Ice Storm of 1998 in Canada or that of 2008 in China.

As for dry snow, soft rime and hoar frost, adhesion strength is normally low, so that these accretions can be easily blown off by wind or shaken off from conductors or earth wires. In the case of hoar frost, while an increase in ice load is usually small, it can cause high corona power losses on HV and UHV lines (Farzaneh, 2019) and can lead to flashover on insulator surfaces (Farzaneh & Chisholm, 2009).

Atmospheric icing (or simply, "icing") refers to freezing and sticking of water in various forms (solid, liquid, vapour) on the surface of an exposed object to the atmosphere (Farzaneh, 2008, 2019). Therefore, icing describes both ice and snow growth on an exposed structure such as power network equipment.

Icing can be classified into the following categories:

- precipitation icing such as freezing rain and snow
- in-cloud icing involving the freezing of supercooled water droplets in a cloud or fog upon their impact with an exposed structure

Table 1.1 Main characteristics of different types of ice (CIGRE TF B2.16.03, 2006)

Ice type	Density (kg/m^3)	Description	Adhesion strength
Glaze ice	700–900	Translucid solid ice, sometimes with icicles underneath the ground wires/ conductors. The density may vary with the content of air bubbles.	Very strong adhesion and difficult to knock off
Hard rime	300–700	Homogenous opaque structure with inclusions of air bubbles. Pennant (triangular flag)-shaped against the wind on stiff objects, tending to circular on flexible wires and conductors	Strong adhesion and difficult to knock off, even with a hammer
Soft rime	150–300	Granular structure, "feather-like" or "cauliflower-like". Pennant-shaped also on flexible ground wires/ conductors	Can be removed by hand
Wet snow	100–850	Various shapes and structures are possible, mainly dependent on wind speed and torsional stiffness of the conductor. The liquid water content depends on temperature. Accreted material can be subject to further modification (freeze–thaw) upon temperature variation	Adhesive strength can be very high in a range of LWC between 20 and 40%
Dry snow	50–100	Very light pack of regular snow. Various shapes and structures are possible	Very easy to remove by shaking the ground wires/ conductors
Hoar frost	<100	Crystal structure (needle-like)	Low adhesion, easily blown off

- hoar frost resulting from condensation of water vapour in the air on a structure with a temperature below 0 °C.

Figure 1.1 from CIGRE TF B2.11.06 (2007-06) shows the role of wind speed as well as ambient temperature in the type of accretion.

Soft rime, hard rime, and glaze (grown in a dry regime) due to supercooled cloud droplets are all classified as "in-cloud icing" (Farzaneh, 2008). In-cloud icing usually occurs at temperatures between 0 and –10 °C with droplet diameters in Fig. 1.2 of less than 40 μm.

Figure 1.3 shows photographs of the accretion types identified in Fig. 1.1.

1.4.1 Glaze

Glaze has a density higher than 700 kg/m^3 (Smith, 2007). It grows in a clear, smooth structure with no air bubbles. It is usually formed from freezing

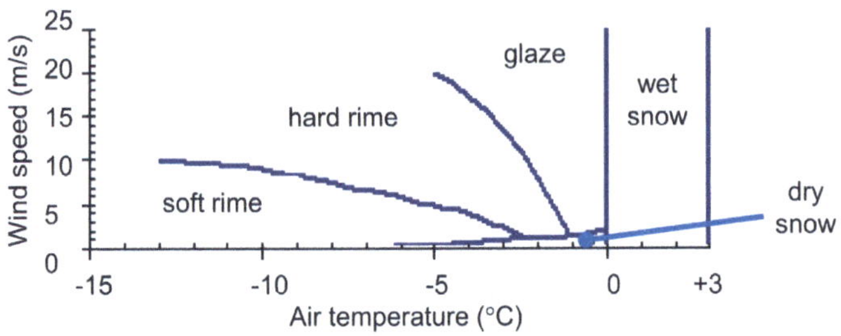

Fig. 1.1 Relation between accretion type and some meteorological conditions (CIGRE TF B2.11.06, 2007-06)

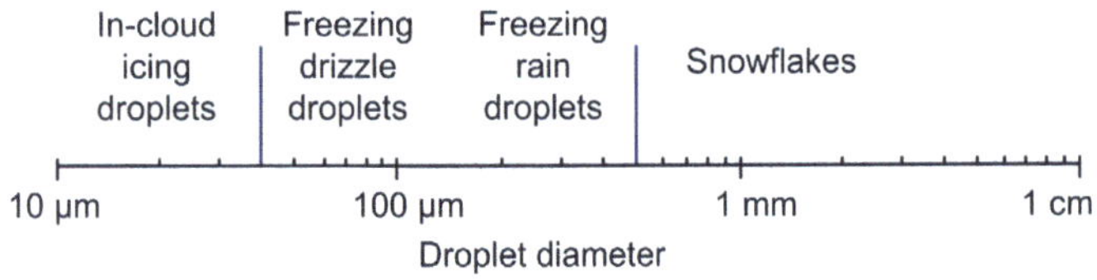

Fig. 1.2 Size spectrum of different droplets in ice accretion processes (CIGRE TF B2.11.06, 2007-06)

precipitation, rain or drizzle, or from clouds with large liquid water content (LWC) and real drop size. The freezing rate of droplets is less than the impingement rate, which causes part of the drop to splash or flow around the conductor before freezing. While glaze contains no air bubbles as such, in strong wind situations, it grows in irregular shapes incorporating pockets of air.

1.4.2 Hard and Soft Rime

Rime has a density of 150–700 kg/m^3 (ISO, 2017). It is usually classified as hard rime and soft rime. Hard rime is a white ice that forms when the water droplets in clouds or in cold fog freeze to the outer surfaces of an exposed object. It has a density ranging from 300 to 700 kg/m^3 and tends to grow in a layered structure with clear ice mixed with ice containing air bubbles (Fikke et al., 2007). In this case, the freezing rate of the droplets is equal to the impingement rate. It is often seen on trees at mountains and ridges in winter when low hanging clouds cause freezing fog. Hard rime formations are difficult to shake off. Soft rime has a density of 150–300 kg/m^3. It has a granular opaque structure, containing large air quantity. Therefore, it shows very low adhesion to surfaces and very low internal cohesion. Soft rime is a white ice deposition that forms when the water droplets in light freezing fog or mist freeze to the outer surfaces of objects, with calm or light wind. The fog water droplets freeze usually to the windward side of tree branches, wires,

Fig. 1.3 Appearance of different types of ice. **a** Glaze; **b** hard rime; **c** soft rime; **d** wet snow; **e** dry snow; **f** hoar frost (Fikke, 2005; Yin et al., 2016)

or any other solid objects in a triangular or pennant shape pointed into the wind. The granular structure results from the rate of freezing of individual droplets freezing completely before another one impinges on the surface. They are fragile and can be easily shaken off from objects. Factors that favour soft rime formation are small drop size, low liquid water content, high degree of supercooling, and fast dissipation of latent heat of fusion.

1.4.3 Dry and Wet Snow

Snow consists of an agglomeration of ice crystals, liquid water, and air, in varying relative quantities. Therefore, snow shows a wide range of physical characteristics that are dependent on composition and ambient parameters. Snowflakes present a wide variety of morphologies, form simple hexagonal prisms to the dendritic form, thus widening the range of physical properties that snow can display (Newell et al., 1968-04). The limit between dry and wet snow has been set by different authors to be at –1 or –2 °C (Farzaneh, 2008; Poots, 1996) or at –3 °C (Colbeck & Ackley, 1982). In fact, wet snow is a complex system made up of ice crystals, liquid water, and air; the amount of each varies over a wide range, so that the physical properties of snow can be very different (Colbeck & Ackley, 1982).

1.4.4 Hoar Frost

Hoar frost has a density less than 100 kg/m^3 and is deposited as interlocking ice crystals formed by direct sublimation of freezing water vapour in the air on an exposed structure. It is featherlike in appearance and occasionally forms in large volume but with a relatively low weight on exposed objects. Normally, hoar frost does not constitute a significant loading problem. However, it is an efficient collector of supercooled fog or cloud droplets and at subfreezing temperatures with light winds, fog conditions gradually become soft rime of significant volume and weight. Furthermore, its presence on the overhead transmission lines can still cause significant energy losses due to corona discharge (Lahti et al., 1997).

1.5 Overview of Atmospheric Icing and Its Effects on Overhead Power Networks

Atmospheric icing events can occur in many parts of the world including North America, Europe, Asia, and South Africa. Such events are recurrent in some countries (USA, Canada, Iceland, etc.) occasionally causing serious socioeconomic impacts, while they are less frequent in other countries (Sweden, Spain, South Africa, etc.) but can nonetheless cause major damage to overhead power lines.

In this chapter, some examples of important icing events in many parts of the world are presented, showing the importance of finding solutions in the interests of safe operation and reliability of networks against icing. These reports are drawn from detailed reports in Farzaneh (2008), CIGRE TB 179 (CIGRE TF 22.06.01, 2001), TB 291 (CIGRE WG B2.16, 2006), TB 438 (CIGRE WG B2.29, 2010),

TB 631 (CIGRE WG B2.44, 2015), IEC TR 60826 (IEC, 2017), ISO 12494 (ISO, 2017). Related parameters such as the weight of ice, density, conductivity, and liquid water content are also reported.

Icing has obvious direct, detrimental consequences to the power network equipment when conductors are overloaded and sag near the ground. There are other, less obvious detrimental consequences related to the effects of heavy ice loads, low temperature, high tension, and aeolian vibration on subsequent ability of the line to carry current while maintaining satisfactory ground clearance in summer under peak loads. The effects of winter conditions on conductor creep are mentioned to contribute to a comprehensive process for overhead line design and operation.

1.5.1 Direct Impact of Icing Events in Canada (1988)

In January 1998, a major ice storm caused catastrophic losses of power over eastern Canada and the north-eastern United States. It was an unprecedented event in Quebec history. In fact, from January 4 to 10, freezing rain fell for 80 h and left as much as 110 mm of ice in the Montreal region and the St-Lawrence valley. This was more than twice the precipitation of the second largest storm which hit Montreal in 1961. Figure 1.4 shows a map of freezing rain accumulations over the entire affected region.

The precipitations from three consecutive storms fell mainly as freezing rain and ice pellets but also as rain and snow. A review of the meteorological conditions is found in (NOAA National Weather Service, 2008) At the peak of the storm, 1.4 million people were without of power. Emergency centres had to be opened to accommodate people. The storm left the power grid with unprecedented damage: 3000 km of power lines, 4000 power transformers, and 1000 steel towers were destroyed or damaged.

Table 1.2 lists the damage recorded for the transmission system. The Canadian Armed Forces deployed over 15,000 troops, their largest operation since the Korean War. Line crews from Canadian and American electric utilities came to help with restoration of the power grid (Henson et al., 2007; Hydro-Quebec, 1988; Jones, 2003-01).

Table 1.3 shows the number of people deployed during the refurbishing of the distribution system. The power outage lasted for up to 4 weeks in some regions. According to Statistics Canada, the damage caused by this event was evaluated at 6.4 billion dollars for the affected Quebec regions.

1.5.2 Direct Impact of Icing Events in UK (1990)

Severe icing events that cause major disruption in the UK occur about every 16–18 years (1947, 1963, 1981, 1990, and 2010). The wet snow event of 8 December 1990 caused severe disruption at all voltage levels from low voltage up

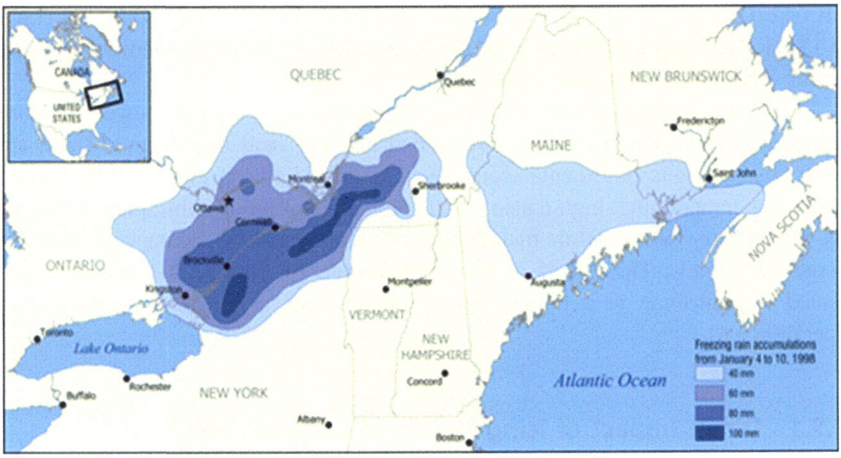

Fig. 1.4 Freezing rain accumulation (CIGRE WG B2.44, 2015). *Source* Environment Canada

Table 1.2 Damage recorded for the transmission system

Voltage level (kV)	Total lines	Broken and/or collapsed towers
735	10	149
315	12	58
230	13	306
120	67	1091 (steel and wooden poles)
49	14	1500 (wooden poles)
Total	116	3104

Table 1.3 Staff affected at the refurbishing of the distribution system by region

	Montreal	Laurentides	Richelieu
Hydro-Quebec line workers	170	225	640
Private line workers		100	190
Canadian and American line workers	30	50	545
Private pruners	25	125	85
Canadian and American pruners	65	125	5

to the 400 kV lines of the National Grid. The North Yorkshire, Cleveland, and Nottinghamshire areas suffered from wet snow accretion for a period of 30 h from Friday, 7 December 1990 to Saturday night, 8 December 1990 as the storm moved from North to South. Over 1 million customers were cut off, some being without power for up to 9 days. Snowfall affected Scotland and North Wales on 7 December and intensified as a weather front became virtually stationary on a line approximately NNE-SSW through Nottingham and Birmingham. As the temperature hovered just above 0 /°C, a rainfall equivalent of 30 mm of precipitation came down as wet snow in 14 h with winds of around 15–25 m/s. Lines running East-West were subject to the greatest snow thicknesses with radial ice thicknesses of up to 200 mm on single small conductors which could rotate. Twin Lynx conductors, which could not rotate, accreted ice horizontally, leading to many galloping incidents and line failures. Detailed reports of network damage have been provided by East Midlands Electricity, Northern Electric, Yorkshire Electricity, and Midlands Electricity. These were the names of the utilities at the time of the event. A report was also produced by the UK Met Office with details

A low-pressure area and accompanying cold front moved SE across the country on 7–8 December and then the front became occluded and slow moving. The strongest winds occurred in the Tyne and Cleveland areas, whereas the worst snowfall was observed in the East Midlands. Measurements of ice load indicated levels of up to 3 tonnes on a single MV conductor span. Three weather maps for this event are found in CIGRE WG B2.44 (2015). The first, at 16:00 on 7 December, confirmed that snow had already been recorded at the EA Technology snow accretion test site in the South of Scotland. By 19:00, snow was already falling in NE England as the local temperature dropped from 3 to 1 °C. By 21:00, 24 mm of (rain equivalent) wet snow had fallen in the Northern Electric area in the previous 12 h and a further 17 mm (rain equivalent) was to fall over the next 7 h.

The snow reached Nottingham by 23:00, and the local temperature dropped from 1 °C to around 0 °C. Winds gusted to 28 m/s. By 9:00 on 8 December, 19 mm (rain equivalent) of snow had fallen in the Nottingham area with snow depth on the ground of 1.3 m. The worst area for precipitation was in Yorkshire with 61 mm rain equivalent. By 11:00, the snow was widespread from Anglesey in North Wales to the Wash on the English East coast and from Shetland in the North to the Isle of Wight off the English South coast. By 17:00, the snow accumulation rate had tapered off and temperatures ranged from 0 to 2 °C.

About 40% of the customers in the East Midlands area (800,000 out of 2.1 million customers) lost power and ice shedding caused further problems with over-stressed poles breaking. 8000 incidents were recorded. The UK National Grid had 800 incidents on their 275- and 400-kV lines, resulting in power loss to 25 of East Midland's 132 kV substations. The main transmission line problems were clashing, conductor burning due to contact caused by excessive sags and major galloping on 132, 275 and 400 kV lines. A total of 724 wood poles were broken, and the 132 kV line suffered 30 times its normal annual fault level. Conductor failures occurred on Dingo and Wolf ACSR conductors up to 150 mm^2.

In the Northern Electric area, the weather load standard for overhead lines is 30 m/s with 15 mm radial ice. The winds gusted to 30 m/s, and radial ice thickness was 25 mm. Although much less than the Midlands area, there were over 1000 conductor breakages and 354 broken poles.

Radial ice thicknesses of up to 62 mm were measured in the Yorkshire Electricity area with 41 failures on 33/66 -kV lines and earthwire breakages on the 132 -kV tower lines. Similar problems were reported in the Midlands area with one noticeable feature—on the 132 kV system, failures of twin Lynx conductor were

10 times more than the single conductor failure rate (after taking circuit length into consideration). Most wood pole conductor failures (93%) were at mid-span due to clashing of overloaded lines.

The causes of this widespread event were the excessive ice build-up on (mostly) smaller conductors, and the faults caused by icing across twin conductors (leading to galloping and clashing). The excessive sags meant that phases could contact each other, leading to arcing and subsequent conductor failure. Reduction of accretion levels would not only reduce sag (and so less clashing) but also reduce structure loads.

1.5.3 Direct Impact of Icing Events in South Africa

While people in the Northern Hemisphere do not visualize South Africa as a region where ice is a concern, operating experience shows that ice loading incidents, while infrequent, have the potential to severely disrupt the supply of electricity in parts of this country (Marais, 2004).

Figure 1.5 shows a total of 31 ice loading incidents were recorded in different areas of the country in the period 1995–2004. The incidents were compiled from extensive interviews with regional transmission and distribution engineers and field staff, and the information is mainly limited to the decade indicated. While incidence of icing in the high-lying areas in Lesotho is known, the frequency of events as well as the intensity of ice loading are not as severe as found in the northern hemisphere.

Ice loading incidents, while infrequent, have the potential to severely disrupt the supply of electricity. Extreme cold weather has the potential to affect a large geographical area, in contrast to more localized extreme wind conditions such as those reported in KwaZulu Natal in 2001.

1.5.3.1 *Icing Event, KwaZulu Natal, 2001*

Heavy snow over the Drakensberg escarpment on 13 September 2001 resulted initially in many interruptions of supply on the distribution system and later in the tripping of transmission lines feeding KwaZulu Natal. The line trips were caused by broken earthwires and crossarms due to severe ice loading (Cretchley, 2001). Over the next two days, 400 kV lines and a 275 kV line tripped at various times. These caused the power system to oscillate, eventually resulting in a system voltage collapse in KwaZulu Natal.

The whole KwaZulu Natal load of 3343 MW and 1000 MW of pumping load at Drakensberg were shut down. The system frequency jumped to 50.84 Hz. The total loss equalled 6.81 system minutes.

1.5.3.2 Cross-arm Failure

A phase crossarm failed on a suspension tower of one of the 400 kV lines. It was evident that the crossarm was subjected to high vertical loads, as well as a longitudinal load in the direction of the longer 600-m span (Fig. 1.6). The failure of only

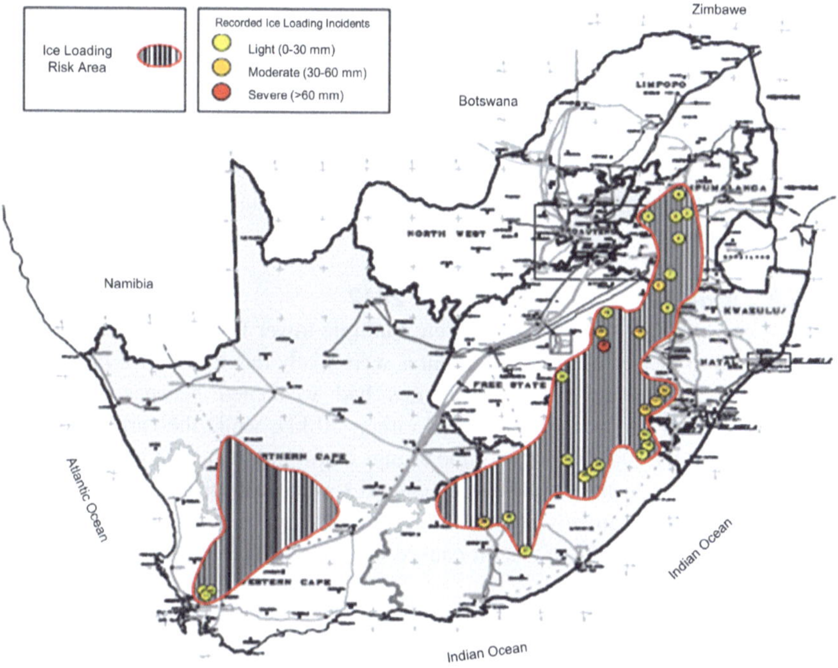

Fig. 1.5 Ice loading incidents on overhead lines in South Africa from 1995 to 2004

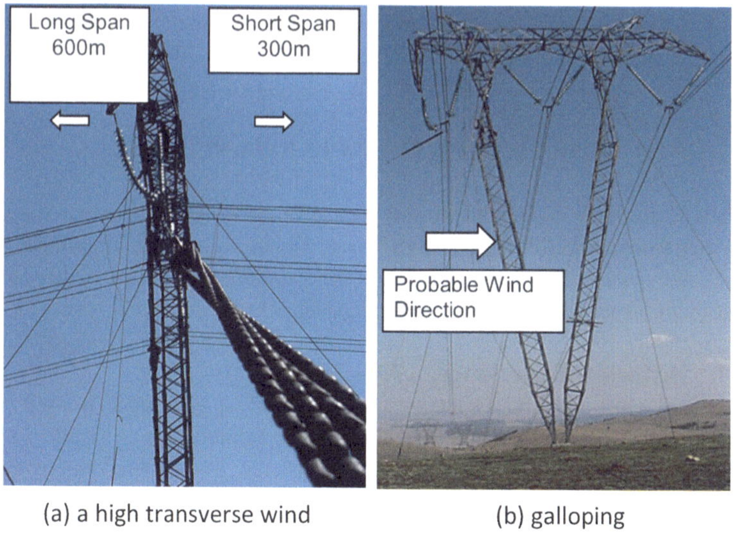

Fig. 1.6 Failure of one of the crossarms on a 400 kV suspension tower

one crossarm would indicate the presence of either (a) a high transverse wind or (b) galloping—which typically results in more variable loading.

Ice loading is causally related to conductor size. The largest conductors in use by Eskom are 36-mm diameter Bersfort ACSR conductors, 1355 kcmil (686 mm^2) aluminium area. These would have attracted a considerable amount of ice build-up. This by itself could have resulted in sufficient vertical load to collapse the crossarm. Calculations show that anything more than 10 mm of radial ice could cause failure of the crossarm.

1.5.3.3 *Icing Event, KwaZulu Natal, 2012*

In the 7 August 2012 incident, there were multiple tower failures. Generally, there did not seem to be any wind, so the failures were likely due to high ice build-up. It was evident that some of the earthwires had weakened because of ageing. The OPGW fitted had a low mechanical rating, 50 kN, while the earthwire rating was approximately 100 kN. Some of the crimps on fittings also gave way. When the earthwire failed, then these side forces caused tower earth peaks to collapse, the tower crossarms were already heavily loaded with ice on the conductor and the peak or earthwire falling on the crossarm caused the crossarms to collapse.

1.5.3.4 *Snow/Ice Events, Southern Cape: 2002–2009*

In the winter of 2002, a 400 kV transmission line in the Southern Cape had a sustained outage for a total of 31 h 27 m due to two permanent faults (22 h 21 m and 9 h 6 m), believed to be caused by the icing of insulator strings at tower 259. During the winter of 2007, the line was locked out for 7 h 28 m. In 2009, four other line faults occurred in the space of 5 h. One of them was a permanent fault which lasted for 8 h 40 m, and icing was found to be present (Swinny, 2011). Figure 1.7 shows some of the accumulation on the insulators in the 2009 event.

1.5.4 Direct Impact of Icing Events in Iceland

Iceland is a northern country with complex geography and a winter temperature which often fluctuates around 0 °C at the coast. Hence, icing occurs regularly on the overhead line system. Wet snow icing is most common, but in-cloud icing can also be a problem, especially in highland areas, over 300 m above sea level.

1.5.4.1 Historical Experience, 1960–2012

After serious icing incidents in the years 1991–1995, a systematic work began in the distribution system (11–33 kV) to improve its reliability with regard to icing. The power companies have collected and registered all known icing events on overhead lines into a database which contains information from many decades. The number of broken poles/towers exceeded 600 in 1972, 1974, and 1991 as shown in Fig. 1.8.

This database was used to identify distribution lines that were most sensitive to icing incidents, and they were prioritized when overhead lines were replaced with

Fig. 1.7 Snow/ice accumulation on a strain insulator arrangement

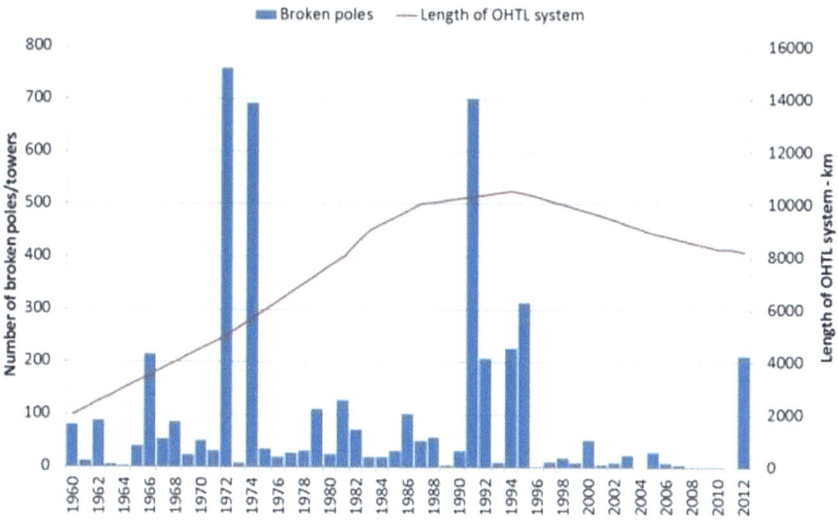

Fig. 1.8 Number of broken poles in icing events and length of overhead lines in Iceland. A reduction of broken poles from 1995 to 2012 is partly explained by the fact that distribution lines in the most severe icing areas were replaced by underground cables

underground cables. This has been very effective in reducing icing damage in the distribution system in the last years. The database of operating experience is also used when preparing specifications of new transmission line.

1.5.4.2 *Wet Snow Event, 2012*

An example of a recent wet snow icing incident which affected the 66- and 132-kV transmission line system is one that occurred on 10 September 2012 in Northern Iceland. Estimated ice load and description of tower failures are shown in Fig. 1.9.

There was a strong wind from NNW, with heavy fall of wet snow, especially in areas 100–200 m above sea level in the western part and 200–300 m in the eastern part. In the eastern part, ice load was measured with a load cell on one tower which recorded an ice load of 14.5 kg/m when the conductor broke (Fig. 1.10).

However, the data show that ice accretion continued for two or three hours after the conductor break in Fig. 1.10. Based on observations and simulated data, mean winds (10 min) are likely to have peaked at the site early in the night at 27 m/s, with gusts exceeding 40 m/s. Many towers were broken or damaged (Fig. 1.11), which caused loss of power in three villages in the Northeast of Iceland for three days, while the line was being repaired.

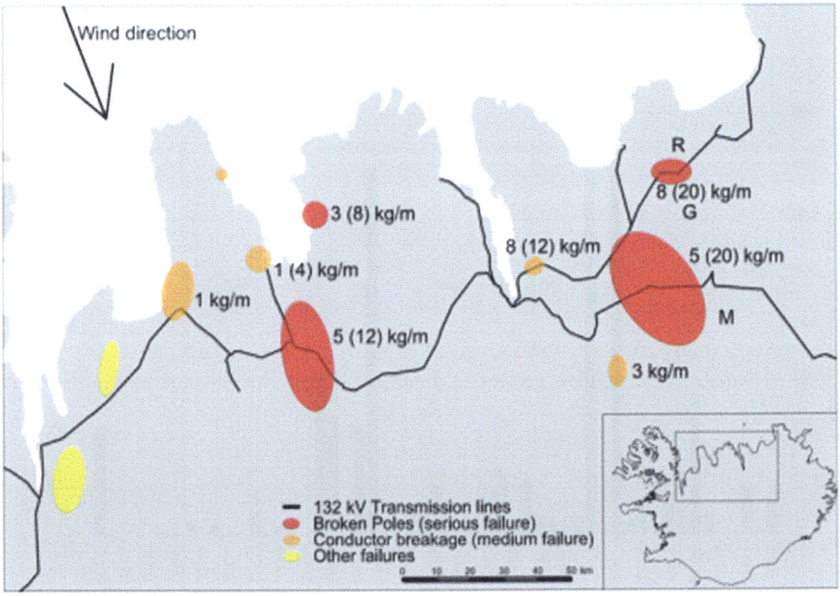

Fig. 1.9 Areas affected by the icing event of 10 September 2012. Overview of measured/estimated wet snow loading as well as main regions of failures. Values show median (max) loading on 132 kV transmission lines

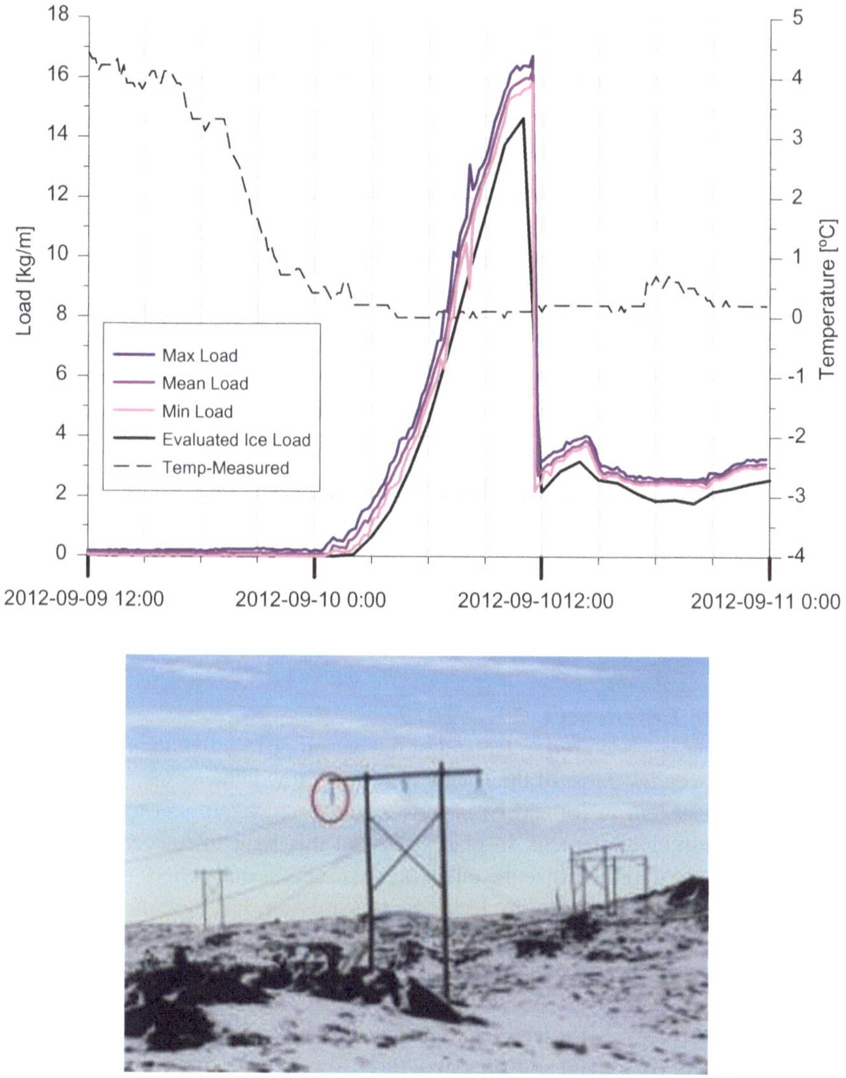

Fig. 1.10 Measurement with a load cell in a 132 kV suspension tower

Diesel generators were used to supply electricity for domestic use, but the fishing industry was without power. Although many towers were broken in another 132 kV line in the area, the resulting loss of power was minor, because the so-called n-1 rule allowing for loss of any single asset in redundant grid design made it possible to supply power by other transmission lines.

Fig. 1.11 Broken 132 kV OHGL. Inset: Ice ellipse diameters, 200 × 300 mm

1.5.4.3 *Icing Experiences, 2014–2019*

Fewer icing hazards occurred in Iceland in the period 2015–2019 than generally in the two prior decades. Most of the events were wet snow icing as seen in Fig. 1.12.

The power system in Iceland is, in most cases, N-1 system and tolerant of the loss of any individual element. In places without this fault tolerance, local diesel generators are used to minimize the effect on the users. In this period, the following icing events caused failures in the Icelandic transmission line system. In all cases, the towers or poles are made of wood; no failures were in the 220 kV system, where steel towers are used.

- On 7–8 December 2015, icing caused failures on lines in East-Iceland and the Westfjords. In Skriðdalur, East-Iceland, three towers were broken in the 132 kV system, and close to Kópasker, Northeast-Iceland, three towers broke in a 66 kV line, which caused the lines to be out of operation for one to two days. Disruption in the power supply to the end users was shorter. In the Westfjords, severe icing caused the breakdown of 13 towers in the main 66/132 kV line (Fig. 1.13), which was out of service for a week, but other lines and diesel generators provided electricity for most of the users. In all cases, the maximum diameter of the icing was 10–15 cm (5–12 kg/m).
- On 4–5 February 2016, a 17 cm icing diameter (16 kg/m) caused a severe failure on the main 132 kV line in Öræfi (guyed towers), Southeast-Iceland (Fig. 1.14).

Fig. 1.12 Wet snow icing in Prestbakkalína 1, 132 kV, Southeast-Iceland on 1–5 February 2014

Conductors were broken, and two towers damaged. The line outage lasted for three days, but only one hour for the users.

- Rime icing was observed on conductors and towers in April 2017, as shown in Figs. 1.15 and 1.16.
- On 23–24 December 2016, some farms close to Egilsstaðir, East-Iceland, were without power for several hours after icing hazard on distribution lines (icing diameter 10–12 cm, 5–8 kg/m).
- On 12–13 May 2017, a 10 cm icing diameter (5 kg/m) on a 66 kV line caused one-day interruption in part of the Westfjords (broken conductor and damaged mast). Diesel generators supplied power for most of the users.
- In the period 2015–2019, icing was as usual, detected and measured in many of the test spans in Iceland. As most of them are in the highlands, on exposed places, the icing was mainly in-cloud icing.

1.5.5 Direct Impact of Icing Events in Sweden (1999)

On 10–12 February 1999, the Swedish 130–400 kV network was affected by several interruptions (of up to 6 h) in the southwest part of Sweden. Ice accretion on insulators near the coast caused flashovers, which led to several 130–400 kV lines

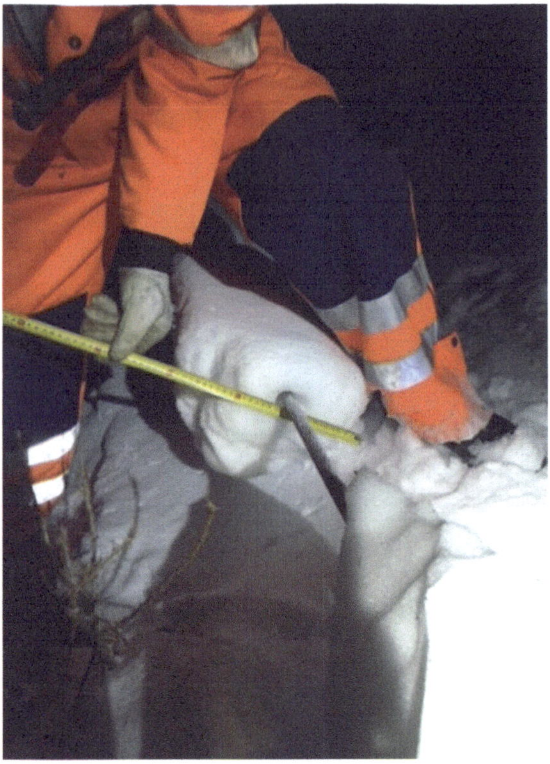

Fig. 1.13 Wet snow icing on Breiðadalslína 1, 66 kV, in Dýrafjörður in the Westfjords, 8 December 2015

being out of operation all at the same time (Svenska Kraftnät, 1999). This type of event is very rare in Sweden. An example of ice accretion on a 130 kV insulator (on the 9th of February) is shown in Fig. 1.17. This insulator did not flash over.

The probable cause of the interruptions was a very favourable weather condition for ice accretion, with extensive west winds and low temperatures although on the day of failure there was clear sunshine. More precisely, during 1–5 February, 92 % of the wind direction was from the sea and the wind speed was 1–17 m/s (during that time the average temperature was 2.6 °C). Salt monitoring at the 400 kV substation in the southwest of Sweden showed high level of salts, requiring automatic washing of substation insulators. On 6 February 1999, the temperature went down to about –1 °C, and on 7 February, the intensive snowfall came (14 cm of snow recorded at the ground). Insulators were covered by snow and the snowfall stopped during 8 February with the average temperature of –3 °C. During the following days, the temperature went down further, in average to –12.1 °C during 9–10 February (with fog and cloudy weather), and to –10.6 °C on average during 11 February (clear weather). Thus, the probable cause of the event was that when the snow melted from the sunshine, the water layer on the ice captured the salt from it and, as a result, the conductivity of the water became very high. Finally, a flashover occurred. This explanation was confirmed by conductivity measurements

Fig. 1.14 Prestbakkalína 1, 132 kV, in Öræfi South Iceland. February 2016

Fig. 1.15 Rime icing on Bolungarvíkurlína 1, 66 kV, at Hestakleif in the Westfjords, in April 2017

Fig. 1.16 Detail of rime icing on structure in Fig. 1.15

Fig. 1.17 Example of ice accretion on a 130-kV insulator string in Sweden (1999-02-09)

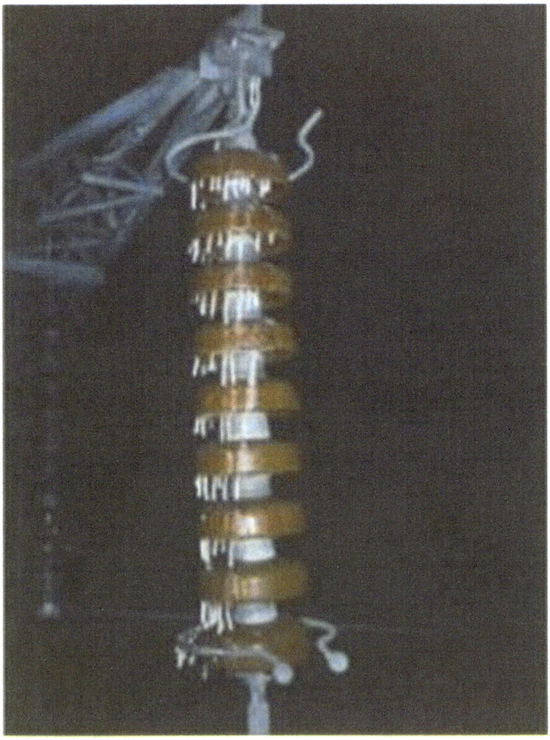

Fig. 1.18 Removal of ice for inspection at a 400 kV line (1999-02-12)

of melted icicles, which reached 680 μS/cm. The removal of ice for inspection is shown in Fig. 1.18 taken on 12 February on a 400 kV line. In the period of one hour (between 11:00 and 12:00, which indicated possible melting due to the sun), four 400 kV lines fell out of operation. After ground and helicopter inspections, ten sky lifts crews were sent in the field and most of the ice-accreted insulators were cleaned. No interruption occurred during the whole event.

The most frequent types of insulators used in the Swedish grid are cap-and-pin porcelain or glass types. These two insulator types have long lifetime. One type of insulator which is only moderately used in Sweden is the composite insulator. It is intended for contaminated environments. The incidents of February 1999 did not require any serious countermeasures.

1.5.6 Direct Impact of Wet Snow Event in Japan (2005)

On 22 December 2005, an outage occurred in the Northern part of Niigata Prefecture around 8:10 where as many as 650,000 households were affected (Onodera et al., 2007–10). Power distribution was restored at 15:10 on 23 December, 31 h after the occurrence of the outage. The system collapse was the result of ground faults and short circuits that occurred simultaneously at many locations on transmission lines under rarely encountered and most severe weather conditions. This

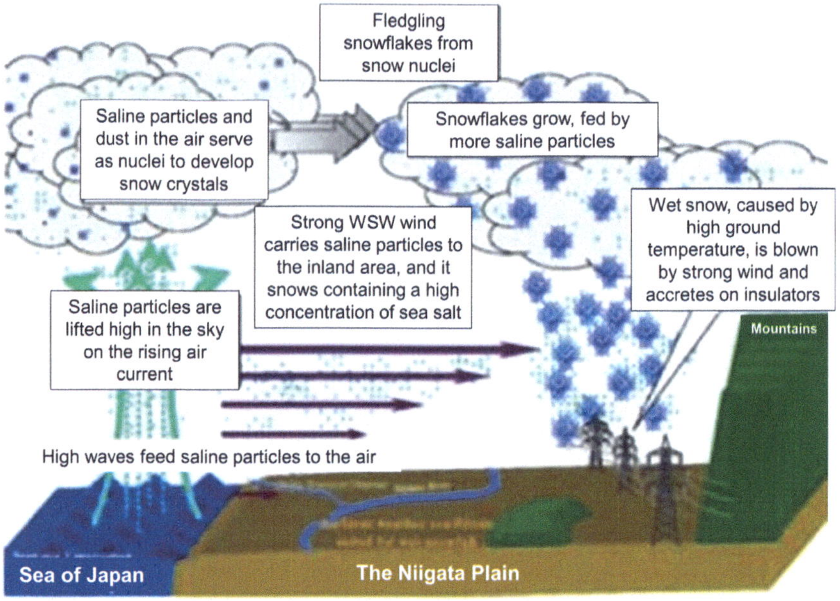

Fig. 1.19 Process of snowfall illustrating sea salt contribution to electrical conductivity for 22 December 2005, Niigata Storm

snow event caused unprecedented damage to the power system by reduction in insulation by wet snow polluted with sea salt and by short-circuit faults due to galloping. The rare and severe weather conditions were reported in several papers that are summarized comprehensively in Yaji and Homma (2019-01).

A schematic representation of the event is presented in Fig. 1.19. The progression of the event can be explained as follows. With a large-scale depression going over the area, it began to snow around 3:00 on 22 December. Figure 1.20, showing the meteorological parameters for these two days, characterizes the event with strong winds with 10-min average velocity of more than 10 m/s between 6:00 to 8:00 as the depression passed.

Ambient temperature of 0–2 °C, which is the range favourable for wet snow accretion, was observed in Fig. 1.20 between 3:00 and 17:00 on 22 December, a continuous exposure duration of 14 h when wet snow could very easily accrete on power lines. The cumulative precipitation was the largest recorded in the period 1975–2005 at Niigata Meteorological Observatory. There was enormous precipitation of snow in a temperature range suited for wet snow while strong winds blew for a long time.

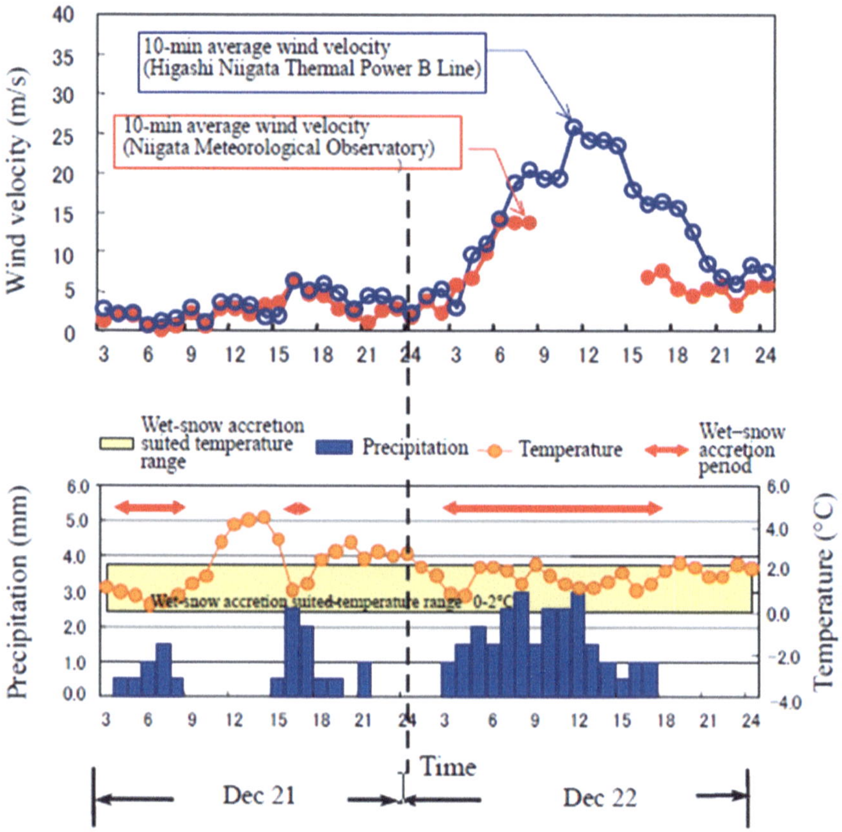

Fig. 1.20 Meteorological conditions at the time of power outage

1.5.7 Direct Impact of Icing Event in China (2008)

In early 2008, the Central and South parts of China encountered the most serious freezing weather in the previous 50 years (Meteorological Service Center of China, 2008). Figure 1.21a shows that the average temperature of seventeen of the thirty-one provinces of China was under 0 °C during the icing disaster. According to the statistics of China Electric Power Research Institute (CEPRI), 36,740 lines were not supplying electricity, 2018 transformers shut down, and 8381 electric towers collapsed at voltages ranging from 110 to 500 kV.

Figure 1.21b shows that the worst-hit areas were in Central China, such as the Hunan Province. The electric equipment of Hunan Power Grid was seriously damaged. Table 1.4 shows the statistics of damaged lines, from 110 to 500 kV, of the Hunan Power Grid.

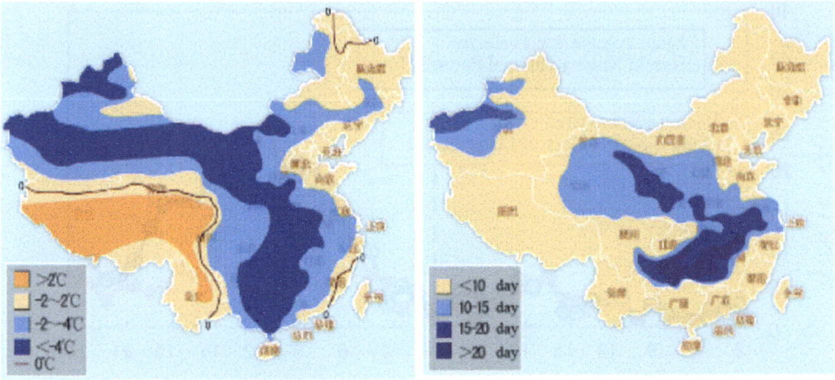

(a) Average temperature (b) Freezing days

Fig. 1.21 Average temperature and freezing days from 20 January to 2 February 2008, China

Table 1.4 Statistics of damaged lines in Hunan Power Grid

Voltage level (kV)	Total # of affected lines	# Of line flashovers	# Of outages	# Of broken conductors	# Of collapsed towers
500	33	28	89	14	174
220	246	>100	Not known	42	185
110	Not known	>100	>2000	61	247

Table 1.5 Statistics of substation outages in Hunan Power Grid

Voltage level (kV)	# Of affected substations	# Of substation outages	% Of affected substations with outages
500	12	6	50
220	85	32	37.6
110	389	87	22.4
35	433	181	41.8

Table 1.5 shows the substation outage statistics. During the disaster, there were many icing flashovers, collapsed towers, and seriously damaged substations. There were consequent outages, and it was even reported that a city of Hunan Province was out of power supply for more than two weeks. The direct economic losses of China in the icing disaster were estimated at more than 3.5 billion US dollars.

1.5.8 Direct Impact of Icing Event in Catalonia, Spain (2010)

Catalonia, a region on the Mediterranean south of the Spain–France border, is not a region where one expects icing problems in the summer months. However, this region receives a "tropical snowstorm" once every 4–5 years.

On the day of 3 August 2010, the meteorological situation described in the diagram at the left Fig. 1.22 produced a heavy snow fall mainly in the coast area known as Costa Brava, as shown in Fig. 1.22 (right). All this area has a height above sea level ranging from 0 to 500 m.

The main features of the snow fall in the most affected areas were:

- Wet snow precipitation between –1 and 2 °C, therefore it was sticky
- Snow fell for over one hour at a rate of 1 mm/min (consisting of huge snowflakes)

Low temperatures caused tension forces in the ground wires of 220 kV towers, leading to damage, but these lines were otherwise only slightly affected. The 25-, 45-, and 66 kV lines coloured yellow in Fig. 1.23 collapsed under snow weight. For the 110- and 132-kV lines, 33 towers collapsed from accretion of wet, sticky snow with examples shown in Fig. 1.24.

The first day after the event, 200,000 customers were left without electricity. Even three days later, there were 20,000 customers in the same situation. Electrical companies had to supply electricity by means of autonomous diesel generators otherwise the number of customers disconnected would have not declined. It took about 6 weeks to repair the damage done by the storm, including closing an expressway to facilitate tower replacement as shown in Fig. 1.25.

As according to Spanish standards such as NNA/2017 Clause 4.5.2, used for constructing overhead transmission lines, snow or ice is not expected at locations that are less than 500 m above sea level. Power line towers at these low altitudes were not designed to support extra loads resulting from the accumulation of wet snow falling so quickly over the conductors. The Spanish design standards have not changed since 2010 regarding the minimum amount of ice to be considered in overhead lines placed below 500 m above sea level, even though the utility which suffered this catastrophic damage decided to design all their new overhead lines accordingly. If the lines are at an elevation of 500–1000 m above sea level, and converting 1 daN/m = 10 N/m, they consider a minimum of ice load per unit length as a function of conductor diameter:

$$\text{Weight of ice or snow } W = 18\,\text{N/m}\sqrt{(d)} \tag{1.1}$$

For zones with altitude >1000 m above sea level, the design ice weight is doubled:

$$\text{Weight of ice or snow } W = 36\,\text{N/m}\sqrt{(d)} \tag{1.2}$$

Fig. 1.22 Meteorological situation 2010-08-03. The blue arrow shows the entrance direction of cold air from central Europe over the Mediterranean; the overall result is shown in the right picture

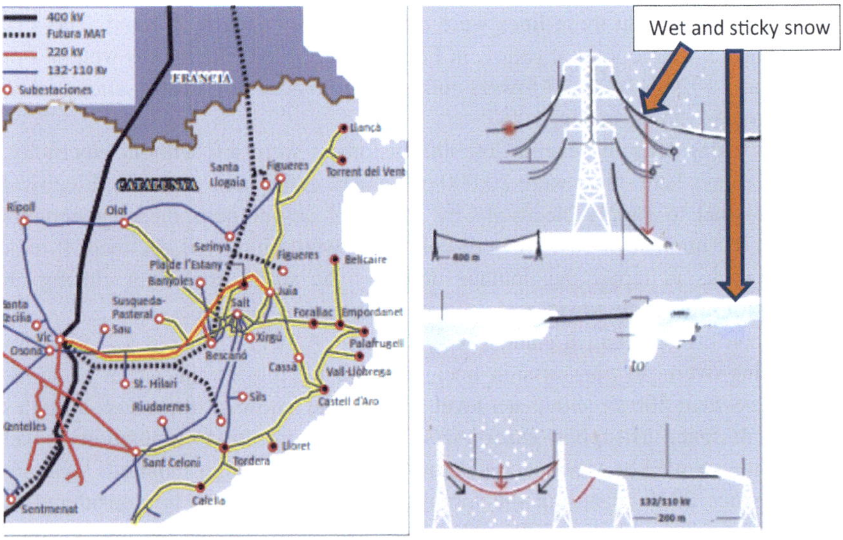

Fig. 1.23 HV lines affected by icing event of 2010-08-03

where d is the diameter of the conductor in millimetres and W is in Newton per linear metre. For a typical 795 kcmil 26/7 aluminium conductor, steel-reinforced (ACSR) Drake with 402.8 mm^2 aluminium cross section, $d = 27.36$ mm, weight of ice or snow is $W = 94$ N/m and the weight of the conductor itself is 1.627 kg/m (15.9 N/m).

Fig. 1.24 Examples of collapsed 110 kV towers

Fig. 1.25 Example of repair work on a 110 kV tower

1.5.9 Direct Impact of Icing Events in Norway (1961–2014)

1.5.9.1 *Heavy Ice Accretion, 1961*

Probably the world's largest ice load on an electric power line was observed in Norway in 1961. Figure 1.26 shows an example of the accretion that measured 1.4 m × 0.95 m and weighed 305 kg/m. This is probably the clearest example on how topography and exposure influence the icing conditions on a power line. This line was supplying a radio and TV transmitter 1412 m above sea level. It appeared to be impossible to maintain this line no matter how short the spans would be, how strong the (wood) poles and the steel conductors were. It was built on the top of a mountain ridge parallel to the coast and therefore with maximum exposure to the humid south-westerly winds from the Atlantic Ocean. It was only after a new line was built on the leeward side of the mountains that the power supply to this radio and TV transmitter became stable.

Figure 1.27 shows an example of collapsed towers due to wet snow accretion in combination with strong winds, about 1100 m above sea level.

Topography influences the icing differently depending on the type of icing. Freezing rain occurs mainly in basins and depressions where cold air can be trapped while warm air with precipitation may intrude the air aloft (temperature inversion). In-cloud icing occurs only above the cloud base, but the cloud base varies significantly with topography. A mountain (ridge) only 50–100 m higher in the upwind direction may be sufficient to reduce this icing to a minimum. However, wet snow may occur at all altitudes and on the leeward side of mountains and ridges (Fikke, 2006-12).

Fig. 1.26 In-cloud icing on a 22 kV line feeding a TV tower 1400 m above sea level in south-western Norway. Courtesy of O. Wist

Fig. 1.27 Collapse of 300 kV towers due to wet snow at 1100 m altitude

1.5.9.2 *Polluted Rime Ice Accretion, 1980s and 1990s*

Mostly in the 1980s and 1990s in Norway, the HV insulator properties were affected by polluted rime ice on insulators, as shown in Fig. 1.28. This pollution was transported over long distances from industrial and densely populated areas in continental Europe as well as from the British Isles. High concentrations of sulphates and nitrates resulted in low pH values of droplets in clouds formed over the mountain ranges in Western Norway (Fikke et al., 1982-06, 1994-06).

Fig. 1.28 Example of burn marks on a 420 kV line insulator set due to polluted rime ice accretion

1.5.9.3 *In-cloud Ice Accretion, 2014*

In-cloud icing typically occurs in mountainous areas, and the cloud base can vary significantly with topography. Therefore, in-cloud icing on relatively close transmission line spans can vary considerably. Such an example is shown in Fig. 1.29, where the spans closest to the photographer are under heavy ice load, and a few spans further down there is almost no ice on the conductors and ground wires. The photograph shows the 420 kV Sima-Samnanger power line in Norway and was taken on 1 February 2018 during a visit to an ice monitoring station.

A characteristic of in-cloud icing is that it is very directional and that it forms on the windward side of the conductor or structure. This is very well illustrated in Fig. 1.30.

The directional nature of the in-cloud icing also affects the way ice forms on overhead transmission line (OHTL) conductors and ground wires. In most cases, the maximum ice load is higher on a simplex conductor or wire than on each subconductor of a twin bundle. This can be attributed to the higher torsional stiffness of bundled conductors which are connected by spacers, difference in the ice shedding and the mentioned directional icing accretion profile.

An example of different ice shapes on bundled and single conductors is shown in Fig. 1.31, where on the left is the ice shape of the bundled phase conductor and on the right is a nearly cylindrical ice of the simplex shield wire.

Fig. 1.29 In-cloud icing on the 420 kV transmission line Sima-Samnanger. *Source* Statnett

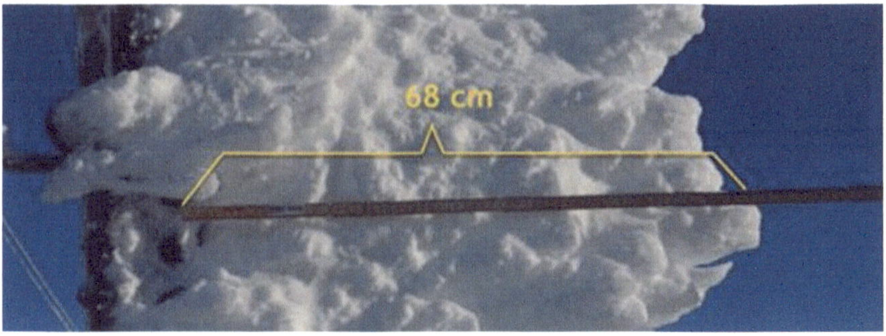

Fig. 1.30 Directional in-cloud icing on a power line structure. *Source* Statnett

However, this is not always the case. Sometimes, the icing load on a subconductor of a bundle is higher than the icing on the single ground wire or conductor. There have been several observed cases which can be attributed to different reasons. One hypothesis for higher accretion on subconductors than on individual conductors of similar diameter is the so-called bundle collapse, where the entire bundle rotates under extreme torsional loads due to heavy icing. The bundle rotation then enables a more cylindrical accretion of ice on the subconductors, which in their deformed state present much greater capture area. One such example of ice accretion on a collapsed bundle is shown in Fig. 1.32 with photographs taken on the Ålvikfjellet ice monitoring station in Norway.

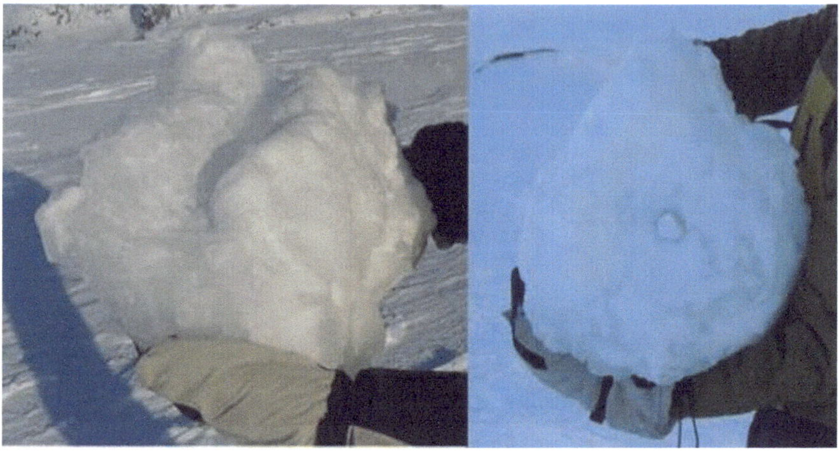

Fig. 1.31 Comparison of ice shapes from the bundled phase conductor and single shield wire. *Source* Statnett

Fig. 1.32 Comparison of icing on single and twin-bundle configurations under "bundle collapse' condition. *Source* Statnett

The largest icing accretion measured recently in Norway was in January 2014 with 76 kg/m ice per metre length of shield wire, see Fig. 1.31. This led to breaking of the Sima-Samnanger transmission line shield wire and partial collapse of towers. This icing event triggered a start of R&D project related to icing (Byrkjedal & Nygaard, 2019; Byrkjedal et al., 2016-09).

Between 2015 and 2019, there were no major icing events in the Norwegian transmission line network. It was found that one span of the Sima-Aurland transmission line had a broken ground wire due to icing two winters in a row, suggesting severe local conditions.

1.5.10 Direct Impact of Icing Events in Italy (2015–2017)

Wet snow is the main meteorological threat in Italy for the entire electricity transmission and distribution network. During this period, three relevant wet snowfalls occurred, also confirming the trend of a gradual increase in these phenomena.

- On 5–6 February 2015 in Emilia and Lombardy regions, a peak of 325,000 users were without electricity for more than eight hours, mainly in lowland areas, with about 1000 MWh of Energy Not Supplied (ENS) with an automatic reimbursement of 33 ML€.
- On 5–6 March 2015 in the Abruzzo region, more than 100,000 users were disconnected.
- On 15–18 January 2017 in Abruzzo and Marche regions, the snow in some areas was three metres deep, and a sleeve load on conductor was up to 15 kg/m. Up to 160,000 users were disconnected more than eight hours, with 35–40 ML€ of automatic reimbursement estimated by ARERA (The Italian Regulatory Authority).

1.5.10.1 Direct Impact of Icing Events in Slovenia

Slovenian transmission network is affected by three different climatic zones (Mediterranean, Alpine, and Continental) and frequently exposed to different levels of ice storms and separately from bora windstorms.

The characteristic of periodical meteorological events during winter in Slovenia is the substantial contrast between the deep cyclone region (low air pressure) above the Eastern Atlantic and partially above the Mediterranean and the distinctive anticyclone (high air pressure) with its centre above Russia. A frontal zone between the two regions develops above the Eastern part of the Alps and the Northern part of Mediterranean. The difference in air pressure between these two may sometimes be as high as 0.1 kPa. Because of this, the cool polar air which is a thin layer close to the ground meets the strong inflow of warm and wet air in the shape of high-altitude winds from the Mediterranean. Therefore, snow falls through a layer of warm air, where it melts, and then cools down again in the lower cold layers of air. Supercooled water drops of rain are in an unstable condition and rapidly turn into ice when they touch objects. The phenomenon of supercooled droplets is therefore the main reason for the occurrence of ice loads on overhead power lines.

During the period from 1997 to 2014, Table 1.6 shows that there were eight significant icing events on the Slovenian power transmission network, affecting

Table 1.6 History of major ice events in Slovenia in the period 1980–2014

Location	Period	Events and comments
Brkini (SW)	5 November 1980	OHL 400 kV Divača-Melina: Destruction of steel construction due to icing on 27 supports OHL 220 kV Divača-Pehlin: Destruction of steel construction due to icing on 29 supports *The ice coating was up to 7 cm thick and because of the damage to the forests some 674,000* m^3 *of timber had to be cut down* (Kern & Zadnik, 1987)
Notranjska (SW)	15–16 November 1984	OHL 110 kV Cerkno-Idrija: Broken crossarms because of icing on 2 supports OHL 110 kV Cerkno-Idrija: Destruction of support Nr 15 because of snow and ice load *Icing damaged about 21,000 ha of forests (nearly 2% of all forest areas) and about 110,000* m^3 *of timber were damaged* (Kern & Zadnik, 1987)
Notranjska (SW)	7–15 January 1997	OHL 400 kV Beričevo-Divača: Damage to conductor clamps and conductors due to icing on several locations OHL 400 kV Beričevo-Divača: Destruction of metal construction because of icing on 7 supports OHL 220 kV Kleče-Divača: Icing on conductors with destruction of one support OHL 400 kV Maribor-Krško: Earth fault because of icing *Icing damaged some 82,000 ha of forest (about 8% of all forest areas), around 870,000* m^3 *of timber were damaged*
Notranjska (SW)	5 December 1998	OHL 400 kV Beričevo-Divača: Earth fault *Extreme sag, caused by additional ice load of the conductor*
Štajerska (Central)	9 February 1999	OHL 400 kV Maribor-Podlog *Earth fault because of icing*
Gorenjska (NW)	14 November 2004	OHL 110 kV Okroglo-Jesenice/Jeklarna II: *The ice load was shaken off the conductors*
Štajerska (NE)	27 January 2009	OHL 2x110 kV Pekre-Vuhred *Destruction of 3 supports because of snow load*
Notranjska (SW)	31 January –9 February 2014 (Fig. 1.33)	OHL 400 kV Beričevo-Divača: *Destruction of 26 supports and heavy damage to 6 supports because of icing* OHL 400 kV Beričevo-Podlog: *3 supports destroyed by icing* OHL 220 kV Kleče-Divača: *21 supports destroyed and 22 supports heavily damaged by icing* OHL 220 kV Beričevo-Podlog: *Heavy damage caused to 1 support by icing* OHL 110 kV Cerkno-Idrija: *Due to icing 8 supports were demolished and 4 supports heavily damaged* OHL 2x110 kV Slovenj Gradec-Velenje: *Due to icing 1 support was demolished and 1 support heavily damaged* *OHL towers: Supports of overhead lines over 93 km were destroyed or damaged, the total cost being estimated to 12 M€* (DELO, 2014)

110, 220, and 400 kV overhead lines including the tower collapse shown in Fig. 1.33.

In 2014, ice storm damage also occurred over 3053 km of the Slovenian low voltage and distribution network, with about 23,000 demolished or damaged supports. Figure 1.34 shows some representative accretion and damage. The total damage to the distribution network was estimated to about 55 M€. Icing also damaged about 500,000 ha of forest, representing half the forest areas in Slovenia. The loss of a timber volume of 7,030,000 m^3 was estimated as 194 M€ (DELO, 2014), about 0.5% of the gross domestic product of Slovenia that year.

Icing also destroyed or heavily damaged the railway infrastructure (electric overhead supply lines with fittings and supports) on the Koper railway line—Ljubljana—where the total freight from the seaport of Koper is being transported in the direction to Central Europe. The railway line was totally closed for several days. A temporary traffic system using diesel locomotives was organized for freight traffic only. Reconstruction of the damaged railway line is estimated to have last for six months at a cost of 20 million Euro. During the 2014 ice storm, also called *The Historic Slovenia Ice Storm*, which particularly struck the hilly region of South-Western Slovenia, some 80,000 households were affected, and the total damage was estimated to nearly 500 M€. The thickness of ice load on conductors of overhead lines in some locations reached 12–15 cm. During this difficult time, many European countries responded generously and sent immediate help in the

Fig. 1.33 Damage to 400 kV OHL, Slovenia, 2014

Fig. 1.34 Damage to LV & Distribution Network in Slovenia, 2014 (DELO, 2014)

form of many diesel power generating units to secure electricity supply to many households and production plants until the damaged public power distribution system was fully restored.

Fortunately, between 2015 and 2019, the Slovenian transmission network did not experience any major icing events (Bakic et al., 2009; Jakl, 2013; Jakl et al., 2003).

1.5.10.2 Indirect Impacts of Winter Conditions

When an overhead conductor heats up in summer peak load and low-wind conditions and sags into a tree, it is convenient for the public, press and utilities to blame the tree. No one blames the prior winter conditions—low temperatures, high ice and wind loads—that changed the conductor over its service life to a state where its sag–temperature relationship and clearance to ground were uncertain.

When a bimetallic conductor such as ACSR is initially strung, it is placed in frictionless pulleys, called travellers. The construction process applies a schedule of increased tension, held for specific time intervals to allow the conductor to creep—mainly through plastic deformation of the aluminium layer.

After a stringing schedule has been completed, the conductor is fixed into clamps at each suspension insulator. The insulators are hanging vertical while the conductor is in the travellers. The stringing temperature is the temperature at which the insulators hang vertical (T_{vert}).

The conductor continues to accumulate creep over its service life. There are several contributors:

- Time
- high conductor temperature
- heavy mechanical load
- high tension at low conductor temperature, compounded by aeolian vibration

The conductor is thought to make a transition from its initial stringing condition, with T_{vert} equal to the stringing temperature, to a final condition in about 20 years. In the final state, T_{vert} has drifted downward as tension in the conductor relaxes. The insulators now hang vertical at the original stringing tension, which is achieved at a lower temperature that contracts and tightens up all the coupled spans in a stringing section between dead-end towers. While T_{vert} is dropping, the "Knee Point Temperature" (KPT) of the ACSR may also drop from an initial value of perhaps 100 °C into the range of 50–75 °C. Operation of a conductor above its KPT adds uncertainty to the equations of state that relate tension, sag, clearance or insulator swing to the average conductor temperature in a stringing section.

An example of the uncertainty in predicting high-temperature sags when operating above the KPT is found in CIGRE WG B2.55 (2019).

Two of the most significant "tree" outages in North America were in Idaho and Ohio, both regions with winter conditions. A strong subsequent focus on condition survey, using LiDAR to precisely establish the clearances in all spans, has guided regulatory policy by FERC in the North American utilities. It is important that this condition survey process be repeated after every significant ice or windstorm. Each severe event may add additional creep to the conductor, reduce the maximum operating temperature and thus reduce the ability of the conductor to carry current (ampacity).

1.6 Global Climatology of Relevant Winter Weather Parameters

The Köppen–Geiger classification of climate considers ambient temperature as well as precipitation. In regions classed as "DfX" (cold, without dry season) in Fig. 1.35, engineers normally consider severe winter conditions. However, regions with humid temperature climate "Cf", coloured green, have also described specific problems.

For planning, design, and operation of wind turbines, the "WIceAtlas" project considers mainly regions with Köppen–Geiger Class D as "Low Temperature Climate".

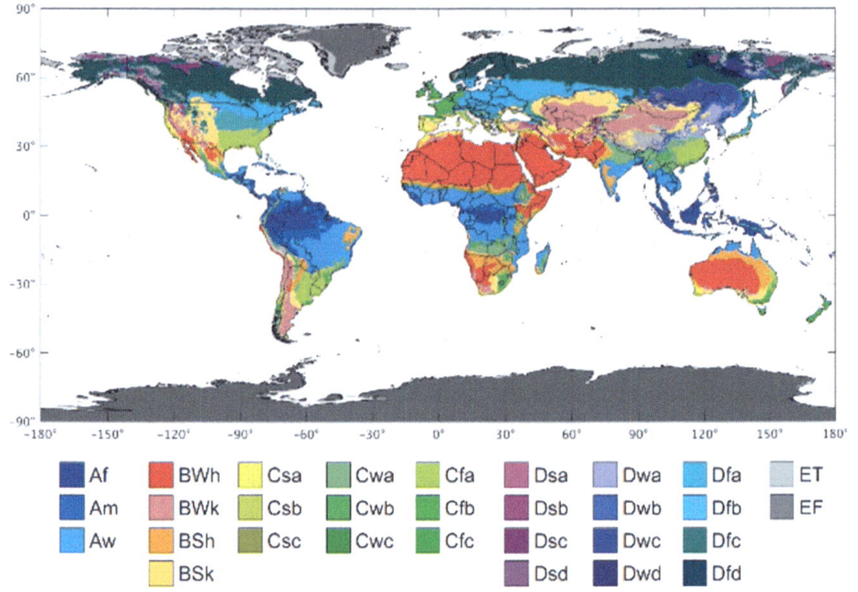

Fig. 1.35 Köppen–Geiger classification of world climates, present-day map (1980–2016) (Beck et al., 2018-10)

1.6.1 Ambient Temperature and Dew Point Under Icing Conditions

One of the difficulties with measuring humidity and propensity to condensation in winter conditions is that the absolute humidity is low, often less than 1 g/m^3 compared to 11 g/m^3 in standard high-voltage laboratory test conditions. Sensors that rely on interaction of water molecules to change the dielectric constant and capacitance do not resolve so well at cold temperatures. It is common in climate data to see a ceiling in the observations of relative humidity for $T_a < -5$ °C. This ceiling, perhaps 80% RH, is an artefact of the sensing technology.

The role of relative humidity is especially important around 0 °C. A close match between ambient temperature (T_a) and dew point temperature (DPT) traps the temperature of any ice layer between these values. If the DPT is several K below T_a, the lower ice temperature has a strong effect on its electrical conductivity and flashover strength.

1.6.2 Precipitation Icing

The International Energy Agency (IEA) sets five classes for wind energy projects in winter conditions, shown in Table 1.7.

Table 1.7 IEA ice classification

IEA ice class	Duration of meteorological icing (% of year)	Duration of instrumental icing (% of year)	Production loss (% of annual electric power)
0	None		
1	0–0.5	<1.5	0–0.5
2	0.5–3	1–9	0.5–5
3	3–5	6–15	3–12
4	5–10	10–30	10–25
5	>10	>20	>20

The lowest elevation considered in the WIceAtlas results (VTT Technical Research Centre of Finland Ltd., 2015) is 50 m above ground. The world icing map at this elevation, overlaid with the area of "Low Temperature Climate" appears in Fig. 1.36.

1.6.3 In-cloud Icing

The public WIceAtlas map (VTT Technical Research Centre of Finland Ltd., 2015) uses cloud base height and temperature <0 °C observations to classify in-cloud icing conditions. It consolidates data from >4000 meteorological stations and estimates icing levels at wind turbine height of 150 m above ground level. There are extensive regions in Fig. 1.37 where the (IEA) ice class is 3, 4, or 5.

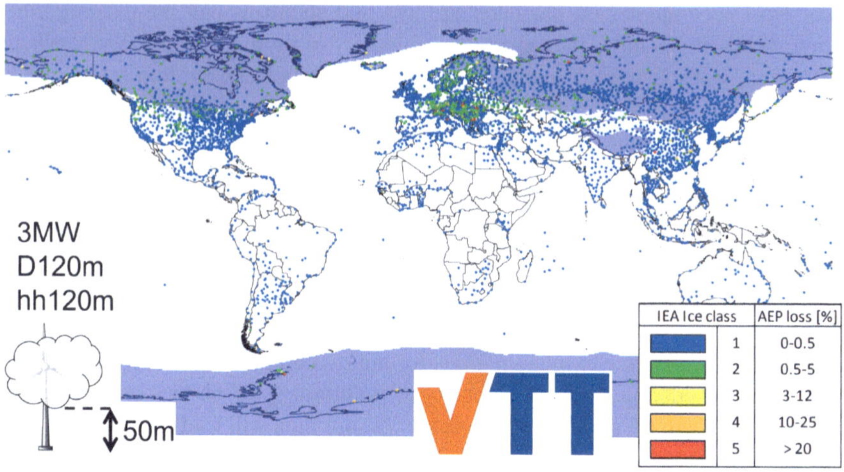

Fig. 1.36 WIceAtlas world icing map, 50 m above ground level, overlaid with regions with low temperature climate, 34-year average (VTT Technical Research Centre of Finland Ltd., 2015). Used with permission

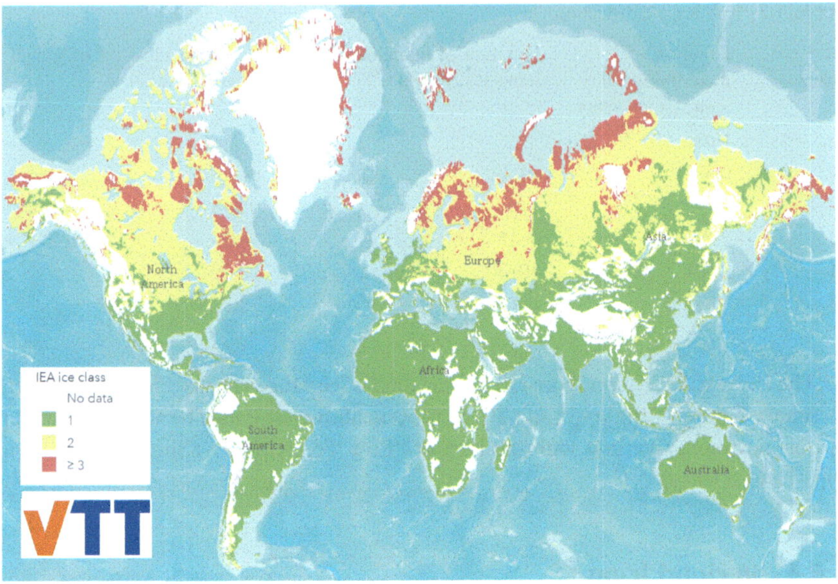

Fig. 1.37 WIceAtlas map of IEA ice class. Scale: Clear—no data; green—Class 1; yellow—class 2; red—class 3, 4, or 5 (VTT Technical Research Centre of Finland Ltd., 2015). Used with permission

1.6.4 Hoar Frost

Hoar frost accumulates on insulator surfaces and stabilizes any pre-existing condition. While it can have an electrical conductivity of more than 300 μS/cm, where 100-μS/cm water is used in simulated high voltage rain tests, the thin frost layer is frozen, and its main role is to stabilize any pre-existing pollution layer. Some countries such as China track the number of fog days per year by month, allowing for identification of possible problem areas. A long-term study result in Fig. 1.38 shows some regions have fog on 30–50% of winter days.

1.6.5 Consecutive Dry Days in Winter

Precipitation is effective in many regions for removing accumulated soluble surface pollution. Considering a constant density of pollution, g/m^3 as well as an average airflow and surface characteristics, greater exposure time will lead to greater deposit of pollution. Areas with more than 90 days without rain in winter season such as

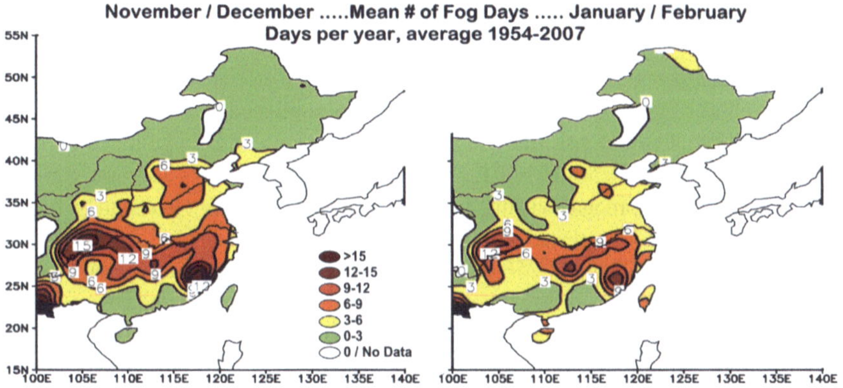

Fig. 1.38 Average number of days with fog in winter season in China, 1954–2007. Data from Yu et al. (2019)

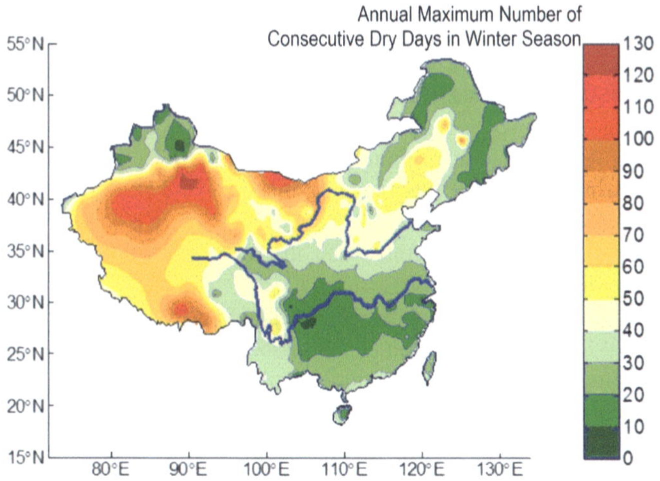

Fig. 1.39 Climatology of consecutive dry days in cold season in China, 1961–1990. Redrawn from Doan et al. (2017)

those indicated in Fig. 1.39 are susceptible to build-up of pollution over an entire winter, and with thaw and/or hoar frost conditions in February or March, these are the regions that experience many wood pole fires and cold fog-type insulator flashovers across the leakage distance.

1.6.6 Synoptic and Exceptional Wind Speeds

Standard codes for wind engineering cover the general needs for overhead power lines as well as other structures. CIGRE WGB2.16 published in 2004 the Technical Brochure (TB) "Current practices regarding frequencies and magnitudes of high intensity winds". TB 256 (CIGRE WG B2.16, 2004-10) focuses on various meteorological phenomena causing high wind speeds. Those effects may hit an overhead line in random places and developments since publication are directed to probabilistic design. The TB 256 does mention icing in its recommendations for wind loading pressure on towers and conductors, increasing the overloading coefficients from 1.2 to 1.4 for wind pressure on ice-covered conductors.

Generally, in winter conditions, it seems that the synoptic wind speeds rather than those high intensity wind gusts exceeding 45 m/s are associated with overhead line failures. In Norway, high-intensity windstorms with velocities >40m/s are regularly experienced; however, the frequency of overhead line failures due to high-intensity winds remains very low and most overhead line failures are due to combinations of wind and ice. The situation is similar in Iceland, with failures every decade normally in association with normal synoptic wind velocity of 37 m/s with ice diameter of 7–10 mm (CIGRE WG B2.16, 2004-10).

1.6.7 Persistence of Ice Accretion

Persistence of ice accretion is monitored with the IEA Wind Recommended Practices for wind energy projects in cold climates:

- Incubation time is the time from when the icing starts to when the cup-type anemometer freezes up.
- Meteorological icing is the duration of the period when ice is accreting.
- Instrumental icing is the duration of the period when the anemometer is frozen.
- Recovery time is the duration from the end of the ice storm to when the ice falls off the anemometer and it returns to normal operation.

These are imprecise measures for the persistence of ice accretion on conductors and towers, which have different sizes and shapes. Specialized sensors such as vibrating-probe ice detectors may be used to measure meteorological in combination with rotating cup or propeller anemometers of various sizes. These heat up and melt any accretion of 0.05 mm thickness, but as no ice can accumulate just after a heat cycle, the sensors can underestimate the actual level of ice accretion.

1.6.8 Occurrence of Multiple (Ice-on-Ice) Events from Consecutive Storms

The 1998 ice storm described in Quebec was a sequence of three separate ice storms. Prior to this event, there was a focus on establishing the ice layer for 50-year return period from a single ice storm. After this event, there was an appreciation that it was much more probable to have two or three, modest ice storms, each accumulating on top of the previous accumulation.

The 2013 North American Storm Complex led to 25 mm of radial accumulation in Toronto, Canada, and up to 30 mm at nearby locations. There were 1 million people without electricity over the Christmas season, mostly from the loss of about 20% of the tree canopy, falling on and dragging down the distribution system below. One report was that the distribution utility (Toronto Hydro) delivered ten years' worth of tree trimming maintenance work in the week following the events. There were two separate ice storms, one on 20 December and the second two days later. There was a single, 500 kV transmission line insulator flashover and no mechanical problems with transmission lines during this pair of events, leading to accretion near the design levels.

Methods to assess ice-on-ice accretion events are under development as replacement for the Weibull model used for return periods of severe ice accretion events considering only a single storm.

1.6.9 Atmospheric Corrosion Measures

In electrical power transmission systems, many economic losses are related to corrosive processes, especially in line sections that are in coastal, industrial, or road-salt exposure areas. Road salt in winter conditions acts like marine and industrial pollutants to initiate a corrosive process on all metallic components of electrical networks, such as conductors, fittings, metallic insulator parts, signs, and support structures of transmission lines. In most cases, these components are constructed with hot-dip galvanized steel and/or aluminium alloys with resistance to corrosion. Environmental conditions play an important role in the rate of degradation of metallic components, affecting both transmission line components exposed to air and those in the ground, such as metal footings of self-supported towers.

Durability of the corrosion protection system for galvanized steel and aluminium alloys is determined by the corrosivity of the atmosphere; that is, the more aggressive the environmental conditions, the shorter the protection time. "Durability" in this context, is usually defined as "the expected life of the corrosion protection system until the first major intervention as an act of maintenance". Therefore, it is imperative and advised that utilities address this problem right from the project planning stage. The initial investments made to mitigate or avoid corrosion and ensure the operability of assets over time are often captured under Capital Expenditure (CAPEX), whereas costs of special maintenance activities over time are often covered under Operating Expenses (OPEX).

There are techniques used by corrosion investigators to measure the corrosivity of the atmospheres in which the power network equipment is installed. The measured levels of corrosivity cannot be extrapolated to the entire transmission line. The most critical geographical areas have very particular climatic conditions with some common parameters. Some of the common climatic parameters are high relative humidity, high concentrations of chlorides and/or sulphur dioxides, and low or infrequent rainfall. In these circumstances, accelerated corrosion deterioration of the metallic components of the assets can cause considerable economic losses and eventually power supply interruptions. To mitigate this natural process, it is necessary to apply anticorrosive coatings and/or use stainless materials.

The main objective of this chapter is to present in detail the methodology adopted by some utility companies to identify corrosion rates of equipment components installed in aggressive environments and, in addition, to present some procedures used to mitigate the effect of corrosion.

1.6.9.1 Corrosion Process

The corrosion process can result from chemical and/or electrochemical reactions. In the case of chemical reactions, elements are added or removed from a chemical species. In contrast, electrochemical reactions are chemical reactions in which not only elements are added or removed from a chemical species, but also at least one species undergoes a change in the number of valence electrons (Kelly et al., 2002).

In fact, there are four requirements for degradation by corrosion (Kelly, et al., 2002):

- an anode (where oxidation of the metal occurs)
- a cathode (where reduction of a different species occurs)
- an electrolytic path for ionic conduction between the two reaction sites and
- an electrical path for electron conduction between the reaction sites.

The corrosion of steel arises from its unstable thermodynamic nature. It is manufactured from iron, which is made in a blast furnace by reducing ores such as hematite (Fe_2O_3) with carbon in the form of coke (Bayliss & Deacon, 2004):

$$2Fe_2O_3 + 3C \rightarrow 4Fe + 3CO_2 \tag{1.3}$$

Although the previous reaction is completed at 1200 °C, the final products, iron and eventually steel, are unstable. Consequently, when steel is exposed to moisture and oxygen, it tends to revert to its original form (Bayliss & Deacon, 2004):

$$Fe + O_2 + H_2O \rightarrow Fe_2O_3 \cdot H_2O(\text{iron})(\text{rust}) \tag{1.4}$$

In this case, the result is a hydrated iron oxide (rust), similar in composition to hematite. This explains why steel tends to rust in most situations and the process can be seen as a natural reversion to the original ore from which it was formed (Bayliss & Deacon, 2004).

Hot-dip galvanized steel, commonly used for most tower components, is coated with a thin layer of zinc which helps retard corrosion. However, under favourable conditions, corrosion will occur over time. Corrosion degradation of these elements can result from atmospheric corrosion or soil corrosion (for tower feet).

1.6.9.2 *Atmospheric Corrosion*

Atmospheric corrosion of metals occurs mainly due to the influence of agents such as oxygen, humidity, and atmospheric pollutants such as sulphur dioxide (SO_2), chlorides (Cl^-), nitrogen oxides (NO_x) and carbon dioxide. (CO_2). For atmospheric corrosion to affect power lines, the atmosphere must contain pollutants such as those described above.

The most common atmospheric pollutant is sulphur dioxide. Sulphur dioxide generally comes from natural sources, such as volcanoes, and from artificial sources, such as the industry and motor vehicle exhaust. In the case of chlorides, the main source is the sea, and wind plays two roles: it agitates the waves, increasing mass density and then may transport the aerosol inland. Chloride deposition is more-or-less fixed, while SO_2 emissions in Europe have dropped 70% in the period 1990–2010 and continue to decline with improving environmental concerns.

ISO 9223:2012 (ISO, 2012) establishes a five-level classification system for the corrosivity of atmospheric environments that includes estimates of the rate of thickness loss per year for steel, copper, aluminium, and zinc (Table 1.8).

The ISO methodology to determine the corrosivity of a geographic corridor could be considered. Measurements of average yearly temperature (°C), SO_2 and Cl^- deposition (mg/m^2/day) and average yearly relative humidity (%) at evaluation points along any planned new power line corridor should be taken and the results assessed against the various categories of corrosivity.

Even though the classification of atmospheres into four types not as rigorous as the ISO 9223 approach, it turns out to be very practical for the electricity sector, especially considering the large areas of territory that cover power transmission lines. According to ISO (2012), the environment can be divided into four pollution categories:

- rural (inland): dry environment with little or no pollution
- urban: polluted by exhaust, smoke, and soot
- industrial: highly polluted by industry smoke and precipitate
- marine: on and by the sea, with high humidity and chlorides.

Table 1.8 Atmospheric corrosivity classifications according to ISO 9223 (ISO, 2012)

Category	Corrosivity
C1	Very low
C2	Low
C3	Medium
C4	High
C5	Very high

Rural Atmosphere

The rural atmosphere concerns outdoor regions suffering little influence from polluting air sources, i.e. industrial gases or salts in suspension (aggressive pollutants in the air). The relative humidity of the air could present low or high values. It can comprise land areas rich in forests or used for agriculture and animal husbandry.

In terms of corrosivity, it can be said that it is very low due to the absence or low presence of aggressive pollutants.

Urban Atmosphere

This kind of atmosphere is found in cities, and its main characteristic is that most of the pollution comes from automobile exhaust gases. The corrosivity of this type of atmosphere can be increased by the presence of industries within cities or in close vicinity of the same.

In urban atmospheres, there is a presence of pollutants that in contact with high relative humidity and condensation can generate "acid rain". The large quantity of chemicals emitted in the urban environments in the form of atmospheric pollution starts to react with contained water vapour, forming a series of acidic components that fall in together with rain. The impact of acid rain includes a gradual destruction of buildings and public assets, damage to forests and soils, increased acidity of lakes and water bodies and the consequent contamination and defilement of river fauna and damage to human health (including but not limited to dryness of the skin, respiratory diseases in case of inhalation of pollutants that generate acidity in the rains).

Industrial Atmosphere

Industrial atmosphere is typical from regions where there is a dangerous amount of exhaust from motor vehicles and impurities in the form of oxides of sulphur (SO_2, SO_3), acid soot among others. It causes acid rain, deposition of solid particles, release of CO_2 into the atmosphere, alteration of the wind regime changing characteristics of directed rain and changes in temperature distribution.

Marine Atmosphere

Marine atmosphere is characterized by the presence of salt particles originating from the sea. The corrosivity of this atmosphere is largely determined by the distance from the coast. When the distance is smaller, the concentration of salt particles is increased, and in consequence, the corrosivity is increased.

1.6.9.3 Classification: Atmosphere Corrosivity

For utility companies, it is vital to identify the corrosivity of the atmospheres where their assets will be installed, especially those that will be installed outdoors.

Knowing the corrosivity of the atmosphere will allow to define strategies for the design of the materials and equipment, as well as to define the activities and frequencies of maintenance of the equipment. Carrying out corrosivity studies of the atmospheres in the power sector is less common than it should be; very few companies take the precaution of performing these studies and considering the results in the life cycle analysis.

Some utilities consider three levels of corrosivity based on their maintenance experiences and defined environmental criteria, following a model such as Table 1.9.

This simple classification allows a broader definition of geographical sectors and requirements for renewal of the line components over time. A detailed breakdown of individual component lifetimes is found in CIGRE WG B2.69 (2021-06). The range of observed overhead line component lifetime has a rather narrow range, as summarized in Table 1.10.

Proximity to a busy, elevated motorway with winter road salting fits into the corrosivity classification system of Table 1.9 as "HIGH", based on an equivalent sea distance of about 1 km; no rainfall; relative humidity >75%; average temperature below 0 °C and snow cover, equivalent to desert. The only aspect that does not contribute to "HIGH" ranking is average temperature. The presence of salt or sulphur dioxide on surfaces can form liquid water monolayers that result in high corrosion rates in temperatures well below 0 °C.

1.6.9.4 Classification: Time of Wetness

Guides such as ISO 9223 (ISO, 2012) classify regions with high rate of metal loss using three inputs: chloride deposition rate (related to proximity to the sea), sulphate deposition rate (linked to many sources of man-made pollution), and time of wetting, expressed as number of days per year. Chloride and sulphate deposition expressed in mg/m^2/day are classed in categories S1/S2/S3 and P0/P1/P2/P3/P4, respectively. The most recent version of ISO 9223 estimates the time of wetting from the average annual temperature and relative humidity, which are readily obtained from global climate monitoring systems (Chisholm et al., 2016-05). The global map in Fig. 1.40 shows that there are few land regions with the highest, T5 time-of-wetness classification.

Table 1.9 Usual classification of atmosphere corrosivity (CIGRE WG B2.69, 2021-06)

High	Medium	Low
Sea distance: 0–15 km	**Sea distance**: 15–50 km	**Sea distance**: >50 km
Rainfall: 0–800 mm	**Rainfall**: 800–1200 mm	**Rainfall**: >1200 mm
Relative humidity: >75%	**Relative humidity**: 70–75%	**Relative humidity**: <70%
Annual average temperature: >27 °C	**Annual average temperature**: 24–27 °C	**Annual average temperature**: <24 °C
Vegetable cover: Low vegetation, grass, desert	**Vegetable cover**: Bush bushes, medium vegetation	**Vegetable cover**: Thick forest, high vegetation

Table 1.10 Range of overhead line component life expectancy, according to the atmospheres corrosivity, adapted from (CIGRE WG B2.69, 2021-06)

Atmospheric corrosivity	High		Medium		Low	
Life expectancy of metal components (yr.)	9–11	15	20–22	25–27	35	>46

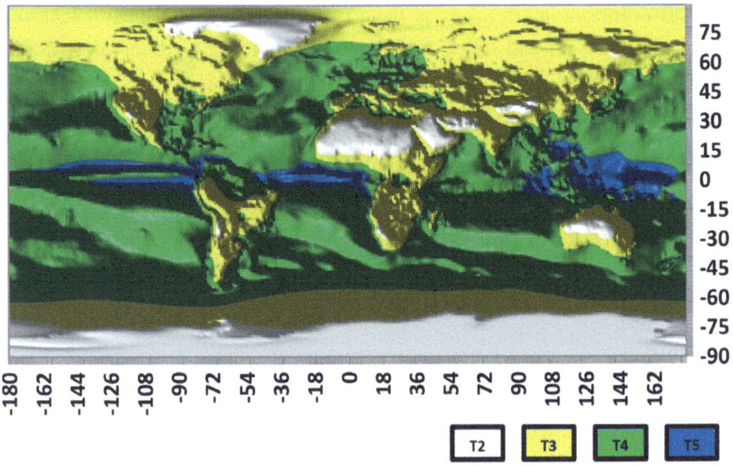

Fig. 1.40 Global map for time of wetness (h per year) in ISO 9223, estimated from average annual temperature and relative humidity, from NASA satellite data, July 1983–June 2005 (Chisholm et al., 2016-05)

1.6.9.5 *Classification: State of Corrosion of Overhead Line Components*

Evaluation of the corrosion condition of a component aims to determine—with an acceptable margin of uncertainty—what the residual life of the component is, and when to schedule preventive maintenance. The evaluation must describe the physical characteristics of the corrosive attack, as well as the percentage of loss of functionality or wear of the element.

There are many ways to evaluate the corrosion condition of the metal components. In South America, for example, several companies have successfully implemented a practice that is part of a methodology called Comprehensive Corrosion Management (CCM), in which it uses photographic patterns with four levels of corrosion damage for each of the components. This has allowed these companies to reliably operate their transmission lines in areas with determined corrosivity.

The procedure fulfils two fundamental objectives:

1. Ease of understanding for maintenance personnel who perform field inspections and objectively report the condition of components affected by corrosion.

2. Clearly describes the level of corrosion of the components inspected, allowing utilities to assess the risk of failure of their assets affected by corrosion and to plan necessary corrective and/or preventive actions.

Table 1.11 summarizes of photographic patterns used by field experience to assess the corrosion condition of the structural elements of towers. These photographic patterns are followed by brief text descriptions of the main characteristics that are present in each grade of corrosion.

Grade 0

Superficial white corrosion of the zinc coating, with no effect on the mechanical integrity of the asset.

Grade 1

The corrosion is superficial; zinc coating could show a general dark brown colour, but there is no corrosion of steel. Generalized corrosion can be present on two of the four faces of the profile. Due to the oxidation of the zinc coating and atmospheric pollutants, the corrosion products appear beige in colour (coffee with milk). Iron oxides can also be presented as small reddish-brown dots. Generally, the affected faces of the profiles are the surfaces perpendicular to the direction of the prevailing winds. The other two faces can maintain the characteristic grey colour of zinc because is not hit by winds with pollutants.

Grade 2

These profiles can present a general reddish-brown colour on the surfaces of the most affected faces; they can also show corrosion products in a laminar form (exfoliation); some of the faces of the profiles or part of their surface may show little or no damage. The main characteristic of this degree of corrosion is the wear of the steel that can occur in the entire profile, also partially somewhere on its surface. Steel wear can be up to 20% of the original profile thickness; oxides in sheet form are evidence of steel wear.

Profiles can have a general reddish-brown colour on most of the affected surfaces, and corrosion products in laminar form (exfoliation) may also be present. Some of the profile faces or parts of the face surfaces may have little or no damage. Main characteristic is that steel corrosion can occur in the entire profile, also partially on the surface. Corrosion extent can represent up to 20% of the original profile thickness; oxides in laminar form testify to the wear of the steel.

Grade 3

The physical characteristics of the corrosion products of tower profiles with this degree of damage are very similar to Grade 2. However, the fundamental difference is that the percentage of corrosion of the steel is greater than 20% of the original thickness of the profile. Sometimes, the corrosive processes are so severe that they can perforate or break the profiles.

Table 1.11 Photographic patterns summary table of the corrosion degrees in structures (CIGRE WG B2.69, 2021-06)

Grade 0			
Grade 1			
Grade 2			
Grade 3	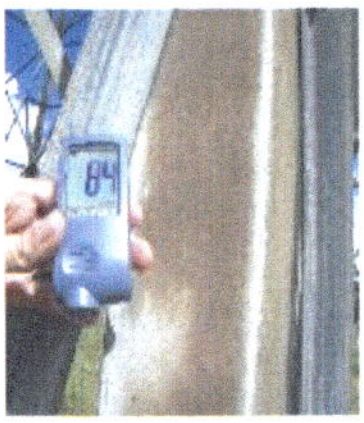		
DEFECT CONDITION			

Defect Condition

This is the most critical condition that can be found in the structural elements of towers affected by corrosion. While similar visually to Grade 3, the difference is that this condition only applies to the main elements of the towers. That is, for elements with a high mechanical load that, when broken, can cause the collapse of towers or the interruption of power transport service. The main characteristic of the defect condition is that the profiles are in an extremely critical condition and their breakage is imminent.

1.6.9.6 *Procedures to Prevent or Mitigate Corrosion of the Components of Overhead Lines*

Imprecise knowledge of corrosion phenomena in many utility companies means that sometimes the failure of assets due to corrosion is understood as a normal and unavoidable situation, leading to ineffective maintenance activities and sometimes to very high associated costs.

Construction is the most critical stage of the asset life cycle in areas with high level of corrosivity, and especially in the elaboration of the project CAPEX and OPEX. It is therefore necessary to know the characteristics of corrosivity of the areas in which the components of lines and substations will be installed. With this knowledge special designs are defined, specific materials and characteristics are determined, maintenance strategies are considered, as well as the actual intervals for the replacement of equipment damaged by corrosion. Purchasing departments also play a decisive role at this stage, since the base of the reference values used for the construction of project budgets depends largely on it. The design and construction stage are also a critical instance where lack of caution or knowledge about the corrosion phenomenon can generate future problems in the components. The care and contributions made in the assembly of projects largely determine the good performance of assets over time. The importance of the maintenance stage in the asset life cycle is almost on par with the construction stage and is where the level of knowledge of the corrosion phenomena becomes more relevant. A wrong field or a misinterpretation of the information received can lead to unscheduled emergency maintenance with significant costs or even a catastrophic failure in the system.

With the renovation and final disposal stage, the asset life cycle is closed and the design and construction phases are validated. The final stage of the life cycle of assets must be reached with a plan foreseen from the engineering phase, and it must be visualized and guaranteed that any asset or component thereof is disposed of properly and in an environmentally friendly manner at the end of its useful life.

1.6.10 Man-Made and Natural Pollution as $PM_{2.5}$ Aerosol Density

In 2005, the use of satellite data for monitoring dust was validated by comparing results from several sources. In the previous decade, researchers such as van Donkelaar et al. (2011, 2016, 2019) calibrated satellite measurements of the optical properties of air against reference air pollution measurements at ground level. The ground-truth data combined models of the relation of aerosol density to wind speed over the oceans, and measurements of man-made pollution in many cities around the world. Over the observation period 1998–2018, satellite estimates of fine particulate matter with size <2.5 μm ($PM_{2.5}$) matched surface observations at almost

4000 sites, mainly in China, Europe, and North America, with a Pearson linear regression coefficient of $R^2 = 0.81$, slope of 0.90 and negligible offset (Hammer et al., 2020), suggesting that predictions for remote regions without $PM_{2.5}$ sensors can be used to estimate local pollution levels.

Scientific data, averaged by year and separated into total $PM_{2.5}$, SO_4^{2+}, NO_3^-, NH_4^+, organic matter (OM), black carbon (BC), inorganic dust, and sea salt (SS) are found in open websites such as Atmospheric Composition Analysis Group (2020). A sample dataset with 0.01° × 0.01° latitude/longitude resolution has about 3.4 GB of data and can be plotted with open-source viewers such as *Panoply* as described in Chisholm et al. (2019-10-23). The average $PM_{2.5}$ density for 2018 in Fig. 1.41 shows, for example, increased levels of aerosol in northern British Columbia, on the west coast of Canada. This was associated with significant forest fire activity from the "worst fire season on record" with 1.35 million hectares burned.

1.6.11 Pollution Deposition Velocity (relating $PM_{2.5}$ to Rate of Increase of SDD)

In addition to direct correlation between $PM_{2.5}$ and electrical conductivity of precipitation, there is an environmental link between this parameter and the rate of accumulation of pollution on structure and insulator surfaces. The deposition flux is the product of a vertical "deposition velocity" V_d and the ambient concentration of

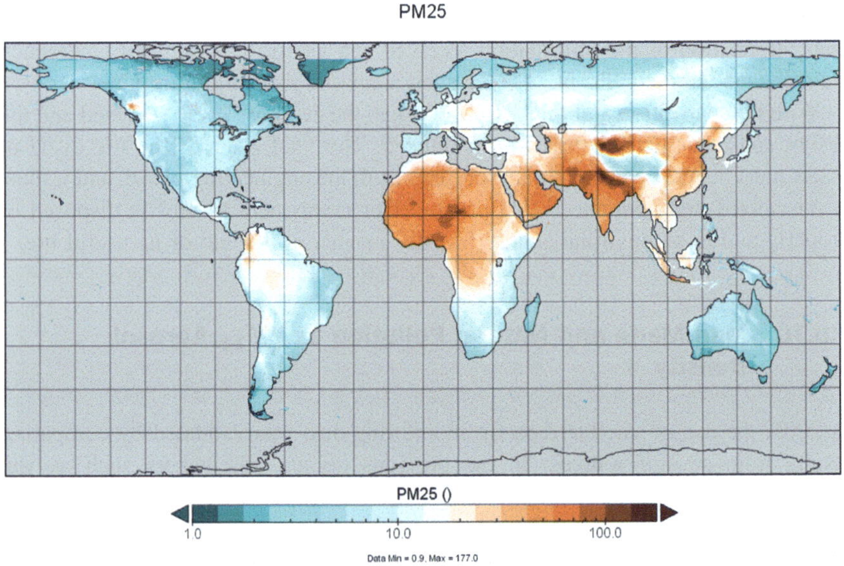

Fig. 1.41 Annual average density of aerosol particulate matter $PM_{2.5}$, 2018, units μg/m³. Data from Atmospheric Composition Analysis Group (2020)

the pollution. Deposition velocity tracks particle settling velocity when the particle size is greater than about 1 μm, so V_d increases from 0.03 cm/s for 2.5 μm d_p to 50 cm/s for d_p = 100 μm, considering low wind speeds of 0.1–0.3 m/s. V_d scales with wind speed as well, so establishing the link between $PM_{2.5}$ and rate of increase of pollution on a surface requires integrating the wind exposure over the same observation period. Natural rain washing will usually reset the surface accumulation process, but one important aspect of winter conditions is that, as described in Fig. 1.39, winter is a time when there can be extended periods without rain.

Studies of deposition velocity focus mainly on pollution accumulation on plants, tree leaves and water surfaces rather than smooth surfaces such as insulators.

1.6.12 Annual and Winter Season Lightning Parameters

A simple criterion defines "winter lightning" as lightning that occurs when the temperature is ≤ -10 °C at the 700 hPa level, 3 km above mean sea level. When the temperature exceeds –10 °C at this height, more typical charge separation of summer lightning, with positive charge above negative charge, is observed. Figure 1.42 shows the regions with relatively high winter lightning stroke density, but this does not generally exceed 0.1 strokes/km^2 per year.

What distinguishes winter lightning from summer lightning is the high fraction of positive flashes, which sometimes deliver exceptionally high levels of charge. CIGRE TB 578 (CIGRE WG C4.409, 2014-06) discusses the important effects of winter lightning on wind turbine blades. The median level of charge for a first negative return stroke is $Q_{\text{-first}} \approx 5$ C, and the level for a median positive stroke is $Q_+ \approx 85$ C. Records at three wind turbine sites in Japan suggest winter lightning strokes have Q > 300 C combining data from positive and negative polarities.

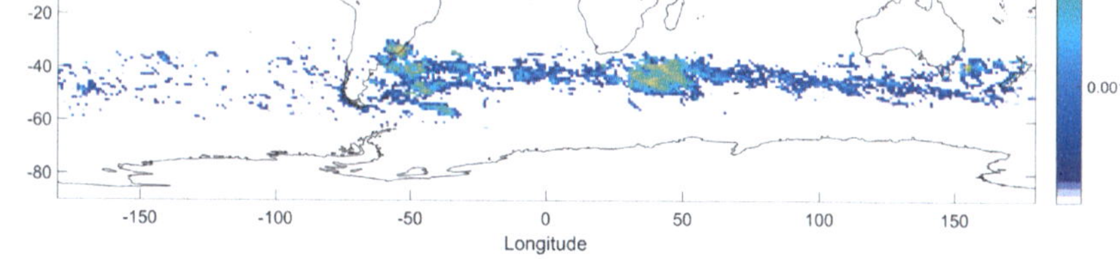

Fig. 1.42 Global distribution if winter lighting stroke density (strokes/(km^2 yr)) for the period 2009–2013 (Montanyà et al., 2016)

Energetic winter lightning storms in the Mediterranean Sea and the North Atlantic Ocean are also observed in Fig. 1.42.

1.7 Anticipated Effects of Climate Change on Winter Weather

The role of rising atmospheric CO_2 concentration, from 340 to 410 ppm over the period 1980–2020, tracks the global land and ocean temperature anomaly which rose from 0.1 to 0.9 K in the same period. The incidence of forest fires in USA and Canada has become highly variable. For example, in British Columbia, 4 million hectares were burned in the entire period 2008–2020, but 63% of the area was consumed in two years (2017 and 2018). Figure 1.43 shows a similar trend to large year-to-year variations in forest fire activity in the USA. Many utilities are sensitive to the temporary bankruptcy of Pacific Gas & Electric starting in 2019 because of forest fire liability. The positive feedback of forest fires producing CO_2 that leads to higher temperatures, increased lightning activity and drought, and thus more forest fires is also appreciated.

Like forest fire activity, atmospheric icing has a great variability. Figure 1.44 shows annual maxima (and running 5-year mean values) from the longest homogeneous time series of ice measurements in the world, on the top of Studnice Mountain (800 m above sea level) in the Czech Republic. This graph from CIGRE TF 22.06.01 (2001) represents continuous measurements from 1940 to 1999.

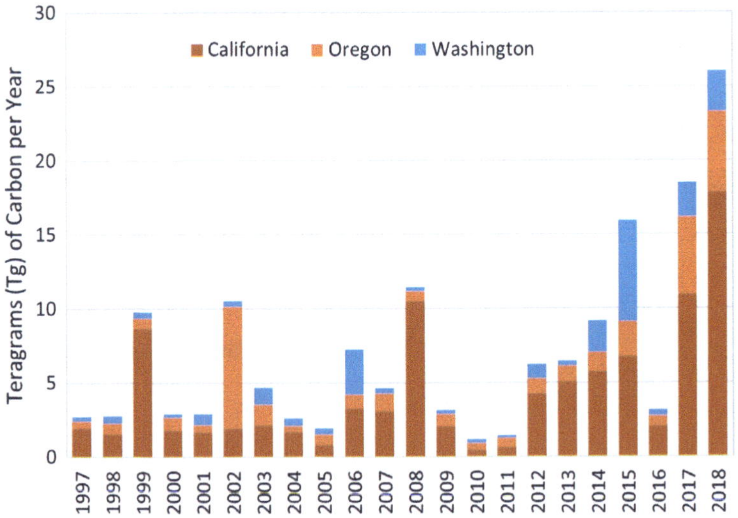

Fig. 1.43 Carbon emissions from fire, 1997–2018. *Data source* Global Fire Emissions Database

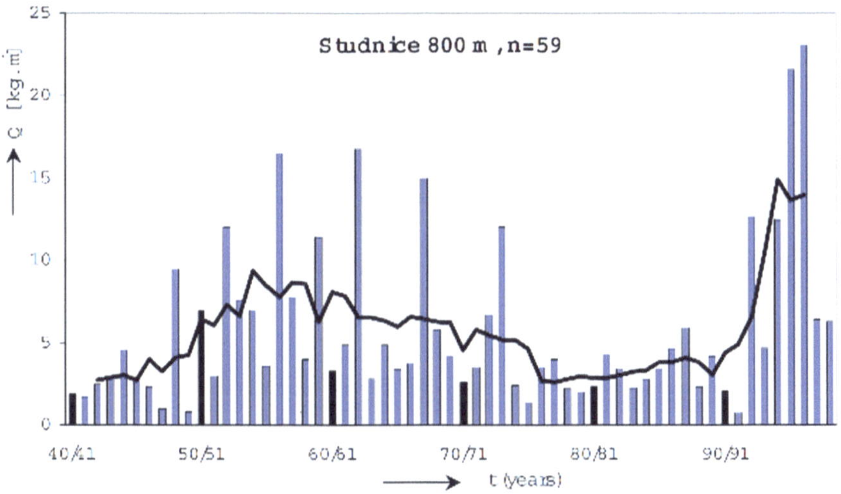

Fig. 1.44 Ice load measurements from Studnice 1940–1999 (CIGRE TF 22.06.01, 2001)

Figure 1.44 clearly shows that the design ice loads based on data from the 1950 to 1960s would be quite different from those selected from the 1970s and 1980s. The most important conclusion from Fig. 1.44 is that it is more important than ever to monitor ice accretions and ice loads for power utilities in all countries where damage cost due to ice loads is high. Furthermore, it is interesting to notice the increase in five-year average ice load the last decade, 1990–2000. It is a reasonable hypothesis that this dramatic increase of ice accretion is related to the parallel increase of atmospheric temperature in the same decade. This underlines the importance of careful awareness regarding ice accretions in the future, especially in the light of possible global warming which will lead to higher atmospheric humidity. The predictions of climate change, covering the expected end of life of new transmission lines today, are illustrated by comparing Fig. 1.35 (1980–2016) with Fig. 1.45, the future climate map suggested for 2071–2100, when today's new transmission lines should still be in service.

1.8 Concluding Remarks

Several major icing events were reviewed. These examples indicate the extent of damage that has occurred in the past during severe icing events, and which have the potential of reoccurring in the future. In addition to these events, there is a range of icing events that occurs more frequently causing operational problems and damage to overhead line conductors. These events took place in different environments and countries from Scandinavia to Southern Africa. Some problems associated with

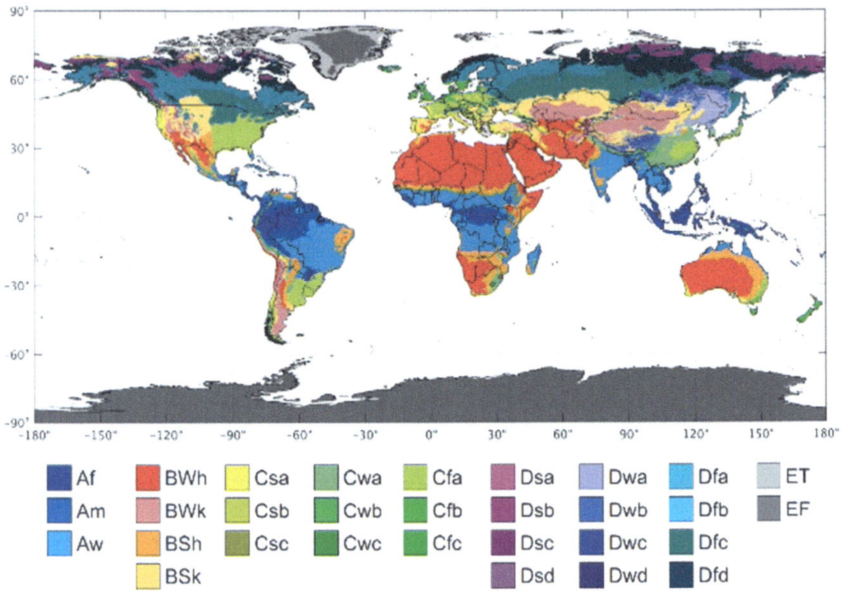

Fig. 1.45 Köppen–Geiger classification of future world climates (2071–2100) (Beck et al., 2018-10)

such events, both from mechanical and electrical point of view, are not yet fully understood or resolved. Significant progress has been made recently in understanding different anti-icing and de-icing techniques and their pros and cons, but all the existing methods have their own limitations. Limited trials of these techniques have also been done in the field, but these trials are not sufficient to indicate whether such techniques can be effective in minimizing the damage or not.

Regions that experience severe winter weather most years have, through trial and error, taught their overhead line engineers to specify designs that are strong enough to withstand most conditions, and resilient enough to avoid cascade failures in the eventuality of a tower collapse or broken conductor. There are niche reliability problems in winter weather that have been addressed by CIGRE in considerable detail. For example, the 180 reports of conductor galloping in TB 322 (CIGRE TF B2.11.06, 2007-06) complement the photographs and descriptions of tower and conductor problems provided in this chapter. Winter pollution issues include long-duration exposure to pollution flux with no washing during long periods without rain. This leads to rethinking of site pollution severity estimates, especially for regions that have fog in the early months of spring. Local pollution levels also affect the electrical conductivity of accreted ice and snow, leading to occasional widespread disasters from multiple line-to-ground flashovers in a short period.

It would be convenient for some utilities to rely on the estimated effects of climate change in Fig. 1.45 to abandon their concerns with winter conditions. However, there have been many unexpected winter disasters in countries not normally associated with ice and snow. Combined with an appreciation that there will be more uncertainty in the weather parameters, appropriate overhead line design methods should continue to consider appropriate winter conditions for all regions affected by icing events in Table 1.12.

Table 1.12 Impact of icing events from CIGRE WG B2.44 (2015)

Country	Line voltage (kV)	Tower/ pole supports	Conductors	Insulators	Type of events	Date	Cost
Canada	735	149	Earth wire failure		Glaze icing	Jan 1998	6.4 Billion CA$
	315	58	Earth wires failure				
	230	306	Earth wires failure				
	120	1091					
	49	1500					
Sweden	130-400			Flashover in polluted ice	Ice accretion	Feb 1999	
UK	132		Twin Lynx conductors failure		Wet snow	Dec1990	
	LV—66	1078	Small Cu, Wolf and Dingo failure (1000 in total)				
Catalonia (Spain)	25–66				Wet snow	Aug 2010	
	132	33					
	220		Earth wires failure				
Norway	22	Several			Wet snow	1961	
	420			Polluted rime ice	Ice	1980/90	
Iceland	11-33	760			Wet snow	1972	
	11-33	690			Wet snow	1974	
	11-33	700			Wet snow	1991	
	11-33	310			Wet snow	1995	
	132	205		Broken	Wet snow	Sept. 2012	
Slovenia	400	27			Glaze ice	Nov 1980	
	220	29			Glaze ice	Nov 1980	

(continued)

Table 1.12 (continued)

Country	Line voltage (kV)	Tower/ pole supports	Conductors	Insulators	Type of events	Date	Cost
	220	3			Wet snow	Nov. 1984	
	400	7	Several broken		Glaze ice	Jan1997	
	220	1				Jan1997	
	400		High sag		Glaze ice	Dec 1998	
	400		High sag		Glaze ice	Feb1999	
	110	3			Snow	Jan2009	
China	500	174	14		Ice	Jan/Feb 2008	3.5 Billion US$
	220	185	42				
	110	247	61				
Japan				Sea salt flashovers	Wet snow	Dec2005	
South Africa	275		Earth wires broken		Wet snow	Sep 2001	
	400	1					
	275/400	Multiple	Earth wires broken		Ice	Aug 2012	
	400			Icing on insulators		Winter 2002	
	400			Icing on insulators		Winter 2007	

2 Dimensioning for Winter Conditions in Overhead Line Design

Many CIGRE Working Groups have advised on the basic requirements and design principles for overhead power lines, including (CIGRE WG 22.06, 1995-06) on tower top geometry and (CIGRE TF B2.11.06, 2007-06) with specific advice for “galloping”. Tower top geometry makes provisions for galloping and clearances between conductors in midspan, as well as ensuring sufficient distance for live-line maintenance. All three provisions are more generous in areas with winter conditions.

The same meteorological data are ideally used for establishing structure mechanical strength, insulator swing angles, electrical characteristics and required clearances. CIGRE TB 48 proposes two cases when considering tower top geometry that considers swing of suspension insulators:

- Still air or moderate wind, applying to conductor positions during 99% of the operation period. Clearances should withstand lightning or switching surge impulse voltages with probability of 90%.
- High wind conditions. For a wind velocity with return period of 50 years, TB 48 recommends design with a probability of flashover of AC voltage of 1% per year.

Wind acting at right angle to the line direction causes the conductors and suspension insulators to swing. If wind acts steadily along the line and over a long period of time, compared to the skipping rope or “galloping” fundamental frequency, the calculation of swing angle is simple. Ice accretion adds to the effective cross-sectional area of the conductor, adds mass, and changes the drag coefficient, all introducing some complications. However, the wind does not act in a uniform way, and peak wind velocity does not cause swing angles that are statically equivalent to a locally observed peak wind velocity.

The mass of the conductor must be accelerated first and moved into a displaced position, before wind force is transmitted to a support. Peak wind velocity of short duration thus does not affect either the swing angle or the force acting on the tower. It is the mean value of wind velocity, acting over a sufficiently long time, that is

M. Farzaneh and W. A. Chisholm, *Techniques for Protecting Overhead Lines in Winter Conditions*, Compact Studies, https://doi.org/10.1007/978-3-030-87455-1_2

relevant to the swing angle. CIGRE TB 48 also describes how the ratio of wind span to weight span affects the results.

The probability of exceeding a wind velocity is described by a Gumbel distribution with mean value of yearly maximum wind velocity and standard deviation, considering many years of observations. Corrections for conductor height above ground and terrain roughness are introduced.

Conductors can be characterized aerodynamically as bluff bodies. The airflow in proximity of their surface separates and forms vortices shed downstream in their wake. Conductors are flexible structures with almost no internal or self-damping and will become prone to aeroelastic instabilities consecutive to this flow separation. The aeroelastic instabilities of major concern are galloping and aeolian vibrations (CIGRE TF B2.11.06, 2007-06). Both instabilities are important when overhead lines are operated in winter conditions.

2.1 Tower Head Clearances for Galloping

Galloping is a spectacular large-amplitude oscillation of single or bundled conductors, caused by the action of wind on ice or wet snow accretion. The conductor motion is characterized by amplitudes that approach or exceed the conductor sag. The frequency of oscillation may reach 3 Hz, depending on the amplitude and the number of loops within the span. (CIGRE TF B2.11.06, 2007-06).

Galloping has caused various kinds of structural damage in overhead lines. Some types of damage result directly from the large forces that galloping waves or loops apply to supports. For example, crossarms have failed on both wooden and metal structures. Conductor ties on pin-type insulators have been broken, and support hardware has failed. In other cases, cotter pins have been damaged, and insulator strings have consequently uncoupled.

So far, galloping has only damaged transmission line surge arresters mounted on the conductor. There is a possibility that the pendulum frequency of the arrester can match a galloping frequency. Jumpers at dead-end towers have been thrown up onto crossarms, and the long ground connections of non-gapped line arresters (NGLA) may also cause electrical problems if they break during galloping.

Dynamic shock loading occurs when a steep-fronted galloping wave is reflected at a tower. Repeated reflections can cause aeolian vibration damper weights to "droop", and in extreme cases, damper messenger cables can break, dropping the damper weights. Dynamic loads have also caused loosening of crossarm and bracing bolts in tower structures and loosening of wood poles in the ground.

There are three strategies to protect overhead lines from galloping in winter conditions (CIGRE TF B2.11.06, 2007-06):

- Prevent the formation of ice or remove it, using systems described in Chaps. 4 and 5.
- Interfere with the galloping mechanism, for example, by changing the torsional frequency ("detuning") under icing conditions so that it does not match any galloping frequency.
- Make the lines tolerant of galloping by providing increased phase clearances or controlling the modes of galloping with interphase ties along the spans.

Each span has a different galloping frequency. There is some coupling of energy at lower frequencies with single-loop galloping as shown in the mode shapes in Fig. 2.1.

The cross section of the conductor motion during galloping traces a somewhat elliptical path as shown with numerical simulation in Fig. 2.2. This figure is viewed along the span from midspan. In this case, the peak-to-peak displacement is about 10.5 m vertical and 3 m horizontal. Staggering the phases, on a double circuit line using longer crossarms for the middle phase than top or bottom phase, can provide adequate electrical clearance between conductor trajectories and thus protect the overhead lines from phase-to-phase faults during galloping.

The orbit shape of a galloping eight-conductor bundle traced a figure-8 pattern with horizontal displacement of 4.9 m and vertical displacement of 2.6 m over a 15-s interval in wet snow conditions with a wind speed of 17 m/s.

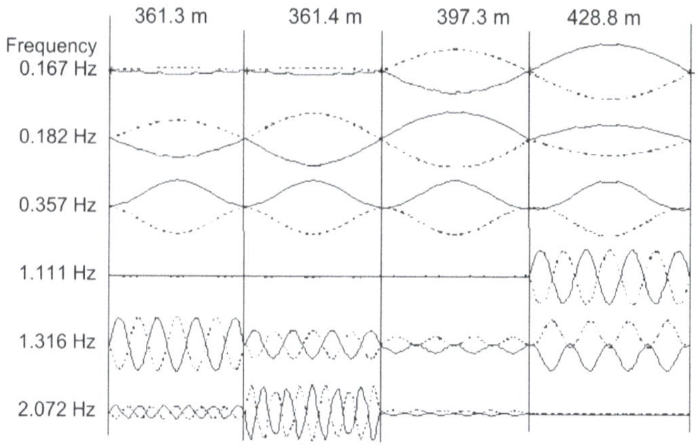

Fig. 2.1 First six possible eigenmode shapes and frequencies for four-span galloping test line (CIGRE TF B2.11.06, 2007-06)

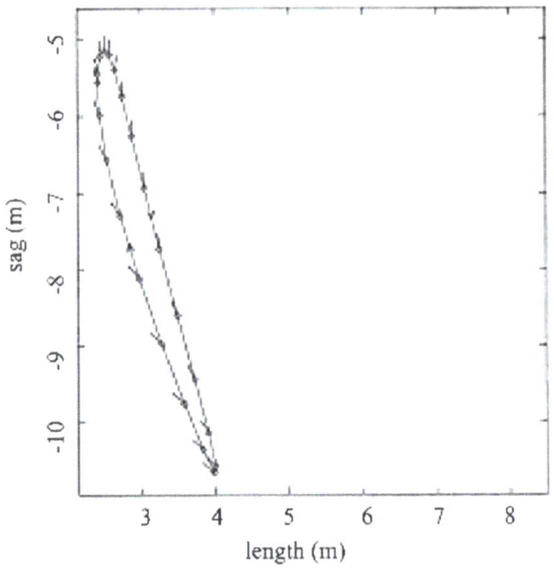

Fig. 2.2 Typical galloping ellipse in vertical plane at midspan, adapted from CIGRE TF B2.11.06 (2007-06)

2.2 Tower Head Clearances for Passive Mitigation of Ice Accretion and Sleet Jump

In some countries, where only a relatively small number or sections of double circuit lines are affected by atmospheric icing, several passive methods are used to minimize the impact of icing and shedding events, including

- Installation of longer middle crossarms
- Increases in spacing of crossarms or circuits
- Installation of interphase spacers
- Modification of phase subconductor bundling, for example, decreasing the bundle size to mitigate rime icing in river valleys
- Installation of separate structures for each phase
- Rearrangement of the phasing configuration between circuits so that some phase-to-phase contacts occur between conductors at nearly the same potential.

In some cases, a combination of these methods has been used. Some of these options may involve increased right-of-way (servitude) width, tower modification and strengthening, special operational guidelines, increased loss, or other disadvantages.

Taking an example of passive icing mitigation methods from New Zealand, the biggest effect of ice accretion is

- Increased conductor sag while the ice is on the conductor
- Occasional ice dropping (sleet jump) resulting in unwanted phase clashing causing outages.

An obvious method for double circuit lines, and most appropriately initiated before the line is built, involves creating a horizontal separation between the phases within a circuit by using longer middle crossarms. This allows uneven sags of iced/un-iced conductors in the phases to be accommodated. This is not feasible in all countries. In Austria, for example, overhead line regulations require a horizontal rather than vertical circuit configuration to provide mitigation of sag and sleet jump effects.

A similar method has also been used on existing lines where there was provision to use a redundant ground wire peak or similarly add an upper section to the tower to increase the separation between the top and middle crossarms to again accommodate uneven conductor sag profiles. The towers would either need to have additional capacity or utilization available, or sufficiently strengthened to accommodate increasing the tower loading.

2.3 Midspan Clearances for Passive Mitigation of Ice Accretion and Sleet Jump

Midspan clearance management can achieve passive anti-icing with the use of interphase spacers. These are articulated, stiff rod insulators which reduce the potential effect of ice build-up by maintaining conductor separation both during icing events and during ice unloading. Interphase spacers are usually composite long rod insulators of suitable length to match the dimensions of the phases and electrically rated to accommodate the phase-to-phase voltage. The disadvantage with this approach is that it introduces another component to the line, with its own possible failure contributions such as conductor fretting under the many additional clamps involved. Interphase spacers also require inspection and maintenance of the insulators themselves and will only provide mitigation if installed in the spans where icing is most likely to occur. However, interphase spacers do ensure adequate conductor separation. In 2010, there were about 500 interphase spacers installed on four affected lines in New Zealand. A further 40 units have been installed on specific spans to mitigate against similar clearance issues related to high winds. Many views of overhead lines in other regions, such as the 115 kV circuit in Fig. 2.3, also show applications of midspan spacers from roadways. This is unfortunate because, in winter conditions, these roads are maintained with salt, with

Fig. 2.3 115-kV transmission line with interphase spacers, viewed from six-lane expressway, Kitchener, Ontario, Canada 43.42°N, 80.42°W

application rates of 65 kg per km of lane with every pass. Overhead lines next to these roads are thus exposed to a marine environment, affecting winter corrosion rate and insulator performance.

Interphase spacers can also mitigate some effects of galloping by introducing nodes that raise the resonant frequency and reduce amplitude (CIGRE TF B2.11.06, 2007-06).

2.4 Conductor Sag and Clearance to Ground

Areas with winter conditions may have a wide range of conductor temperature, from perhaps an ambient of -40 °C and no electrical load up to a summer rating more than 100 °C in an ambient of +40 °C. This means that there is a wide range of conductor tension, as the aluminium and steel of the phase conductors expand and contract at rates of 0.23%/K for (61% IACS) aluminium and 0.115%/K for galvanized steel, where 1 K is a temperature difference of 1 C°. At low temperature and high tension, conductors are more susceptible to aeolian vibration.

The temperature coefficient for resistivity of aluminium is normally taken as 0.403%/K, and nonlinearity of this coefficient below 0 °C leads to roughly 2% error in conductor resistance estimates at –50 °C, which can be neglected in thermal rating calculations.

Wind tunnel tests of ACSR heat transfer at –10 °C suggest that the power-law expressions used for convective heat transfer, with exponent of 0.6, stated to be valid at ambient above 16 C° (IEEE Standard 783-2012), understate the relation between Nusselt number Nu and Reynolds number Re for most conductors at higher wind speeds. Indeed, a linear relation between Nu and Re with different slope for each of four ACSR types (Condor 795 kcmil, Drake 795 kcmil, Carillon 1029 kcmil, and Géant 2152 kcmil) was reported in (Péter, 2006). These observations of heat transfer coefficient at –10 °C, plotted in Fig. 2.4 as a relationship between Nusselt and Reynolds numbers, are lower than those predicted by the CIGRE thermal rating model in TB 207 (CIGRE WG 22.12, 2002).

Most of the interest in thermal rating calculations is for weather conditions with low wind speed. This range of low Reynolds number is shown in Fig. 2.4 as the region where IEEE Std. 738 parameter $q_{c1} > q_{c2}$. The trends suggests that additional wind tunnel heat transfer test data, in the low wind speed range below 2.5 m/s, are needed in freezing conditions before utilities can make confident use of ambient adjusted rating values in winter conditions.

The simple heat balance used for thermal rating calculations is expanded to include ice accretion effects in Sect. 4.2.

In addition to uncertainty about conductor temperature related to Nu estimates, and the clearance-temperature relation at extreme temperatures—high or low (CIGRE WG B2.55, 2019), safe operating clearances from conductors to ground in winter conditions should consider the typical and extreme levels of snow depth. In some regions, the only physical access to portions of transmission lines for routine maintenance is over frozen muskeg or bog so arguments of inaccessibility may not hold. For lines that are accessible by snowmobile, tracked work platform or the public, utility policy may consider vertical clearance above snow as the sum of a fixed physical buffer along with an electrical buffer related to the power system

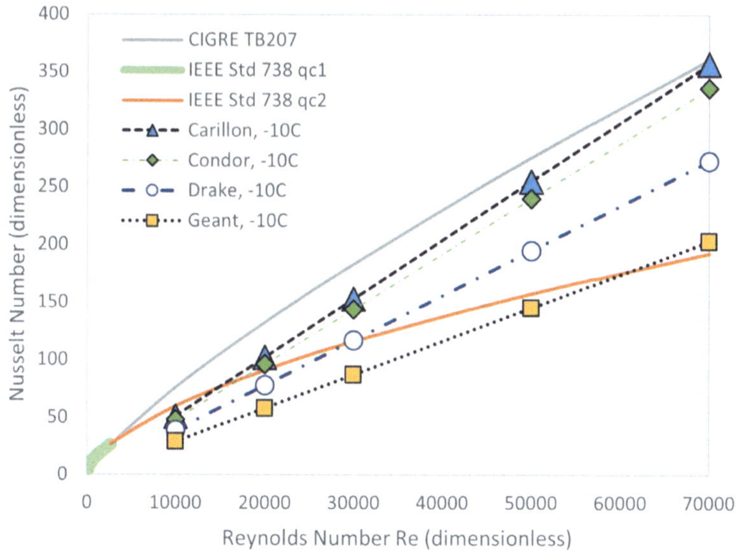

Fig. 2.4 Predicted (CIGRE WG 22.12, 2002; IEEE PES, January 2013) and observed (Péter, 2006) relations between Nusselt and Reynolds numbers for cross flow on dry ACSR conductors at –10 °C

voltage. Typical physical buffers for overhead line clearance in winter conditions may consider a 1.8 m tall person vertically extending a ski pole (2.4 m reach + 1.5 m pole length), or a legal vehicle height of 4 m above the snow surface.

2.5 Electrical Clearances for Insulators

2.5.1 Stress Per Metre of Dry Arc Distance on Vertical Suspension Insulators

Full bridging of vertical suspension insulators by ice, followed by a melting phase, depresses the electrical strength to an extent that leads to line-voltage flashovers. IEEE Standards 1820 and 1863 recommend that electrical stress is reduced below 100 kV ac rms per metre of dry arc distance, to a level as low as 60 kV/m, in regions where the electrical conductivity of the ice is high. This is a modest request for HV transmission lines; for example, fourteen standard 146-mm discs on a 230-kV suspension string give a stress of 68 kV/m considering 5% overvoltage. However, for EHV applications, V-strings are needed for adequate electrical performance in icing conditions.

2.5.2 Stress Per Metre of Leakage Distance

Consensus from the use of IEC 60815 methods is that, for regions with a long period of winter without rain, ending in a season having days with fog, the site pollution severity (SPS) classification should be increased by one level, for example, from light to medium or heavy to very heavy. This reclassification is based on the action of fog as a form of non-soluble deposit, usually greater than 1 mg/cm^2, and thus capable of stabilizing any pre-existing pollution on insulator surfaces.

2.5.3 Stress Per Metre of Live-Line Tool Length

Flashovers on live-line tools in winter conditions have shown that voltage stress should be maintained below 70–75 kV ac rms per metre of tool length with normal handling of clean tools. It is feasible to increase this stress to 100–105 kV/m if special care is taken to ensure the handling of tools in freezing conditions does not smear a thin coating of pollution and ice, transferred from workers' leather glove covers, onto the clean tools.

2.6 Concluding Remarks

Procedures for providing adequate mechanical strength under specific wind and ice loading cases are mandated by standards in many countries. The physical separation of conductors is managed at the tower head and influenced by the type of insulator. Compact lines, stabilizing the horizontal motion of conductors through V-string or post insulators and reducing sag by using shorter span lengths, can reduce the amplitude of galloping activity.

Winter conditions affect several aspects of the line profile. Regions with cold temperature must tolerate high conductor tensions at low temperature and still maintain sufficient tension for safe vertical clearance to ground at the maximum conductor operating temperature. Ice storms cause permanent changes to the conductor, and these changes reduce the thermal rating of the line, expressed as a maximum operating temperature, after every occurrence. The use of joule de-icing methods may also overheat the conductors if not managed carefully. Broken-conductor scenarios cause large dynamic loads. A measure of flexibility in the line design can improve resilience by absorbing these shocks and preventing cascade failures of many towers in a row.

In a stringing section of suspension insulators between rigid dead-end positions, equations of state relate the average conductor temperature in a stringing section to clearances, tensions, sags, and insulator swings. Many utilities have successful operating experience with instrumented dynamic line rating (DLR) systems that measure any one of the mechanical states within a stringing section and infer the other variables, especially ground clearance, at every other point along the section. These systems implicitly assume that the weight of the conductor per unit length is fixed. Operating experience with instrumented DLR can lead to detection of ice accretion. Also, a thorough understanding of the heat balance that underlies all DLR systems is needed to follow the processes of ice prevention and ice melting using Joule (heating) methods.

Ice accretion on insulators weakens the electrical strength. Modest amounts of fog or ice accretion stabilize pre-existing pollution that accumulates after months of freezing weather. High levels of ice or snow accretion can bridge the dry arc distance, causing unique sets of flashover problems in the freeze–thaw transition process.

3 Systems for Monitoring and Predicting Ice Accretion and Shedding

Any operational de-icing system will always depend on a reliable system for monitoring and predicting icing events and their developments. Ice or snow build-up on overhead transmission lines is a problem which can occur over a wide geographical area in cold climates throughout a significant period of the year. Furthermore, it is often local in nature with some areas heavily affected, while adjacent regions are virtually ice-free. This makes direct measurement of icing phenomena on transmission lines very difficult, e.g., for planning of de-icing or other operational measures, as many sensors would be needed. Moreover, ice accretion often occurs in remote and mountainous areas where the supply of low voltage power and lack of telecommunications constitute obstacles to the installation of measuring systems.

Furthermore, using meteorological data from weather stations, radar and satellite observations have not proven to be reliable enough for predicting icing near the ground, especially for rime ice and wet snow (Fikke, et al., 2008). The main reasons for this are that the existence and amount of liquid water in cloud air (rime ice) or the liquid water ratio in snowflakes (wet snow) cannot be estimated directly from such measurements.

3.1 Meteorological Weather Forecasting Models

In recent years, the development of meteorological weather forecasting models of the atmosphere has reached a level of accuracy and details that makes it possible to estimate the amount of ice already accreted or to forecast the expected ice build-up on standard objects. Such a method is described in Nygaard (2009). The most important aspect with this technique is to insert local scale models into a global weather forecasting model by nesting. Since it is not possible to insert abrupt changes in model scale from several tens of kilometres (global models) to the order of hundreds of metres (local scale models) in one step because of stability in the

M. Farzaneh and W. A. Chisholm, *Techniques for Protecting Overhead Lines in Winter Conditions*, Compact Studies, https://doi.org/10.1007/978-3-030-87455-1_3

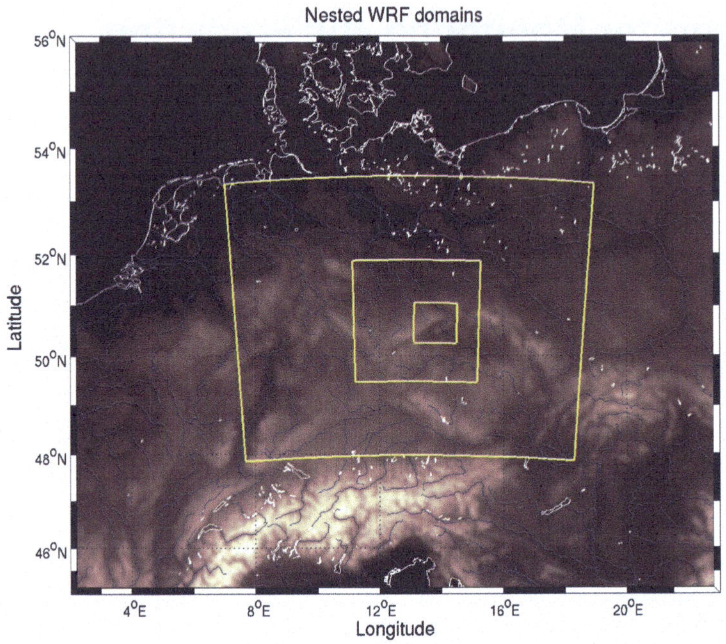

Fig. 3.1 Nested Weather Research and Forecasting Model domains for the Zinnwald test station in Germany. The three domains are indicated by the yellow squares. The grid spacing from outer to inner domain is 12.8 km, 3.2 km, and 0.8 km, respectively (Nygaard, 2009)

models, such transitions are made by nesting of intermediate models, as shown in Fig. 3.1. Here, the nested models have grid spacings of 12.8 km, 3.2 km, and 0.8 km, respectively. For each nested model, the ground surface characteristics are represented with corresponding resolution with the number of vertical layers adjusted accordingly.

Each model domain has its own set of equations for the dynamic and physical processes in the 3D atmosphere. These processes are again forced by the corresponding processes in the outer model at the boundary between the two domains.

For the inner model, any icing parameter can then be calculated for each grid point by applying any already established physical icing model, as for instance shown in CIGRE TF B2.16.03 (2006). Hence, a map of ice occurrence, or accumulated ice loads, may be produced for either a selection of points along one line, or in a 3D (map or Google Earth). An example of such a regional icing map from Norway is shown in Fig. 3.2.

Models of this type can be used for retrospective analyses of historical icing events or as operative forecasting of potential risks in an electrical network. Accordingly, it can be used to initiate any operational system for de-icing of

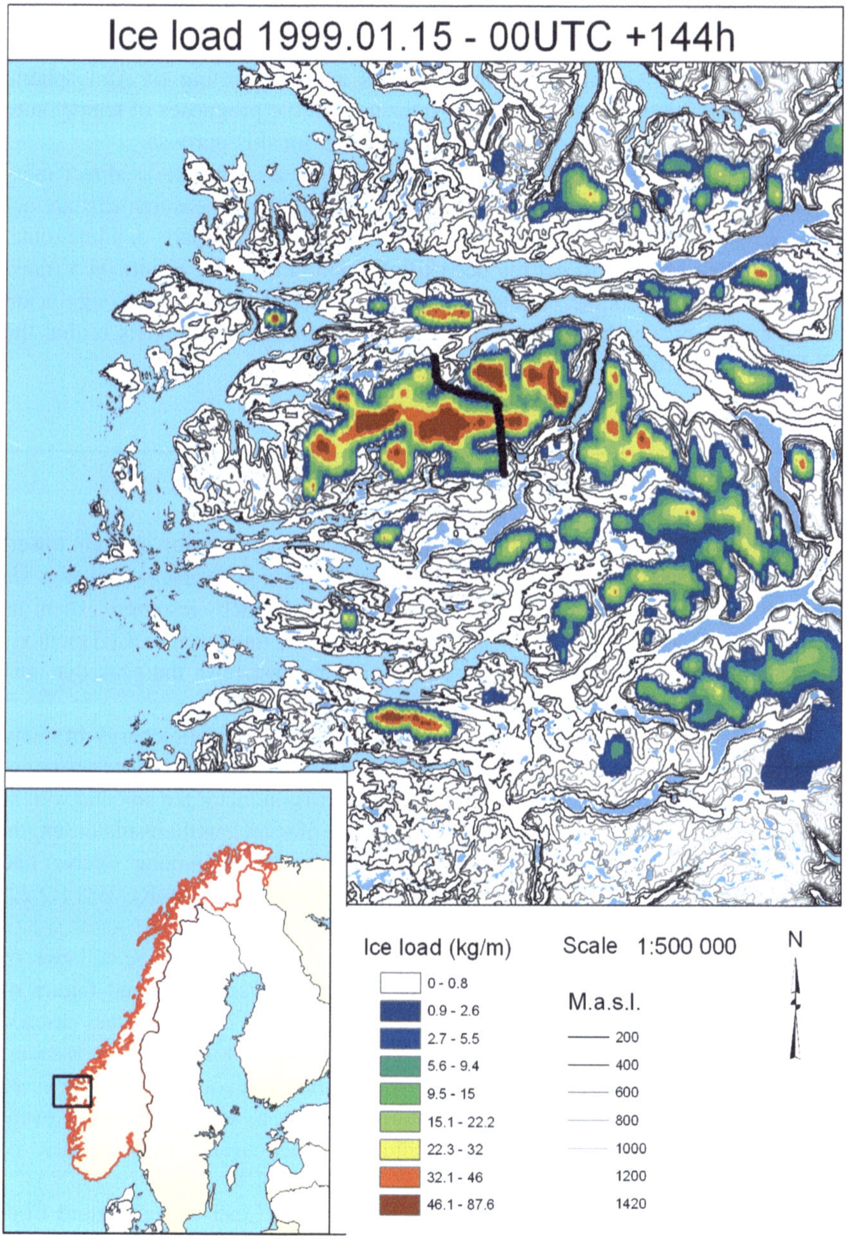

Fig. 3.2 Accumulated ice (colour shadings). The black line indicates a transmission line route, adapted from Nygaard et al. (2007–10)

overhead transmission lines. Such models have already been successfully used for rime ice (Nygaard, 2009). They have also shown promising results for wet snow during the COST Action 727 "Measurements and Forecasting of Atmospheric Icing" (WMO, 2009–12). However, for freezing rain the prognoses of temperature inversions near ground are not yet accurate enough for this purpose.

Local scale weather forecasting models cannot be as accurate as direct measurements. But it is also important to remember that reliable measurements are not always easy or even possible to perform for the most important parts of a line route. Therefore, the best way to get the needed information on icing events on a transmission line is to combine measurements and atmospheric models. A significant feature gained by combining measurements with atmospheric models is that the validity range of the measurements will be enhanced.

3.2 Measurements of Conductor Mechanical State

The mechanical state of overhead conductors—measured by tension at strain tower, sag or clearance within a span or inclination angle of a suspension insulator—is used for thermal rating as described in several CIGRE resources such as CIGRE WG B2.55 (2019). The "instrumented" dynamic line rating (DLR) methods monitor important variables, all inter-related, that also establish the peak dynamic load and residual static load on towers during icing failures.

Strain energy in the conductor is the primary source of kinetic energy in many overhead line mechanical failures. High mechanical tensions at low temperature lead to higher dynamic and static loads. In addition to conductor tension and weight per unit length, CIGRE TB 515 notes the influences of span length, insulator length, and number of suspension spans in the stringing section, among twelve line parameters affecting PDL and RSL due to conductor breakage (CIGRE WG B2.22, 2012-10).

Power line cascades are classical examples of failure by progressive collapse, of the "zipper type" where failure progression is due to overloading and failure of adjacent elements. CIGRE TB 515 (CIGRE WG B2.22, 2012-10) describes cascade failure cases in more detail, including "domino-type" longitudinal cascades, "instability type" buckling failures from torsional load and combined types. There is a distinction made between multiple failures from a widespread icing event (weakness) and progressive collapse failures, which result from a lack of robustness.

Present-day DLR sensors or future "smart structures" can be repurposed from DLR projects to improve structural resiliency in several ways. Broken wires can be detected by a sudden change of state anywhere within a stringing section. In the case of structure failures, all phases suffer a simultaneous and significant reduction in tension but may not break.

3.2.1 Dedicated Test Lines

Instrumentation of transmission lines for direct measurements of ice loadings on live lines would deliver optimum results on ice data. Such measurements would also give data for various unbalanced loads like transversal, longitudinal, and torsional loads. This is done by several utilities, mostly for research purposes. Instrumentation is most relevant to extremely exposed lines, and the results may be difficult to transfer to other lines or regions. The use of external expertise should be considered.

Due to the costs involved, only a few line sections in the world are equipped with the suitable instrumentation, and ice events on many other lines are missed. Such instrumentation is recommended for research purposes rather than for guiding power system operations.

Many dedicated test lines for mechanical shock tests use a helicopter load-release device in series with a tensioned conductor. Results from CIGRE TB 515 in Fig. 3.3 show two resonant frequencies, one at 3 Hz for the transmission tower and a slower response, with return time of t = 3 s, from the adjacent tower.

Transfer of mechanical shock along typical stringing sections, with many suspension towers terminated at each end with tension towers, is of significant interest with studies on ten-span test lines found in TB 515 (CIGRE WG B2.22, 2012-10). On dedicated icing test lines, data from single spans are used to study the level of ice accretion with greater precision, since this varies from span to span depending on line orientation to the wind as well as terrain.

3.2.2 Lines Equipped with Dynamic Line Monitoring

Frequent interrogation of real-time DLR monitoring systems, possibly exploiting built-in alarm functions, is recommended during actual or forecast icing activity. One rationale for this is illustrated using Fig. 3.3. It is feasible to calculate the sag of the conductor in mm as:

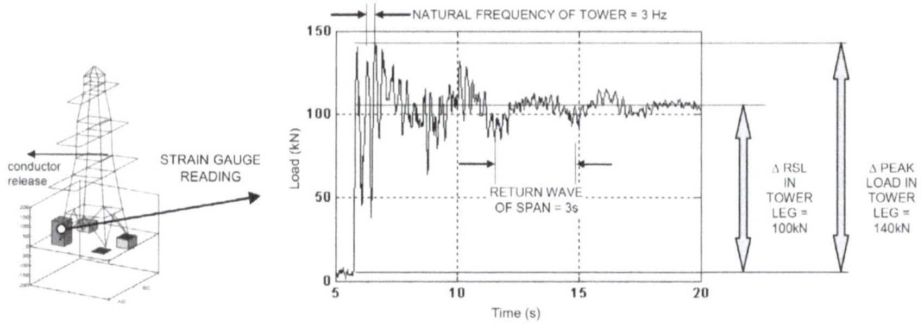

Fig. 3.3 Results of conductor tension release experiment showing peak dynamic load and residual static load (CIGRE WG B2.22, 2012-10)

$$D = 307t^2 \tag{3.1}$$

with the single return wave in Fig. 3.3 using the traditional return wave method. In this case, with t = 3 s, the sag is computed to be 2.76 m. This sag corresponds to a 200-m span of ACSR strung at 20% of rated tensile strength (RTS) using

$$\mathrm{Sag}_{\mathrm{ACSR}} = 0.0014(\mathrm{Span})^2/(\%\ \mathrm{RTS}) \tag{3.2}$$

Equations 3.1 and 3.2 illustrate the strong coupling among travel time of return wave, sag, tension, and conductor weight per unit length (in the 0.0014 factor of Eq. 3.2). This explains why monitoring any deviation in mechanical state with DLR equipment can be used to monitor small changes in conductor weight per unit length (ice accretion) averaged along an entire stringing section. DLR systems also respond to the large changes in state associated with broken conductors.

Some DLR sensors are more robust than others in icing conditions. DLR systems such as insulator inclination, video, LIDAR, or sonar, that inexpensively monitor two or more phases in a span, can be used to distinguish between a single broken conductor and a collapsed tower, provided the sensors remain in service during and after the line collapse. Three-axis inclinometers and load cells in suspension and dead-end insulators all provide some indications of the response of a span or a stringing section to accretion and release of ice if time synchronization among sensors is sufficiently precise to support travelling mechanical wave analysis of the observations are delivered in near real time. Operating experiences suggest that some DLR system components register fault conditions—frozen anemometers, obscured views—under ice and snow accretion. In the worst cases, loss of signal from a monitoring system can provide a binary indication of an icing problem in winter, in the same way that lightning damage to a tower-mounted monitoring system localizes the time and location of a severe flash in summer weather.

3.2.3 Dynamics of Ice Persistence and Release

Low-cost battery-operated wireless systems to monitor three-axis acceleration presently offer frequency response up to 1 kHz, which is sufficient to resolve mechanical travelling wave features such as those in Fig. 3.3. Adequate time synchronization among sensors can be maintained through Network Time Protocol (e.g. in IEEE Standard 2030.101-2018), Precision Time Protocol (IEEE 1588), or Global Positioning System (GPS) receivers. With time-synchronized sensors, displacements and/or accelerations at two or more points can be used to localize the source of mechanical disturbances. Projects to monitor time-synchronized conductor acceleration or displacement at multiple points along a line, in response to broken conductors, bird strikes or vandalism, also offer an interesting additional winter application to establish when ice drops off the spans.

3.2.4 Conductor Slip Loads

Suspension clamps with engineered slip loads are often considered when dimensioning towers and supports for winter conditions. When considering failure modes under severe winter conditions, the friction of the conductor surface affects the performance of several types of fittings:

- Compression dead-end clamps, mid-span joints, and jumper terminals
- Suspension clamps
- Bretelle clamps
- Spacers and spacer dampers.

The surface frictional properties influence the minimum or maximum slip loads achievable inside a conductor fitting. Achievable fitting slip loads can significantly influence the performance and lifetime of an overhead line. Selected properly, controlled slip can reduce peak forces and contribute resilience to an overall line design.

3.2.4.1 Controlled Slippage Suspension Clamps

There are three types of suspension clamps:

- Clamps with specified minimum and maximum slip load
- Clamps with only minimum slip load specified
- Controlled slippage clamps.

With clamps that only have the minimum slip load requirement, the tower and supports are often dimensioned to withstand conductor or shield wire breakage. When, in addition, a maximum slip load is specified that is done to protect the tower or support overload in case of large axial forces in the conductor. An example of tower member overloading during conductor slipping through the suspension clamp is given in Fig. 3.4.

In the case of controlled slippage clamps, even more control is required in case of conductor or shield wire breakage. The slip load force should not exceed a maximum value, but while the conductor is slipping through the clamp over a specified length, the axial force in the conductor is not to fall under a specified minimum. In addition to the minimum and maximum slip loads, it may be required to test the friction coefficient between the suspension clamp and the conductor. Figure 3.5 shows a schematic representation of the test set-up to establish this friction coefficient.

Fig. 3.4 Example of a collapsed tower top after shield wire breakage

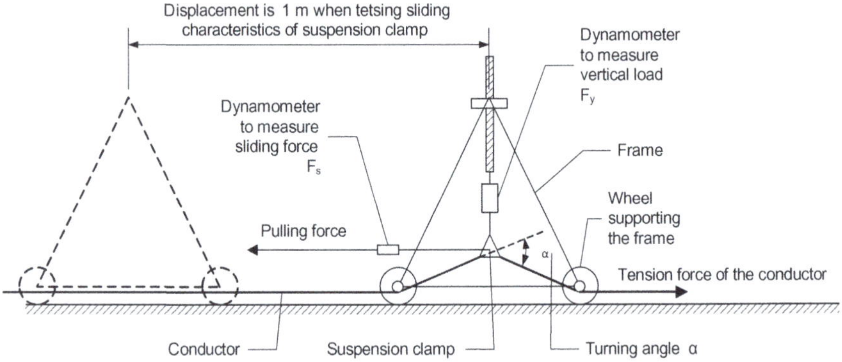

Fig. 3.5 Testing of controlled slippage clamps (IEC, 1997)

According to IEC 61284 (IEC, 1997) the controlled slippage clamp test is done as follows:

1. The conductor is resting in the suspension clamp without the keeper installed. The conductor departure angle α and the vertical force are agreed between the clamp manufacturer and the utility or line designer.
2. The conductor is tensioned on both sides, usually to 20% of the conductor rated tensile strength (RTS) or everyday stress (EDS).
3. The keeper clamping bolts are then tightened according to the manufacturer's installation instructions.

4. The conductor is then released on the side where the clamp puling force is to be applied. The clamp is then pulled, and the sliding force is measured over a length of one metre.
5. The friction coefficient is determined as a ratio of the maximum pulling force and the vertical load force.

3.2.4.2 Spacer Clamp Performance

Spacer clamp can be subject to significant axial or torsional loads. Clamp slipping has been noted as cause of damages on transmission lines, for example, as shown in Fig. 3.6. Spacers should be retested on conductor with changed surface properties, through surface treatment or ageing.

Testing methodology for spacer axial slip loads is described, for example, in IEC 61854 or other relevant standards and specifications.

3.2.4.3 Tension (Dead-End) Clamp Performance

Tension clamps hold the conductors and shield wires in the air. Their failures could result in conductors or shield wires on the ground. In the case of sudden failure, mechanical shock loads transfer to all adjacent towers in a stringing section between dead-end towers. This increases the possibility of cascade failures.

Tensile tests of tension clamps are performed according to IEC 61284 or other industry standards, using load frames like the one pictured in Fig. 3.7.

Fig. 3.6 Example of a torsional load on spacer clamp due to ice loads

Fig. 3.7 Tensile test of a dead-end clamp installed on a coated conductor

3.3 Measurements of Ice Accretion

3.3.1 Need for Field Data

CIGRE Technical Brochure 179 (CIGRE TF 22.06.01, 2001) encouraged utilities to establish appropriate routines to collect ice loading data. In many parts of the world ice loading is the most important parameter influencing the investments and performance of electric overhead lines. Ice loadings influence the life cycle costs of power lines in many ways, such as investment, maintenance cost, repairs after failures, or loss of delivered power during outage periods.

For steel towers, when ice loadings increase, the weight of individual towers increases, and the average span length decreases. This means that the investment increases rapidly with ice loadings. An overestimate in the design ice loadings means that the investment costs of new 420 kV lines may be 5–10% higher than necessary. An underestimate in the design load can, on the other side, imply catastrophic costs regarding maintenance, tower restoration, and compensation for non-delivered energy. For wood pole lines, the similar sensitivity is valid for ice loads of 3–4 kg/m or higher.

Information on ice loading is important when the reliability of electrical networks is to be assessed. Without long-term information about icing, it is more difficult to justify exclusion of events from annual reliability figures using the IEEE 1366 "2.5-β" criterion.

Ice loading data are also crucial where upgrading old lines is considered, including decisions about whether to include overhead ground wires (as opposed to line surge arresters) in areas of heavy icing and modest lightning activity.

A more recent consequence is the potential loss of telecommunication facilities, due to the increased use of optical fibre wires such as OPGW on transmission lines.

3.3.2 Direct Measurements Using Meteorological Data Sources

Ice load measurements on overhead line conductors, combined with other information such as meteorological conditions and load estimates from ice models, are key factors to improve line design criteria. Complementary information from several sources is important, gathered using the strategy flow chart in Fig. 3.8.

The upper right-hand box of Fig. 3.8 represents field data for ice. Such data may come from especially designed measuring racks or test spans. However, the many kilometres of transmission lines that traverse areas with icing exposure represent probably by far the most comprehensive source of information to cover this point. Therefore, ice loads obtained from existing transmission lines represent a major key to improved line design in the future. Such data are in many cases the most relevant when evaluating the possibilities for upgrading of older lines, e.g., installing larger conductors, underbuilt optical cables, increasing the number of sub-conductors (bundles), or removing overhead ground wires. Such issues have been raised already in many regions where the need for increased transmission capacity cannot be met by building new lines. Investments will be greatly affected depending on whether it is possible to utilize existing towers or new towers must be built. Decisions can in most cases only be made in terms of probabilities. This means that owners must take into consideration the probability of failure they are willing to accept or, in other words, how much they are willing to pay for a greater reliability.

It is important to compile primarily good records of major icing events, for instance, with significant damage to the lines. But for statistical purposes, it is also important to record all icing events. The statement of "NO ICE" during an observation period is for the same reason of great value.

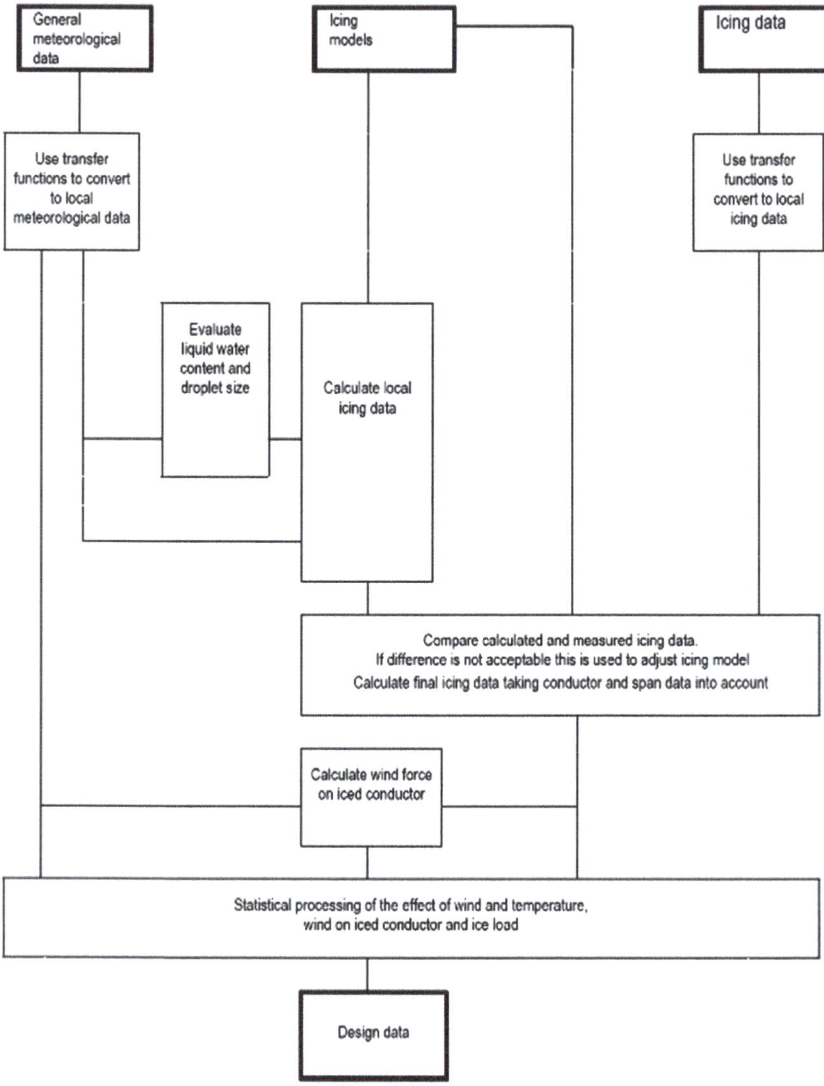

Fig. 3.8 Strategy flow chart for obtaining ice design loads (CIGRE TF 22.06.01, 2001)

3.3.3 Preparations by Utilities and Field Staff

The importance of ice data collection on overhead lines should be explained to the field staff, who should be motivated to be as accurate and thorough as possible. The various icing processes, measurement methods, tools, and safety procedures should be clearly explained. A regular training program for field staff, every year just before the icing season, is strongly recommended. An overview of the lessons learned from the previous year's data or experiences from other utilities should be considered.

Observers are understood to have completed electrical safety awareness training and to be fitted with the usual personal protective equipment (PPE) including high-visibility arc-resistant coveralls, hardhat, gloves, and safety eyewear. Before any measurements are made near energized lines, all appropriate electrical safety requirements must be met, and all necessary regulatory procedures followed.

When observing or making actual measurements, the probability of ice shedding from other conductors should be considered, especially under galloping. The possibility of arc flash exposure should be considered when filming partial discharge activity across energized station post or transmission line insulators.

Training should include sets of observation forms such as those found in CIGRE TF 22.06.01 (2001). The quality of data recorded on these sheets will be substantially improved by providing the staff with appropriate tools, as well as good headquarter routines during the icing season. These topics are dealt with in the next two subclauses. Since the conditions and demands vary significantly between utilities, these sections indicate what is recommended as mandatory (M) and what may be optional (O), according to local needs (Table 3.1).

3.3.4 Observations on Non-instrumented Lines

It is urgent for utilities in regions exposed to icing to implement a program for ice load recordings on existing lines. In the case of failures or serious disturbances, the priority is to re-establish the system without delay. Nevertheless, someone should, in parallel with the restoration work, be appointed to secure the evidence of ice accretion before they are removed or melted. In fact, the more extreme or unusual the event seems to be, the greater is the value and importance of detailed and

Table 3.1 Mandatory and optional equipment for monitoring ice accretion

Mandatory	Optional
Observation forms, pencils, clipboard, and cover	Weighing device
Instructions for data collection	Electrical conductivity metre
Measuring scale	Flexible wire to form around the accretion and preserve the contour shape
Plastic (not electronic) calliper	Suitable equipment for measuring ambient temperature, relative humidity (<0 °C), and wind speed
Compass and GPS (smartphone)	Bag or knapsack to carry equipment
Sampling devices such as knife, hacksaw blade, and plastic hammer	Travel aids—snowshoes, skis, studded boots
Watertight, strong plastic "freezer" bags to store samples	Voice recorder (smartphone)
Tags or marker to write on plastic bags	Spare battery/charge bank for smartphone
Digital still and movie camera with time and date code (smartphone)	

comprehensive data and information from the event. Therefore, motivation and proper training of the people involved are necessary. The grid owners are encouraged to predefine a group of people dedicated to this task.

The weight (ice load) and the diameters (shape) of the icing are parameters that are dependent on each other. If both have been measured, the density can be calculated. It is mostly easier to measure the diameter than the weight, and hence, the diameter measurements have priority. If the weight cannot be measured directly, it may be estimated from the diameter and proper assumption of density (based on ice type or measurements elsewhere in the area).

3.3.5 Weather and Line Location Data

General information about the weather may often be obtained from the Meteorological Service. Records should also capture local observations made in the icing area.

The extent of the icing-affected zone should be marked on maps or described in words, supported with longitude and latitude values from personal GPS equipment. If the affected area is large, or there are distinct variations of accretion within the area, multiple reports should be prepared. Iced sections of the overhead lines or substations should be featured as waypoints in the mapping and reports.

3.3.6 Physical Dimensions

The priority in field observations is to measure the representative diameter (shape) of the icing. Physical dimensions of the ice accrual are important for three purposes:

1. Wind exposed area (drag coefficient)
2. Weight of ice pr. metre (can be estimated from the diameter and the ice type) and
3. If the weight has been measured, the density of ice may be calculated.

If the physical dimensions of a cross section cannot be taken by means of a centimetre scale or a ruler as in Fig. 3.9, it may be useful for instance to cut paper strips or small branches from a tree, equal to the smaller and greater diameter of (an elliptic) ice accretion, for later measurements. It is useful to apply any indirect method that gives as accurate representation as possible, for instance:

- Make sketches—use manual scaling methods by eye.
- Take photographs—use zoom lens.
- Use binoculars (or better: scaled monocular for estimating geometry and dimensions).
- A distance meter (range finder), or a scale fitted to a hot stick, may also be useful to find exact dimensions.

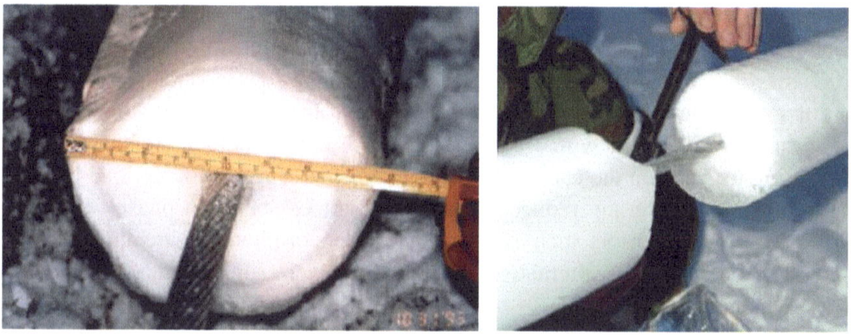

Scale: Greatest diameter is 19.7 cm

Calliper is useful supplementary tool

Fig. 3.9 Examples for measuring diameter of ice accretion (CIGRE TF 22.06.01, 2001)

3.3.7 Weight of Ice on Conductors

For load calculations, it is important to know the average weight of ice per metre of conductor. Choose a representative piece of ice, cut plane ends, and measure the length. Then, the sample can be crushed or melted and kept in a watertight container or plastic bag as shown in Fig. 3.10.

The collected ice sample can be weighed on-site, or the sealed bag can be transported for later analysis. If it is not possible to get an ice sample from the

Fig. 3.10 Collecting ice sample in plastic bag using knife/saw and plastic hammer (CIGRE TF 22.06.01, 2001)

conductor, indirect methods may be used, for instance, by measuring the sags of conductors and earth wires together with ambient air temperature. The additional loads may then be calculated using a conductor design program directly if the line is out of service, or possibly corrected for current for an operating line. Since there is no current in the earth wires, and the ice loads affect them more than phases, measuring the sags on the earth wires will give better estimates of ice load but take care not to aggravate any reduced electrical clearances between phases and earth wires. Noncontact sensors such as laser rangefinders or photographic analysis from side views are modern alternatives to traditional telescoping poles for vertical measurements.

On conductors, it has been common to exploit estimates of ice weight based on ice diameter measurements, as illustrated in Fig. 3.11. This shows a quadratic relation between mass of ice and outer ice diameter.

3.3.8 Density

The vital role of ice density measurements is also clear in Fig. 3.11. Design standards in some countries use the load (mass per unit length), and others use the ice thickness (radial ice) as input. In both cases, the density is needed in case (a) for combined wind load calculations and in case (b) to calculate the vertical loads or tension loads on towers.

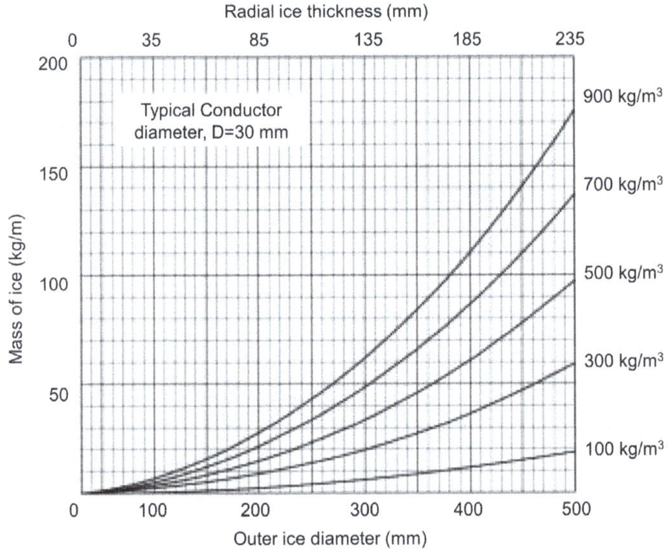

Fig. 3.11 Ice load as a function of diameter, with parameter of ice density (CIGRE TF 22.06.01, 2001)

When the ice load (mass/length) is given, the cross section is needed for the calculation of wind pressure. On the other hand, if the radial ice concept is used, the load must be calculated. In both cases, it is necessary to know the density of the ice. When both the physical dimensions of a (representative) cross section and the weight of the same sample are properly taken, the density of the ice accrual can be calculated as the mass divided by the volume.

Density measurements should preferably be taken from the conductors or earth wires, but samples from guy wires or other nearby structures may also be valuable.

Density measurements are generally not necessary for pure glaze ice since it is close to 900 kg/m^3. However, if air bubbles are visible, then density is lower and should be measured.

Data on density are especially important for model calibrations.

3.3.9 Moulds or 3D Digital Models for Cross Section

Moulds or 3D digital models are useful and informative for several reasons, in particular:

- To get exact volume
- To study drag factors for combined wind and ice loadings
- To study surface structure for modelling purposes, especially relating to wind impacts on wet surfaces, including icicle formation on glaze ice.

Moulds must be taken in two steps according to recommendations in Nigol and Buchan (1981-02):

1. Make a negative mould by putting the original ice sample in a low-temperature-cure RTV rubber. This compound cures at –8 °C in a few hours, and the original can be easily removed by slitting the rubber mould.
2. The empty mould is used to cast a plastic replica by pouring in polyurethane foam. This material expands rapidly and completely fills the interstitial cavities.

A portable 3D scanner application for a smartphone can be used in place of a mould to save time. The ice sample is placed on a reference grid to eliminate background clutter. High-resolution digital photographs are taken from several angles. The scanning tool then generates a 3D digital model that can be viewed and measured.

Cut clean and plane ends of an ice sleeve with a shape as regular as possible, e.g., 50 cm long.

Make an accurate cross-sectional drawing of one of the end planes on this sheet or on a separate sheet by putting the paper directly over the cross section of the ice. If a drawing is not possible due to weather conditions, an imprint may be approximated by folding the paper. Indicate approximate shape on the observation sheet as well.

Measure as accurately as possible:

- The length of the sample
- The diameter of the conductor or wire.

Measure for sections 1 and 2 (end 1 and end 2):

- The greatest (maximum) diameter of the cross section (it is often elliptic)
- The smallest (minimum) diameter of the cross section.

Put the sample in a watertight plastic bag or box (now it may be crushed or melted). Measure the mass (weight) on a scale or the water volume with a gauge.

If the icing on the conductor is very irregular so that measurement of diameter is hard to define, it is important to select a sample which is as representative as possible for the span, to get the weight per metre of conductor.

3.3.10 Number of Observations on Conductors

The representativity of ice data is especially important, but also difficult to ensure. It is not possible to describe general rules other than to keep this in mind when recording such data and take notes on variations that are not clear from the measurements. Photographs are important and helpful, with high-resolution close-ups (with high-power optical zoom lens) showing both the ice accretion and the conductor.

The time window for observing ice accretions on overhead lines may often be less than ideal. In such cases, the priority of observations should stand as follows:

- Photo or sketch
- External diameter and other dimensions accessible for observation
- Weight and length of a representative sample
- Ice type

Record the diameter of icing and other information about the accretion.

- Measure the diameter(s) of a representative sample for each span.
- If the ice varies significantly along the span, take note of the variations.
- If the icing episode is extensive, it is recommended to make several reports according to local variations or variations between different lines.

3.3.11 Observations on Insulators

In cases where electrical flashover under ice or snow conditions is investigated, it is important to obtain samples of ice that has fallen from the insulators. This has a distinctive pattern, and samples from both faulted and unfaulted insulators are needed. The important measures in this case are as follows:

- The weight of accreted ice or snow per cm of dry arc distance
- The electrical conductivity of the melted ice sample, at two different temperatures and interpolated to a reference of 20 °C.

The product of these two values, in units of g/cm and μS/cm respectively, forms an "icing stress product" or "snow stress product" that is used to estimate risk of flashover. A photograph including a scale along the length of the ice sample will help to document the findings. Plastic bags to collect and transfer the ice or snow samples indoors as well as plastic gloves to avoid sample contamination are recommended.

3.3.11.1 Consolidation of Data

The next step is to collect the reports and make them accessible as an information source of icing. There are many ways in which this can be done:

- File the prime sources (or copies of them).
- Publish the main information in electronic reports.
- File the icing data, including picture and movie files, in an electronic folder for each event.

3.4 Measurements of Audible Noise

Audible noise is emitted from high voltage transmission lines because of corona discharge in the highly stressed regions near the conductors, insulators, and fittings. This can lead to the release of an acoustic sound perceived as noise by humans. The level of audible noise depends greatly on the density of the air and weather conditions. In fair weather, a well-designed transmission line has generally a good corona performance under operating conditions, and thus, the audible noise is minor and rarely noticed. However, in the presence of water droplets caused by rain, drizzle, snow, or high humidity and enhanced by the presence of protrusions caused by ice, hoar frost, pollution or damaged strands, audible noise in the form of crackling or humming sounds may be heard in the immediate vicinity of overhead lines. With ageing or weathering of the conductor, the audible noise level is usually reduced. In the case of insulators, audible noise is normally caused by wet, polluted, nicked, or cracked ceramic discs. Audible noise from polymer insulators is normally a sign of imminent internal failure.

Audible noise can be potentially reduced by increasing the hydrophilicity or wettability of the conductor surface using several methods, including sandblasting, surface treatment of the conductor or the application of hydrophilic coatings.

The main purpose of this chapter is to review and summarize the literature relating specifically to audible noise associated with corona discharge from high voltage transmission lines in winter conditions. The general topic is covered fully in TB 838 (CIGRE WG B2.69, 2021-06). This review includes some recent laboratory measurements of audible noise for iced conductors, energized with dc. Based on the results, practical solutions to identify icing from its corona noise may become apparent.

3.4.1 Corona Discharge and Dry Band Arcing

Corona discharge and dry band arcs are the main causes of audible noise on high voltage transmission line conductors, hardware, and insulators.

Corona discharge is initiated by the acceleration of free electrons in the air by an electric field. If the field strength is high enough, electrons will acquire enough energy to ionize neutral molecules in the air (Farzaneh et al., 2012). This luminous phenomenon occurs at the surface of overhead line conductors, insulators, or fittings under both positive and negative polarities. Surface irregularities such as stranding, nicks, scratches, and semiconducting or insulating protrusions are the usual cause for corona discharge. Dry or wet airborne particles in the proximity of overhead line conductors and insulators may cause corona discharges. Apart from these, weather has a pronounced influence on the occurrence and characteristics of corona discharge on transmission lines.

Dry band arc discharge appears between wetted regions on energized, polluted insulator surfaces. The formation of the dry bands is the result of the heating effect of the leakage current passing through the wetted pollution layer (Farzaneh & Chisholm, 2009). Indeed, once the current is interrupted by the formation of dry bands, the entire insulator potential will appear across these spaces. Initial dry bands are a few centimetres in size. The electrical strength of the gap will be of the order of 5 kV/cm and with a typical voltage stress of 10–15 kV per insulator disc. It is likely that a partial arc will form. The current in this arc will be limited by the high resistance of the pollution layer, but it will still have a strong acoustic noise signature and significant ultraviolet emission.

3.4.2 Partial Discharge Noise Emitted by Insulators

Some audible noise is emitted by damaged or cracked porcelain insulator units. Pollution particles and more specifically by dry band arcing under wet conditions generate acoustic noise on all ceramic units. At substations, there is a strong correlation between audible noise and pollution. However, few surveys explore in-service aspects of insulator noise on transmission lines. The common use of a "ping" test excites resonances in the audio frequency range when applied to undamaged porcelain or glass units. These resonances at roughly 2 kHz are also excited by partial discharge (PD) activity. Under wet conditions, the narrow-band audio noise from ceramic insulators is often louder than conductor noise at tower base. Composite insulators do not have audio resonant frequencies and make no sound when they are excited by PD.

3.4.3 Audible Noise from Conductors in Icing Conditions

As described in the previous chapter, different types of atmospheric ice (glaze, dry or wet snow, hard or soft ice, and hoar frost) can be deposited on conductors. The ice protrusions formed on the conductor surface significantly increase the conductor surface roughness and lead to a higher audible noise.

The voltage polarity has a great influence on the icing morphology of glaze ice, while it has little influence on the soft rime ice, as shown in Fig. 3.12 (He et al., 2019).

Figure 3.12 shows ice morphology of glaze and soft rime ice-covered conductors energized with different voltage polarities at a fixed gradient of $E = 22{,}2$ kV/cm during icing:

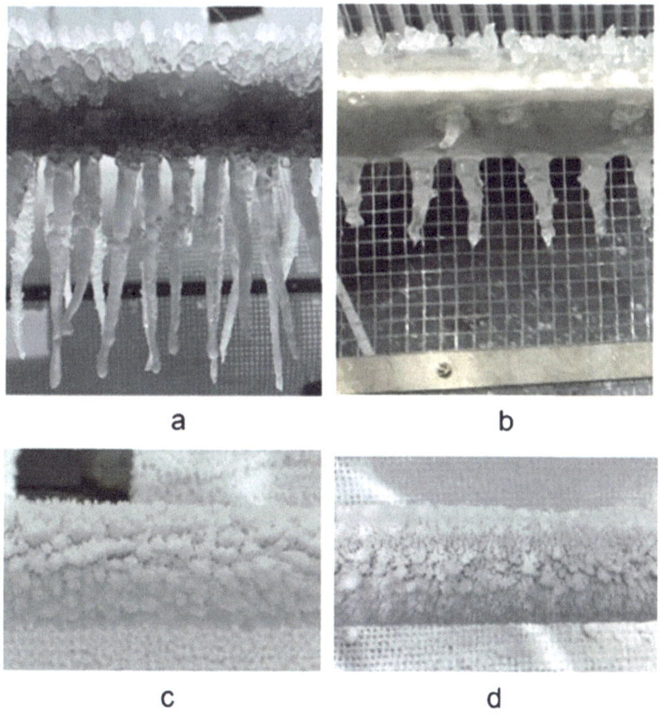

Fig. 3.12 Influence of voltage polarity on glaze and soft rime ice

(a) Glaze ice-covered conductor energized with negative polarity voltage
(b) Glaze ice-covered conductor energized with positive polarity voltage
(c) Soft rime ice-covered conductor energized with negative polarity voltage
(d) Soft rime ice-covered conductor energized with positive polarity voltage

The audible noise levels of different icing categories and voltage polarities are plotted in Fig. 3.13. Figure 3.13 shows the comparison of audible noise levels between a conductor under fair-weather conditions and conductors with glaze and soft rime icing formed under different polarities. Values in shaded area are not meaningful acoustic measurements. The audible noise in glaze icing conditions, for both voltage polarities, is bigger than in fair weather conditions. Once the electric field surpasses 18.0 kV/cm, the audible noise of positive soft rime ice-covered conductor is higher than in fair weather. The fair-weather audible noise levels of DC transmission lines in Fig. 3.13 are calculated by Bonneville Power Administration's (BPA) formula.

The 1/3 octave band spectra of audible noise of different icing categories and voltage polarities are plotted in Fig. 3.14. The applied voltage is around 135 kV (E = 21.9 kV/cm). The prominent 100 Hz components are from the power supply, and they attenuate little even when traversing structures and long distance. In the range from 315 to 20 Hz, all noise is broadband noise, and the curve has an "inversed V" shape (He et al., 2019).

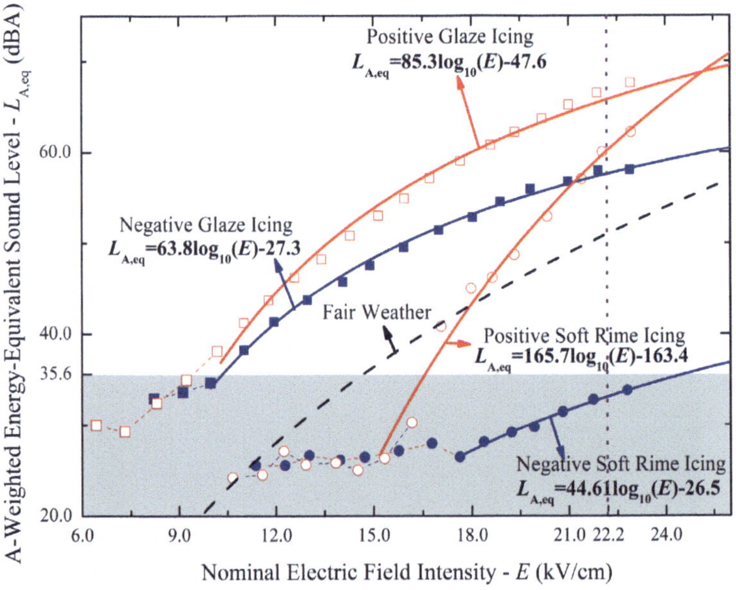

Fig. 3.13 Comparison of audible noise levels for conductors under different conditions

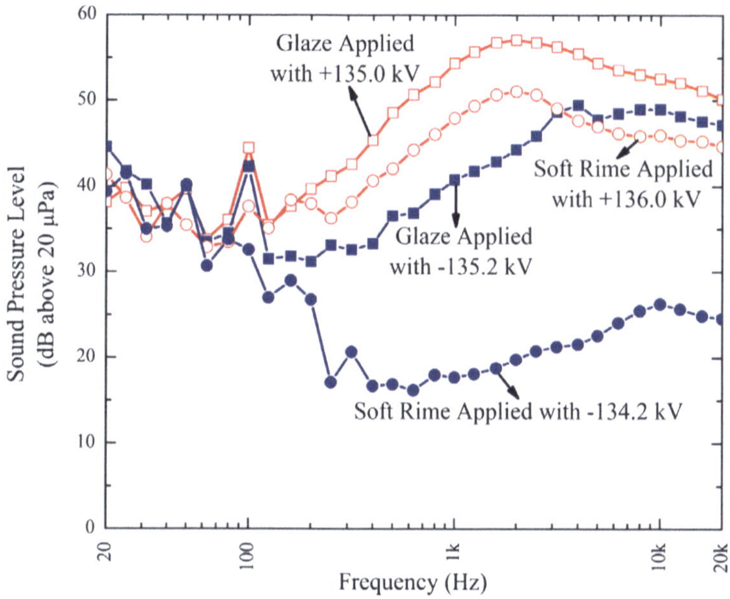

Fig. 3.14 Audible noise 1/3 octave band spectra of different voltage polarities and icing categories

The audible noise levels of the four icing severities are plotted in Fig. 3.15 (with the 35.6 dBA noise level shown by shaded area). The audible noise increases with increased soft rime ice thickness, as expected, and approaches saturation at heavy icing severities (He et al., 2019).

Conductivity of freezing water is also a noteworthy factor that could deteriorate the audible noise of soft rime ice-covered conductor. As Fig. 3.16 shows, audible noise first increases and then approaches saturation with the increase of freezing water conductivity (He et al., 2019). This experiment had the same 35,6 dBA noise floor as Figs. 3.13 and 3.15.

3.5 Attenuation of Power Line Carrier Signals

CIGRE has a rich library of resources describing the use of power line carrier (PLC), including recent digital applications. A PLC system forms a high-frequency transmission path to link sending and receiving ends, through couplers, using energized overhead line phase conductors. These generally work at frequencies of 50–500 kHz, where the surge impedance of the phase is relevant rather than its electrical parameters at 50 or 60 Hz. Uni- and bi-directional PLC systems are used for a range of monitoring, protection, and control functions. In some cases,

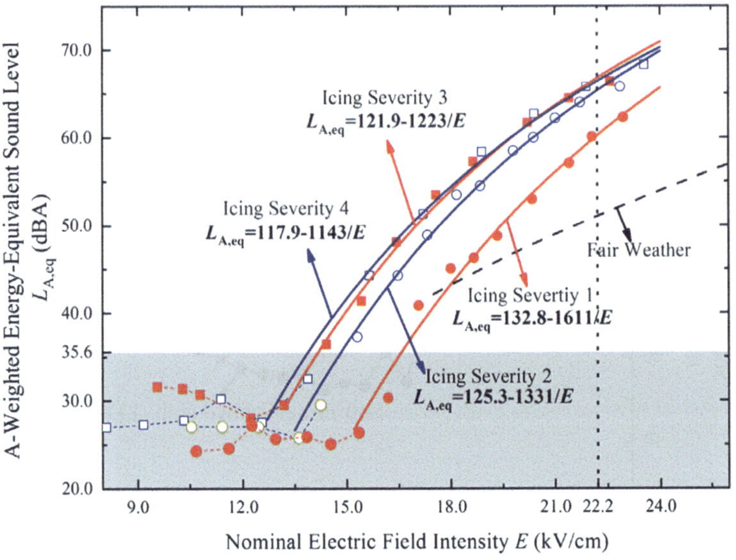

Fig. 3.15 Influence of icing severities on audible noise of soft rime ice-covered conductor

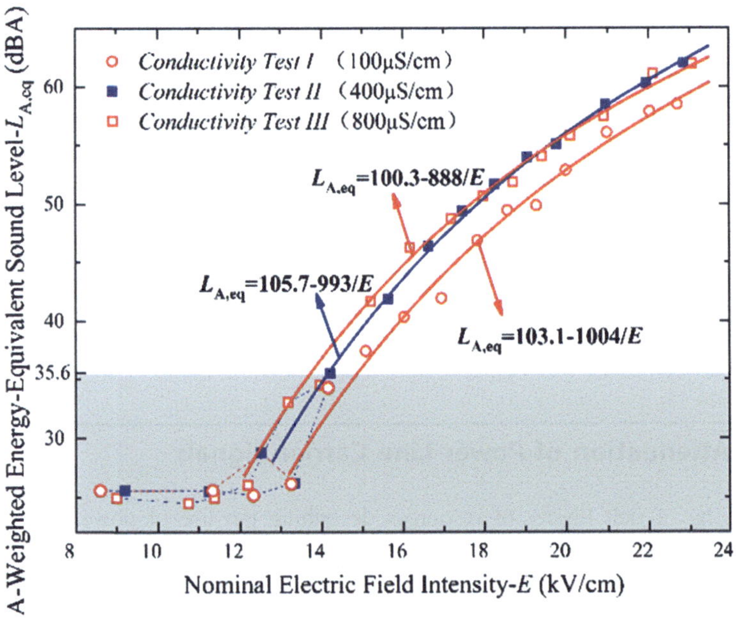

Fig. 3.16 Influence of water conductivity on audible noise of soft rime ice-covered conductors

insulated overhead shield wires provide the transmission path for the PLC signals, either as single conductor over ground (with surge impedance of about 500 Ω) or as a balanced pair configuration (with surge impedance of 900 Ω).

Operators of PLC systems have learned through experience that winter conditions can impair the signal-to-noise ratio. They anticipate and address problems using redundant carrier channels, which also afford the opportunity for testing. According to IEEE PES (2005-06) Std. 643, hoar frost on a line may increase PLC signal attenuation by a factor of 4 or 5 (13 dB). Ice accretion on the conductor has much higher attenuation levels, 18 dB at 50 kHz, and 23 dB at 300 kHz.

Monitoring of signal levels on PLC systems thus provides a simple way to establish the onset and extent of icing activity. This is, of course, not the primary function of the systems, but automated PLC networks may already be reporting signal levels for several frequencies in multichannel receivers. In line rebuilding projects that are replacing PLC with optical fibre ground wires, repurposing the PLC equipment to monitor ice accretion, as well as provide a backup communication channel, may improve the business case to retain existing systems.

3.6 Results from Typical Overhead Line Winter Monitoring Programs

Ice and snow monitoring programs have established relevant design parameters in several countries. Some recent studies, drawn from CIGRE TB 838 (CIGRE WG B2.69, 2021-06), give insight into the expected results from monitoring at other severe winter climate sites.

3.6.1 Results from China

After being struck by the 2008 Chinese Ice Storm, icing problems attracted more and more attention from the government and academia. In late 2008, Xuefeng Mountain Natural Icing Test Base (XMNITB) was started to be built by Chongqing University. After ten years of development, it is now one of the most important natural icing stations in the world, as shown in Fig. 3.17 (Chongqing Daily, 2019).

The XMNITB is sited in the PingshanTang, Xuefeng Mountain and at an altitude of 1400 m. It usually starts to snow in November and melts in March. As a typical microtopography, micrometeorological area, it has as many as 240 foggy days and up to 55 icing/snow days a year with average annual rainfall of 1810 mm. The wind velocity at the Natural Icing Test Base can reach 25 m/s in winter. All these environmental parameters make it a perfect location to study the icing phenomena. Currently, many research activities are performed in XMNITB including:

- Ice accretion on single/bundle conductor
- Ice accretion on wind turbine

Fig. 3.17 Xuefeng Mountain Natural Icing Test Base (XMNITB), China

- Icing performance of composite and porcelain insulators
- Melting ice accreted on conductor by large DC current
- Performance evaluation of anti-icing coating for insulators and conductors
- Measurement of conductor tension and environmental parameters such as temperature, wind velocity, humidity, and water droplet size.

3.6.2 Results from Iceland

Iceland initiated a measuring program on icing on overhead lines in 1972, and it has been ongoing since then (Elíasson & Thorsteinson, 2007; Elíasson et al., 2019-06).

Most of the measurements have been made in specially constructed 80-m test spans, but there are also load cells installed in some operating OHTL. In total, Fig. 3.18 shows there are nearly 60 test spans placed in 40 different locations. Most of them are in higher altitudes, and therefore, the majority of the measured icing events in the test spans are related to in-cloud icing but there are also some wet snow icing events. There are now plans to increase measurements in operating OHTL but reduce the number of specially constructed test spans.

In Fig. 3.18, sites with numbers indicated are test spans, and the first two letters give the year of construction. Sites with yellow boxes letters indicate measurements in operating OHTL.

The measuring site at Hallormsstadahals East Iceland, indicated as (83-1) in Fig. 3.18 has three test spans, two parallel, and one perpendicular to the others (Elíasson et al., 2009). At the same site, there are also measurements ongoing in two

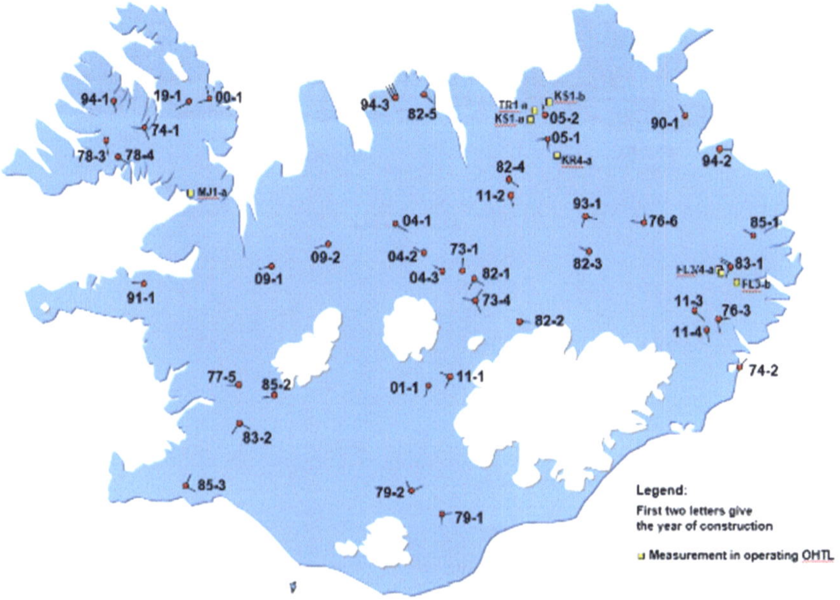

Fig. 3.18 Ice measurement sites in operation in Iceland 2020

spans of two parallel different 400 kV transmission lines, one with a simplex conductor and the other with a duplex conductor. Both transmission lines are equipped with load cells installed with data loggers and video cameras. The site is also equipped with an automatic weather station since 1996, and with these resources, the 400 kV transmission lines are monitored online for winter conditions that lead to reliability issues.

The measuring site at Hallormsstadahals Fig. 3.19 has been used to compare the accumulation of in-cloud icing on different conductors using tension recorders (Thorsteins & Eliasson, 1998). These studies have focused on the influence of conductor size and surface finish on ice loading. One recent test involves a comparison of icing on stranded AAAC conductor vs covered copper conductor. The comparison has been ongoing for almost 5 years and seems not to have been promising for the covered conductor.

Systematic registration of icing events on operating lines has been operating in Iceland since 1977 (Elíasson et al., 2015, 2017, 2019-06). The data from this monitoring program on real lines, combined with observations on the instrumented test spans, form an important icing databank (*Ísingarbanki* in Icelandic) for setting load criteria for new transmission lines in exposed areas. A sample appears in Table 3.2.

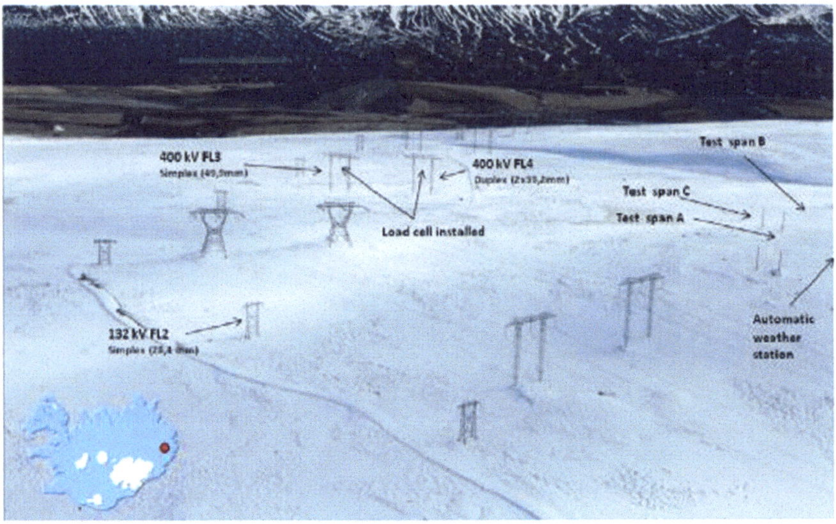

Fig. 3.19 Hallormsstadahals measuring site in East Iceland. Measurements in test spans and operating OHTL

Many of the icing events were observed to have duration of at least one week. Most of the load was associated with accretion of ice or wet snow, with wind pressure adding a median additional force of 1 N/m, increasing to 9 N/m under the worst conditions in Table 3.2.

The *Ísingarbanki* has also supported the use of underground cables for an increasing fraction of the 11-kV distribution system (Elíasson et al., 2019-06).

Table 3.2 Example *Ísingarbanki* data for Mast 168, Kópaskerslína 1, 66/132 kV, 2005–2014

Record number	Date of icing event		Maximum wind + ice load (N/m)	Maximum ice load (N/m)
	From	To		
1	2005 11 30	2005 12 05	16	16
2	2006 12 07	2006 12 17	11	10
3	2007 01 07	2007 01 15	5	4
4	2008 01 11	2008 01 18	6	5
5	2008 11 10	2008 11 16	7	6
6	2010 03 19	2010 03 28	11	10
7	2012 01 21	2012 01 28	12	11
8	2012 09 10	2012 09 12	164	156*
9	2012 11 21	2012 11 22	13	12
10	2013 09 16	2013 09 17	21	19
11	2013 12 24	2014 01 08	70	67

*Catastrophic wet snow icing

3.6.3 Results from Italy

In Italy, outdoor field tests are mainly carried out on wet snow phenomena. An experimental activity started in 2013 with the WILD (Wet snow Ice Laboratory Detection) project, as shown in Fig. 3.20.

The WILD station is operating in the Western Alps of Italy, at an altitude of 950 m. It is the first automated station in Italy devoted to the study and monitoring of the effects of wet snow on the conductors. WILD is able to aggregate completely ice-free unconventional weather measurements, accretion data, and web camera images during wet snow events.

The experiments carried out at the WILD station are as follows:

- Meteorological ice-free measurements: air temperature, wind speed and direction, snow accumulation, and snow water equivalent of precipitation
- Physical measurements and characterization of the snowfall: droplet size, snowflakes terminal velocity, number of particles, and snow density
- Mechanical measurements on conductors exposed to snowfall: weight, size, and diameters of ice sleeve
- Electronic measurements related to the control of the anti-icing current over a conductor
- Pictures of the wet snow event by using four web cameras
- Test of ice-phobic coatings for conductors (6 × 15 m spans and small samples).

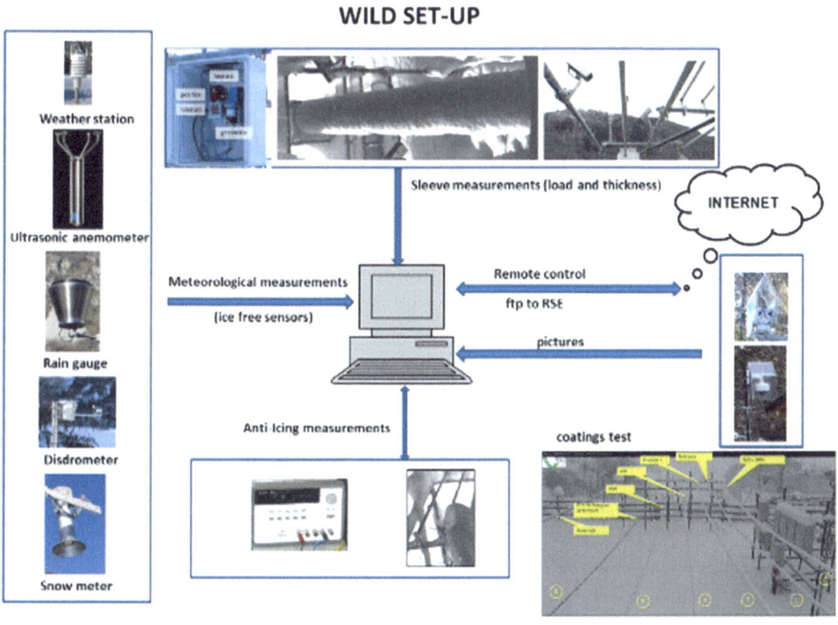

Fig. 3.20 WILD ice monitoring station in Italy

The performances of each coated conductor are assessed qualitatively with respect to the reference uncoated one by analysing the 15′ images from different cameras and correlating the results with the characteristics of the snowfall. Acquired data and images allowed to establish some criteria by which to evaluate the response of the coatings and to retrieve several feedback elements useful for their development:

- The time delay in the first snow film formation over the conductor. Delaying this stage allows to significantly reduce the maximum sleeve load accretion at the end of the event.
- The early shedding of the sleeve from the conductors. Anticipating this stage means to mitigate the sleeve load over the lines during the event.

Applying these criteria to observations made over three years of testing allows to divide the most effective coatings into two groups:

- Coatings able to **delay or inhibit the snow accumulation** at air temperatures below 0 °C and dry snow conditions with very low LWC (superhydrophobic coatings like SIVO, Producer 2).
- Coatings able to **facilitate the sleeve detachment** at temperatures greater than 0 °C in wet snow conditions (moderately hydrophobic coatings like NUSIL, PTFE, ALK, and Producer 1).

3.6.4 Results from Norway

Norway has invested in several long-term icing observation sites. Three of these, Ålvikfjellet, Hardingnuten, and Stølsheimen, were situated as shown in Fig. 3.21 in 2019. All sites are above 1000 m above sea level and are exposed to intense in-cloud icing. The project was to a large degree motivated by an icing event during winter 2013/2014 on the 420 kV line Sima-Samnanger that caused a partial tower collapse and shield wire breaking. Measured icing on the shield wire well exceeded design values for that line, as shown in Fig. 3.22.

The installation of the icing monitoring stations in Fig. 3.21 started as a part of the "Frontlines" project with main aim of developing a toolbox with numerical methods and algorithms for accessing maximum design icing for transmission lines. The Frontlines project was initiated by the utility Statnett and the consultancy Kjeller Vindteknikk in 2014 through an application to the Norwegian Research Council EnergiX program. The project was started in 2015 and completed in 2018.

In addition to the wind and temperature measuring sensors, all three sites were equipped with the following instruments (Nygaard & Honkasaari, 2017):

Fig. 3.21 Location of current ice-monitoring stations in Norway

Fig. 3.22 Icing on the broken shield wire on the 420 kV line Sima-Samnanger. *Source* Statnett

- IceTroll (instrument developed by Kjeller Vindtenikk)—A rotating cylindrical ice load sensor (Fig. 3.23 right) that measures the accumulated ice mass (kg/m) in accordance with the description in ISO12494
- A heated web camera system for reliable ice monitoring (example shown in Fig. 3.23)
- Load tension measurements in 420 and 320 kV overhead lines as well as in test spans, both for single- and twin-bundle conductors

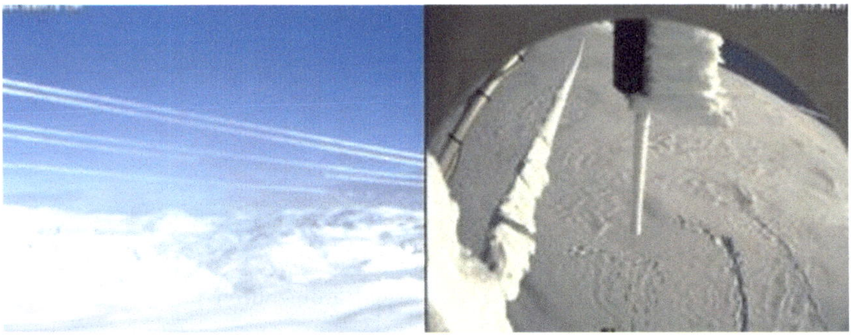

Fig. 3.23 Web camera images from Ålvikfjellet. Left: Three twin-bundle phase conductors of the 420 kV transmission line. Right: Test span (single conductor) and the IceTroll icing sensor

3.6.4.1 Results from Ålvikfjellet Measuring Station

At the Ålvikfjellet site, the first test span is 80 m long and runs parallel with the transmission line span on which the load measuring sensors were installed as shown in Fig. 3.24 (Nygaard & Honkasaari, 2017). The predominant icing mechanism on this site is in-cloud icing. First operational winter for this measuring station was winter 2014/2015.

Measurements on the Ålvikfjellet test span for first winter (2014/2015) showed not to be representative because the test span is only 80 m long, and the conductor span, because of the short length, had a higher torsional stiffness than it would if installed in a much longer transmission line span. After first winter, the conductor was exchanged for smaller shield wire, and swivels were added on both ends of the shield wire attachment to allow shield wire rotation under ice accretion. The swivel is position n.14 (shown in red), and the load cell is position n. 2 on Fig. 3.25.

Experience, photographed in Fig. 3.26, also showed that the position and height of the power supply container that also served as an emergency shelter should be chosen better and that the container should be elevated on a structure to avoid it be snowed in.

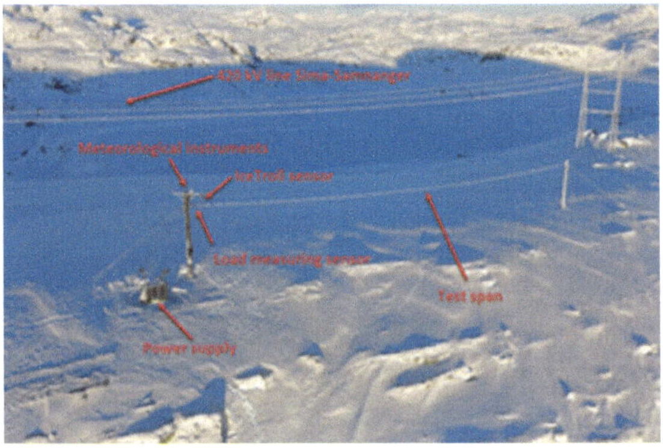

Fig. 3.24 Ålvikfjellet icing monitoring station

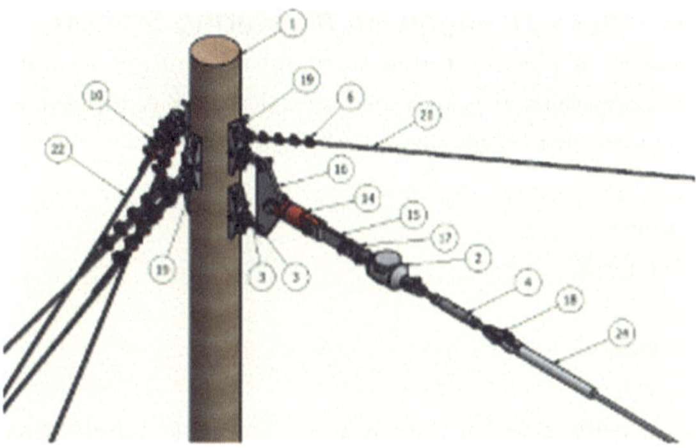

Fig. 3.25 Ålvikfjellet test span—Modified termination of the span. *Source* Statnett

Load sensors installed had signal conditioning, data storage, and data transfer electronics incorporated in them. Initial data transfer from the load sensors was to the container shown in Fig. 3.26 and from there further by modem. First the antenna needed to be reinforced because it was broken by ice and the communication to the container was lost, later the electronics were changed, and the data transfer from the sensor was done using the GSM network.

The camera heating system needed to be optimized to ensure availability of a clear view and at the same time to reduce the power consumption.

Currently, there are new load sensors under development that use GSM and Internet of Things (IoT) technology for data transfer. Prototype testing in under way and trial installations are done.

Also, the two test spans from the Stølsheimen measuring station, mentioned below, have been relocated to the Ålvikfjellet plateau and placed on the possible future route of the Sima-Samnanger line.

Fig. 3.26 Ålvikfjellet test span—Power supply container under snow. *Source* Statnett

3.6.4.2 Results from Hardingnuten Measuring Station

The measuring station at Hardingnuten went into operation in January 2016. The measuring station consisted of a meteorological tower equipped with Ingvaldsen (2019-02-15), Nygaard and Honkasaari (2017):

- Two heated cameras
- Icing sensor (IceTroll)
- 2D wind sensor
- Temperature sensor
- Distrometer
- Data logger with a modem for data transmission and communication
- Diesel generator.

This measuring station did not have a test span, so ice load and ice accretion were measured on spans of two parallel transmission lines. One span had a duplex configuration with two conductors in a bundle, while the other has a simplex conductor. The measuring station is shown on Fig. 3.27, where on the left side is the span of the 420 kV line Kvildal-Rjukan in which is one of the spans in which the icing is measured.

Currently, the camera which is facing the two overhead lines, one simplex 300 kV and one duplex 420 kV OHTL, remains in operation, and the other equipment has been redeployed.

3.6.4.3 Results from Stølsheimen Measuring Station

The Stølsheimen measuring station was built-in connection with a newly built 420 kV line Mongstad-Modalen as shown in Fig. 3.28. The line was energized in fall 2019.

The measuring station has a meteorology tower and two parallel test spans. One test span is strung with a simplex conductor and the other one a duplex.

Fig. 3.27 Hardingnuten measuring station

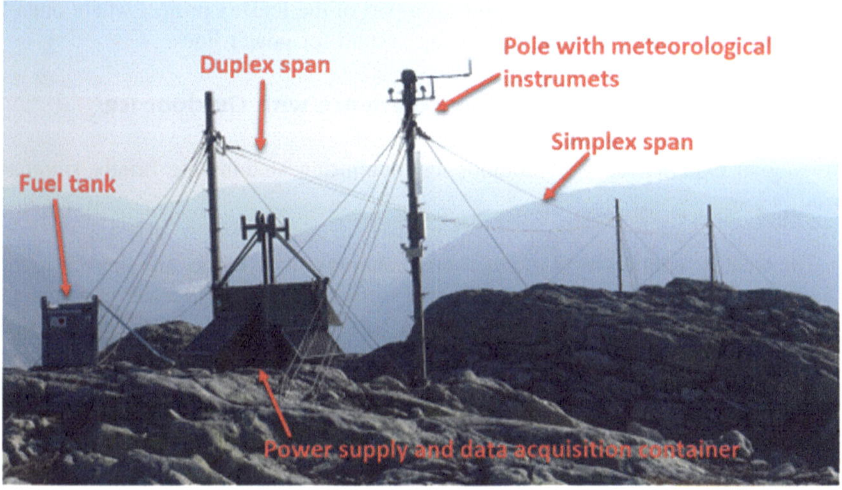

Fig. 3.28 Stølsheimen measuring station

After the 420-kV Mongstad-Modalen project was completed, the Stølsheimen monitoring station was taken down and relocated to the Ålvikfjellet plateau and placed on the possible future route of the Sima-Samnanger line.

3.6.4.4 Description of Other Measurement Stations

Developed prototype ice load sensors are installed on five additional existing transmission lines in Norway:

- Moskog-Høyanger 420 kV duplex
- Salten-Svartisen 420 kV duplex
- Sima–Aurland 420 kV duplex
- Sima-Dagali 420 kV duplex
- Sima-Samnanger 420 kV duplex.

Soon additional installations are planned on the following lines:

- Kvildall-Rjukan 420 kV duplex
- Songa-Vemorktoppen 300 kV simplex.

Each line has a minimum of one span equipped with ice-monitoring sensors. All the spans have V-suspension strings, and one sensor will be installed in each side of the V-string.

On all lines the mid-phase is equipped with sensors. One of the Sima–Aurland line, one of the outer phases is equipped with sensors as well. On that span the mid-phase has a duplex configuration, and the outer phase has a bigger simplex conductor.

Installation of these sensors is done as a part of the IceBox project where one of the goals is developing an ice monitoring system for power lines.

3.6.4.5 Summary of Norwegian Experience with Outdoor Ice-Monitoring Sites

Norwegian experiences with outdoor ice-monitoring stations are briefly summarized below:

- The weather conditions at the measuring stations are extreme, and there is rapid accumulation of rime ice on all unheated surfaces which could result in complete encapsulation of sensors with ice. Heating of sensors is necessary, and the key factors are (Byrkjedal et al., 2019):

  - Enough power, e.g. wind sensors with 150 W maximum heating power
  - Optimized distribution of heating elements on the objects, i.e. heating the right parts
  - Optimized heating control

- It took a couple of seasons develop a reliable enough power supply solution which involves both fuel cells and solar panels.
- Poles at the end of the test span are guyed. Ice build-up on the guy wires can influence the measurement results, and corrections might be necessary.
- Some parts of the equipment or sensor could be completely snowed in if not placed carefully.
- Communication using GSM or IoT technology could be a challenge in remote areas with no or poor coverage.

3.6.5 Results from United Kingdom

The icing performance of two microalloy copper conductors (Cu-95 mm^2 and Cu-150 mm^2) with anti-icing coatings has been evaluated at a natural exposure test site at Deadwater Fell in UK during winter 2013-14. The 190-m test span and a conductor monitoring system detail appear in Fig. 3.29.

Eight icing incidents occurred during the test period. These events resulted in relatively high mechanical loads compared to everyday stress. Expressed in terms of the ultimate tensile strength, UTS, of the conductors the maximum recorded loads were 41.7% UTS for the ACSR, 32.8% UTS for the Cu-150, and 38.5% UTS for the Cu-95 conductors. The increase in tension for each conductor in the various incidents is given in Table 3.3. Note that these loads resulted from a combination of wind and ice.

The time histories of selected tension measurements from Table 3.3 are shown in Figs. 3.30, 3.31, 3.32, and 3.33.

Viewed from the North end of the 190m test span   Cu- 95 mm^2 conductor, vibration monitors, load cells

Fig. 3.29 Deadwater Fell test site

Table 3.3 Measured increases in conductor tension during 2013–2014 icing events, using ice-free condition as reference

Date	Temperature (°C)	Wind speed (m/s)	Wind direction (degrees)	Tension increase (kN)		
				Anti-icing treatment		No treatment
				Cu 95	Cu 150	ACSR 180
2013 12 19	–0.9	15.8	280	1.25	1.41	3.14
2014 01 12	–0.2	10	250	1.82	2.34	6.07
2014 01 15	0.1	8	210	1.59	1.55	3.12
2014 01 28	–0.6	9	200	1.4	1.3	2.8
2014 01 31	–1.8	15	180	1.4	1.54	5.9
2014 02 04	0.1	14	215	0.39	0.44	2.9
2014 02 12	–1.2	20	290	2.17	2.69	3.7
2014 02 14	–1.5	11	300	7.9	7.7	10.6

The long duration of build-up over 24 h and the mixed precipitation, rime ice and snow, combined to make the highest tensions on the reference Cu 150 conductor, approaching 25 kN in Fig. 3.33 compared to peak load of 18–20 kN on the other three occasions. In every incident, the ACSR accreted more load than either of the treated copper conductors. This implied that the anti-icing coating was having some effect although it did not stop icing completely. The results were considered promising and led to further investigations of anti-icing coatings.

3.7 Concluding Remarks

"Smart grid" systems for monitoring overhead power lines have generally focused on real-time sensor systems that monitor a mechanical state of the conductor, such as tension, sag, clearance, or inclination (tilt) of tower or insulator within a stringing

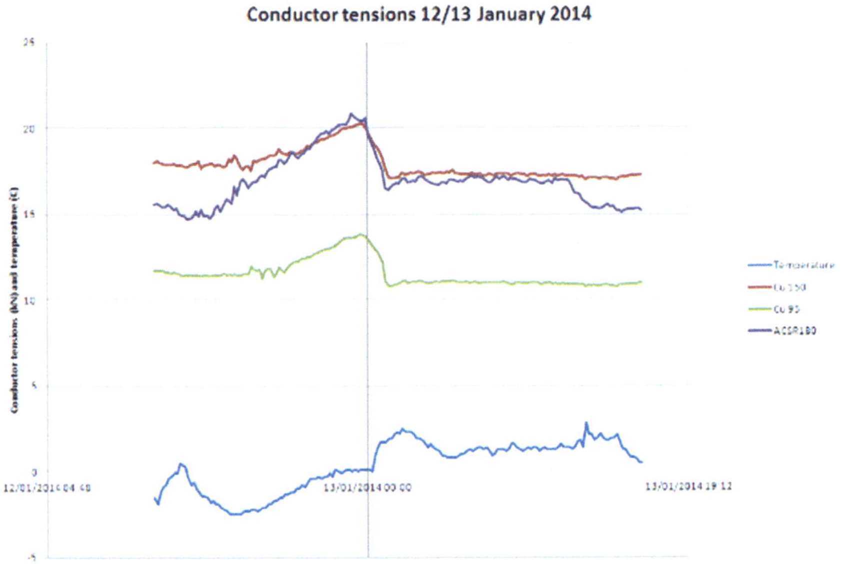

Fig. 3.30 Deadwater Fell observations: Conductor tensions and ambient temperature during rime ice accretion, 12–13 January 2014

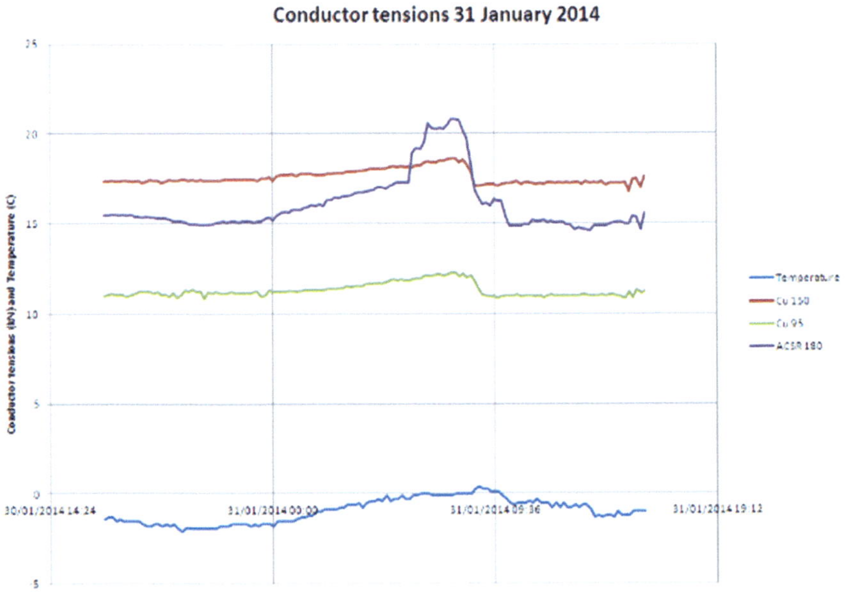

Fig. 3.31 Deadwater Fell observations: Conductor tensions and ambient temperature during rime ice accretion, 31 January 2014

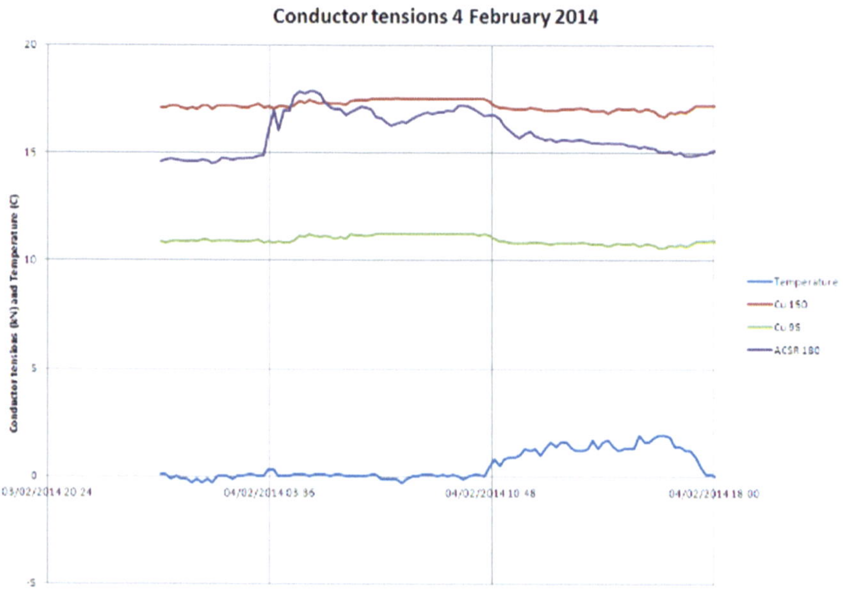

Fig. 3.32 Deadwater Fell observations: Conductor tensions and ambient temperature during wet snow accretion, 4 February 2014

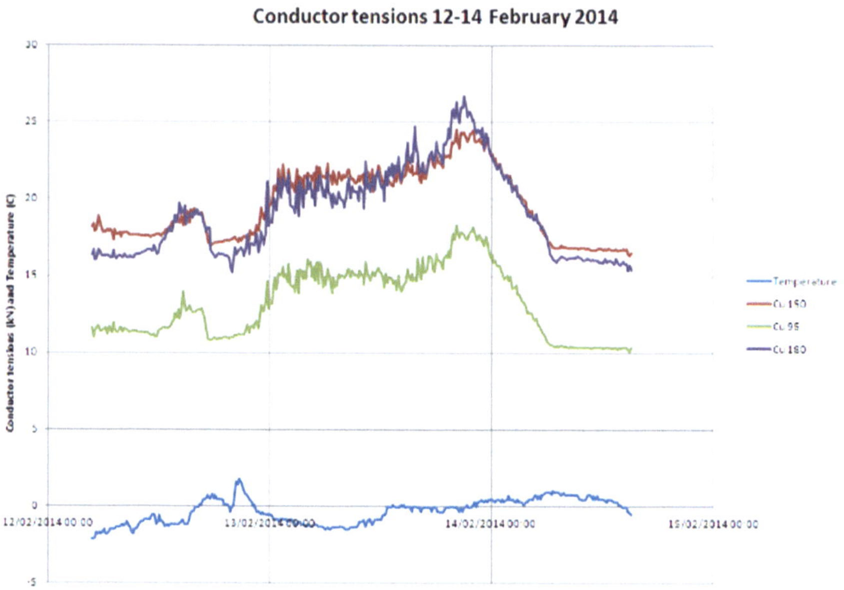

Fig. 3.33 Deadwater Fell observations: Conductor tensions and ambient temperature during combined rime ice / wet snow accretion, 12–14 February 2014

section. One mechanical measurement per stringing section is sufficient to establish the mechanical states—tensions, sags, clearances, and insulator swings, in all other spans, often with excellent precision. With a knowledge of the initial state of the stringing section, expressed as an average temperature at which all suspension insulators are vertical, it is also feasible to predict the near-term ability to carry current (ampacity). Heavy ice and snow accretion, wind loads, and low winter temperatures giving high tensions all affect the creep of the conductor, reduce the temperature at which tension equalization is now achieved, and thus reduce the ampacity of the conductor in future summer operations. Confidence in the results improves when multiple mechanical measures of state are supported by observations of weather parameters—temperature, wind speed, wind direction, relative humidity, and their ten-minute standard deviations.

The dynamic line rating (DLR) instrumentation for summer conditions can also be used to monitor the response of a stringing section to winter conditions. This monitoring can take several forms. The most useful today is probably the progressive impairment of observations, whether through deviation of results suggesting an apparent increase in conductor weight per unit length, loss of image quality in primary sensors or change (increase or rapid decrease to zero) of ten-minute standard deviation. The return of ten-minute standard deviation to normal levels, for example, establishes the duration of instrumented icing after a cup or propeller-type anemometer freezes up. For these reasons, close monitoring of DLR system inputs and outputs should be maintained during predicted or ongoing severe winter conditions.

Remote observations of icing on lines should trigger deployment of teams to collect and observe the ice accretion, using well-known protocols. Some modern technologies, including geo- and time-tagging of photographs, simplify the task of building, analysing, and maintaining an archive of observations. It is just as important to establish the icing levels that do not cause failures as those that make headlines in local newspapers.

The unique characteristics of audible noise from iced, energized conductors as well as attenuation of high-frequency energy in PLC systems could also be the basis of new or repurposed applications to monitor ice accretion and shedding.

Field experiences show that maintaining a measurement system at a remote site in severe winter weather taxes logistics such as adequate power for heated instruments and consideration of snow depth in placement. It is also clear that there are dynamic requirements that move the monitoring equipment after a few years to locations where new projects are in development.

The near future—with permanent, time-synchronized tilt sensors or accelerometers at all structures on a line—should bring increased precision to the monitoring of ice accretion (through slow changes in tilt angle) as well as ice shedding (through rapid changes) and component failures, which cause mechanical travelling waves that can be precisely located from the times of signal arrival. Application of permanent sensors will allow utilities to confirm design assumptions over the long term, rather than relying on projects that have successful monitoring data for two or three years.

4 Systems for Preventing Icing on Existing Overhead Power Line Conductors and Ground Wires

Atmospheric icing and its accretion on power transmission lines may cause considerable damage to power networks in cold climate regions. In this chapter, some of the fundamental mechanical and thermodynamic properties of ice and snow are presented, which will help understanding the process of anti-icing (AI) and de-icing (DI) methods.

AI/DI methods are subdivided into the four following categories:

1. Passive methods based on natural forces or physical geometries
2. Active coatings and devices requiring some electrical energy to be effective
3. Mechanical methods based on breaking down accreted ice
4. Thermal methods based on ice melting.

These methods have been discussed in detail in the CIGRE TB 438 (CIGRE WG B2.29, 2010) and can be briefly summarized as follows.

Passive methods are ones which do not require an external source of energy but rather use natural forces such as wind, gravity or solar radiation, or phase/circuit geometry. Consequently, they can function on both energized and non-energized conductors as well as ground wires. Passive methods include most of the anti-icing methods used to prevent or reduce the formation or development of wet snow and ice on conductors or ground wires. To achieve this, different strategies are used: (i) weakening ice adhesion strength, (ii) preventing freezing of supercooled water droplets on impact, (iii) using a combination of specific devices for limiting the impact of ice overload on conductors, and (iv) exploiting natural forces such as wind, gravity, or solar radiation in order to limit the adverse effects of ice loads on overhead lines. Some of these methods are already effective for wet snow but their efficiency for ice needs to be studied. The coatings discussed in this book are part of the passive methods.

Active coating and devices require some electrical energy to be effective. One of the proposed methods makes use of losses in a specific dielectric coating covering the entire surface of the conductor. By choosing an adequate dielectric coating

M. Farzaneh and W. A. Chisholm, *Techniques for Protecting Overhead Lines in Winter Conditions*, Compact Studies, https://doi.org/10.1007/978-3-030-87455-1_4

among ferroelectric materials, it is possible to maintain the surface conductor temperature above the freezing point to melt the ice/substrate interface. However, this would require the use of a higher frequency, 60 kHz, instead of the normal 50 or 60 Hz service voltage frequency. This condition is problematic as the use of such high frequencies can lead to electromagnetic perturbations and other problems and needs additional investigation. However, most of the proposed methods are not available commercially. In the same area, other methods are based on the use of a ferromagnetic coating for the purpose of sustaining a positive temperature of the energized conductor surface. Instead of absorbing energy from the electric field, the ferromagnetic coating absorbs energy from the magnetic field, which is at a maximum at the surface of the conductor.

Mechanical methods refer to any method involving ice breaking to accelerate ice shedding. In most cases, they can be considered as de-icing methods as they are used to speed up the shedding process after snow packs or ice have formed on conductors or ground wires. It has been demonstrated that mechanical methods require around 100,000 times less energy than thermal methods to force ice shedding. Generally, most of the mechanical methods are based on two strategies. One strategy consists in breaking the ice by scraping it and the second in releasing energy from shock waves, vibrations, or ground wire/conductor twisting to pull or fling off the ice. One of the main advantages of mechanical methods is their ease of application compared to thermal methods. In fact, mechanical methods are preferred for timely and fast intervention to de-ice short critical sections of a power network. However, in the absence of any precise instructions, mechanical methods involving significant bending of ground wires/conductors should be avoided for optical ground wires to prevent damage to the optical fibres.

Thermal methods include all non-natural methods causing the ice to melt in order to force shedding. They consist in the heating of line conductors or ground wires to prevent ice accretion or for de-icing purposes. It is recognized worldwide as the most efficient engineering approach to minimize the consequences of severe ice storms on overhead lines. Some of these methods can be used for anti-icing purposes to prevent supercooled water droplets from freezing during their impact on the conductor surface. In that case, less energy is required for anti-icing than for de-icing. Thermal methods can be divided in two categories: (i) methods based on pure Joule effect, and (ii) methods based on dielectric losses, radiative waves, and external heat sources, which are discussed in detail in this book.

4.1 Mechanical Properties of Ice

The mechanical properties of atmospheric ice depend mainly on temperature, wind velocity, the liquid water content (LWC) of the air, and the volume mean diameter (VMD) of the supercooled droplets during ice accretion. These parameters also depend on the ambient temperature, ice structure, density, grain size, specimen size, loading rate, and failure mode.

Solid ice is a polycrystalline material. Depending on the ambient temperature, loading conditions, and strain level, ice can be interpreted either as elastic, inelastic, and ductile or as almost linear elastic until brittle fracture. At temperatures closer to the melting point, ice undergoes creep deformation, which is divided into four stages: instantaneous deformation, primary creep (transient), secondary creep deformation (steady state), and tertiary creep resulting in failure (Eskandarian, 2005). When ice is subjected to large and sudden loads, its brittle behaviour dominates its ability to creep. The maximum tensile strength occurs in the ductile to brittle transient regime. Ice strength greatly depends on the load type and failure mode. Anyhow, failure of ice in all load types proceeds in three distinctive stages: (1) crack nucleation, (2) crack propagation, and (3) ice fracture. Crack nucleation without pre-existing flaws requires that stresses are locally concentrated at levels matching the theoretical cleavage or cohesive strength. There are several candidate mechanisms for this concentration of stress: dislocation pile up, grain boundary sliding, thermal expansion of extraneous inclusions, and elastic anisotropy of the ice crystals (Ashby & Hallam, 1986).

According to the results of compressive strength (Kermani et al., 2007) and bending strength tests (Kermani et al., 2008), atmospheric ice shows higher strength under compressive loads and increases with decreasing temperature. It also increases and then decreases with increasing strain rate. The bending strength of atmospheric ice has approximately the same value as tensile strength since tensile failure will govern. At the lower strain rates, the bending strength of the ice increases with decreasing temperature, while no temperature effect is observed at the higher strain rates. It is also observed that, depending on the temperature, increasing the strain rate can increase or decrease the ice strength. Ice failure in brittle regime is dominated by crack nucleation and propagation. That is why the de-icing methods require comprehensive information about the fracture toughness of atmospheric ice. The results of fracture toughness of atmospheric ice (Kermani & Farzaneh, 2008) showed that at very cold temperatures (around –20 °C), the fracture toughness of atmospheric ice increases owing to higher strength as demonstrated for freshwater (Liu & Miller, 1979).

De-icing methods by mechanical breaking of the accreted ice will be more effective when the ice failure occurs under tensile stresses in fracture classical opening mode I, in the brittle region of the stress–strain rate curve, at strain rates greater than $10^{-5}\ s^{-1}$. A temperature range between –5 and –10 °C will also allow better results for de-icing because in this range, the ice shows neither ductile behaviour (due to warm temperatures) nor high strength (due to cold temperatures) (Kermani & Farzaneh, 2008).

In summary, the mechanical properties of atmospheric ice and its failure modes under various loading conditions are of interest in many projects including de-icing by mechanical breaking and ice shedding (modelling, prediction, and prevention). Ice shows higher strength at colder temperature. De-icing techniques with rapid loading produce high strain rates that yield better results in ice removal projects. This information may improve the precision of ice shedding models.

In addition to the characteristics of the icing surface, the characteristics of the accumulated ice and snow including their texture are other major factors which influence the adhesion and cohesion strength and consequently the shedding rate. In fact, the cohesive strength of ice and snow is linked to the presence of bubbles or air voids, the density, and the liquid water content. All of these aspects in turn depend mainly on the solidification and deposit process and conditions, possible freeze and thaw cycles, temperature changes, and internal defects or cracks (CIGRE WG B2.29, 2010; Farid et al., 2018; Seifaddini et al., 2020).

A recall of the main factors affecting the characteristics of the accumulated ice or snow that can influence the shedding rate or process is presented below.

4.1.1 Liquid Water Content (LWC)

Liquid water content (LWC) is a very relevant parameter in snow. The LWC in snow accreted matrix controls the mechanical cohesion which keeps snow particles together and adhesion to surfaces (CIGRE WG B2.29, 2010). LWC is influenced by air temperature, relative humidity of air, wind velocity and precipitation intensity. According to a CIGRE report (CIGRE WG B2.16, 2006), snowflakes easily adhere to objects when their liquid water content (LWC) represents between 15 and 40% of the total mass of snowflakes. The LWC of snow sleeves during natural events is reported to be in the range from 20% up to 50%. At the lower limit of the range, snow sleeves of large diameter may grow so that the resulting overload may cause damage to transmission lines. Experimental and modelling studies (Roberge et al., 2007-10-09; Sakamoto et al., 1988) indicate that snow cohesion increases up to a certain LWC threshold (the critical LWC value of 40% is commonly accepted), above which cohesion decreases and snow blocks tend to break up under gravity force and to shed from surfaces.

In a laboratory study (Sojoudi et al., 2015), shear and tensile adhesion of snow as a function of snow characteristics, particularly LWC and density, and of roughness of conductor surface was studied experimentally. The results of this study show that the shear strength increases as the LWC increases, reaching its maximum at 20–25% of LWC, followed by decreasing at higher values of LWC. The shear strength also increases as snow density increases, up to density values ranging between 0.4 and 0.5 g/cm^3; after that, the shear strength decreases at higher values of density. The investigation also showed that the greater the surface roughness, the greater the shear adhesion. However, the opposite is observed for tensile adhesion.

As concerns tensile adhesive strength, the results reported in Sojoudi et al. (2015) showed that it was about 10 times greater compared with shear strength and that it increased with LWC until a maximum occurring between 22 and 32% and then decreased. The tensile force or strength on a smooth surface was about 1.5 times greater than that on a stranded surface and close to that on a rough surface.

4.1.2 Ice Type and Density

The type and density are two important factors in the adhesion strength of ice snow on an exposed surface. Depending on atmospheric conditions, various types of ice and snow with different structures and densities can be formed.

The result of a laboratory study (Rønneberg et al., 2019) showed that the shear adhesion strength of ice on an aluminium surface, measured using a centrifugal device, varied from 0.78 ± 0.10 MPa for precipitation ice, from 0.53 ± 0.12 MPa for cloud ice and from 0.28 ± 0.08 MPa for bulk water ice (ice formed from bulk water). This study also indicated that the ice adhesion strength inversely correlates with the density of ice.

In Hefny et al. (2012), the shear adhesion strength of natural snow to conductor surfaces as a function of its liquid water content (LWC) and density was studied experimentally. The results obtained showed that the adhesion strength increases as the LWC increases, reaching its maximum at 20–25% of LWC, followed by a decrease at higher values of LWC. The results also showed that the adhesion strength increases as the snow density increases up to density values between 0.4 and 0.5 g/cm^3, after that it decreases at higher density values.

4.1.3 Microhomogeneity

Ice cohesion is weakened by the presence of defects in the crystalline structure, voids, or cracks. Some authors find that structural defects can be induced in the ice that accretes on an inhomogeneous substrate (in terms of topography or surficial energy on microscale) (Murase & Fujibayashi, 1997; Murase et al., 1994–12).

Alternatively, a larger inhomogeneity (on macroscopic scale, i.e. mm or cm) can present different friction resistance to ice sliding so that the ice or snow block is subjected to an internal stress, leading to cohesive fracture. On this principle, some authors experimented with inhomogeneous coatings, as for example by applying a spiral Teflon ribbon on conductors, allowing some spacing between spirals (Nakagami, 2007-10).

4.1.4 Shape and Density of the Accreted Ice Sleeve

The shape of the accreted ice sleeve has great influence on the mechanism of detachment. On conductors and ground wires, the shape of ice accretions can be cylindrical, when conductor or wire rotation occurs, or asymmetric, often in a wing (pennant) shape related to the wind blowing direction.

Obviously, an asymmetric shape allows the gravity force to develop a torque that facilitates the shedding. On the contrary, when the accretion is a cylinder or symmetric disc as on insulators, both the surface adhesion strength and the ice cohesion force impede the shedding.

If a symmetrical ice deposit can be fragmented by some combination of mechanical forces, shedding will be easier. For example, sleeve cracking can occur if the sleeve cylinder is subject to uneven torque (due to different adhesion strength on different portions of the surface and to the rotation of the conductor or of the sleeve upon the conductor). On this principle, devices that aim at impeding the conductor rotation (i.e. bundled conductors or anti-rotational devices) or at causing sleeve deformation or fracture [i.e. PTFE tape (Nakagami, 2007–10), snow-resistant rings (Mizoe et al., 2007–10)] can be used, also in combination, to mitigate the accretion phenomena.

4.1.5 Temperature

All the ice properties have a strong dependence on temperature, especially around the melting point. It should be considered that ice shows higher strength at colder temperatures, therefore shedding is easier at higher temperatures. Some anti-icing devices aim at raising the temperature in the ice block or at the adhesion interface to provoke the detachment. These are referred to as "active" coatings or devices, meaning that they need some external energy to be active.

4.2 General Heat Balance

In thermal rating calculations such as Morgan (1991) and CIGRE TB 207, a heat balance consists of solar input, I^2R heating, convection and radiation cooling, and many other factors. Ice accretion changes the heat transfer relation between Nusselt number *Nu* and wind speed.

The mass rate of ice melting dm/dt per unit length of conductor is dominated by the ratio of power input P_i to the latent heat of fusion L_{fus}. Warming the water film from ice temperature to ambient takes a small additional amount of power, based on the specific heat capacity of water c_{pw}.

Heating of line conductors or ground wires to prevent ice accretion or for de-icing purposes is recognized worldwide as an efficient engineering approach to minimize the consequences of severe ice storms on overhead lines (Motlis, 2002). Thermal methods include all methods causing the ice to melt to promote shedding. Less energy is required for anti-icing than for de-icing (Prud'Homme et al., 2005b), and thermal methods for de-icing are discussed separately in Sect. 5.4.

4.2.1 Currents Needed for Anti-icing effects

Thermal methods have been proven for anti-icing purposes to prevent supercooled water droplets from freezing during their impact on the conductor surface. Field test experience in 1913 was that 300 A of current would prevent accumulation of ice on a 300-kcmil (150 mm^2) conductor at ambient temperature of –3 °C and wind speed of 3.2 km/h (Shealy et al., 1952-08). Specialists in overhead conductor thermal rating will recognize the "1 Amp per kcmil" guideline normally used to establish the summer limit current, considering warm temperature, high solar input, and low wind speed. The currents used to prevent ice accretion on overhead conductors must be selected with similar thermal rating considerations in mind, including the possibility of overheating line joints and hardware.

4.2.2 Steady-State Heat Balance Including Freezing Precipitation

The thermal behaviour of bare overhead conductors is described in CIGRE WG 22.12 (2002), CIGRÉ Working Group 22.12 (1992) considering the Joule, magnetic, solar and corona effects as heat gains, and convective, radiative, and evaporative cooling as heat losses. Morgan (1991) also considers the heat transfer to stranded aluminium conductors caused by rain. However, this heat balance must be extended further to account for icing conditions. An appropriate heat balance equation is reported in Péter et al. (2007). More precisely, its form applied to a power line conductor exposed to a two-phase air/water spray flow is as follows:

$$Q_j + Q_f + Q_k = Q_c + Q_e + Q_w + Q_r \qquad (4.1)$$

where Q_j is the rate of heat generation, in watts, by Joule effect, Q_f is the combined heat flow rate due to stagnation and friction effects in the boundary layer, Q_k is the kinetic energy rate of the impinging water droplets, Q_c, Q_e, and Q_w are the heat loss rates due to convection, evaporation, and due to impinging water droplets, respectively, and Q_r is the rate of radiative heat exchange. The terms in Eq. 4.1 can be parameterized as presented in Péter et al. (2007).

The convective term is expressed by

$$Q_c = hA(T_s - T_a) \qquad (4.2)$$

where h is the overall heat transfer coefficient (W/m^{2}K), T_s (°C) is the conductor surface temperature, T_a (°C) is the air temperature, and A (m^2) is the exposed conductor surface. The heat loss rate due to water layer evaporation from the conductor surface is expressed in Eq. 4.3.

$$Q_e = 0.622 \cdot h \cdot l_e \cdot A_w \cdot (p_{sat,s} - p_{sat,a})/(c_p \cdot p_a) \qquad (4.3)$$

where l_e (J/kg) is the latent heat of evaporation, A_w (m^2) is the wetted surface of the conductor, $p_{sat,s}$ (Pa), $p_{sat,a}$ (Pa), and c_p are the saturation vapour pressures at the surface, air temperature over water, and specific heat of air, respectively, and where

p_a (Pa) is the atmospheric pressure. The wetted surface is estimated as equal to half of the total surface area in the calculations.

The rate of heat taken away by the impinging water droplets is formulated as

$$Q_w = \kappa_1 \cdot \kappa_2 \cdot W \cdot v \cdot E \cdot D_c \cdot L \cdot c_w \cdot (T_s - T_a) \tag{4.4}$$

where κ_1 and κ_2 are correction factors, W (kg/m^3) is the liquid water content (LWC) of the air, v (m/s) is the air velocity of the free stream, E is the collection efficiency, D_c (m) is the conductor external diameter, and c_w (J/kg·K) is the specific heat of liquid water. In Eq. 4.4, the water droplet temperature may be assumed to be the same as air, because of their small impact velocity (Makkonen, 2000). However, some larger water droplets could be supercooled. Furthermore, it is assumed in Eq. 4.4 that the air water droplet collection area is equal to the longitudinal projected area ($D_c \cdot L$) of the conductor. The collection efficiency, E, is the ratio of the water content in the free stream to the water content collected after impinging on the conductor surface. E is always smaller than unity because the small water droplets following air streamlines may bypass the conductor (Makkonen, 2000). The leeward side of the conductor, though less exposed to droplet strikes, nevertheless gets wetted by runback water from the front side. The influence of the runback water flow towards the leeward surface of the conductor, as well the effect of supercooled water droplets, is taken into consideration with coefficient κ_1. As it means additional heat loss, this coefficient should be set to $\kappa_1 \geq 1$ (Péter et al., 2007).

The non-uniform water film formed on the stranded surface by the impinging droplets cannot reach a perfect thermal equilibrium with the conductor. Let T_s be the water film temperature near to the conductor surface, and T_a be a good approximation of the temperature at the water–air interface in Fig. 4.1. In theory, the average water film temperature, T_{av}, should be used instead of T_s in Eq. 4.4. However, since the calculation of T_{av} is not feasible due to strong variations in water layer thickness, it is assumed to be the same as the surface temperature of the conductor. This effect is considered in the correction factor κ_2, which should be set less than unity ($\kappa_2 \leq 1$) (Péter et al., 2007). In the setting of this term, the following factors are also taken into consideration: a slightly higher conductor temperature than 0 °C, the displacement of the water layer separation point, as well as the fact that the leeward side of the conductor is sometimes exposed to impinging water droplets in the turbulent wake.

It is a challenge to find the appropriate values of the collection efficiency, E, the overall heat transfer coefficient, h, as well as the κ_1 and κ_2 correction factors for stranded conductors. The relevant literature offers several methods to calculate the collection efficiency (Langmuir & Blodgett, 1945; Makkonen, 2000; Péter, 2006). As for the overall heat transfer coefficient, as well as the κ_1 and κ_2 correction factors, they can be estimated from the experimental results obtained in Péter (2006). Other related references are CIGRÉ Working Group 22.12 (1992, 1997), IEC Standard 60865 (1993), and Morgan (1991).

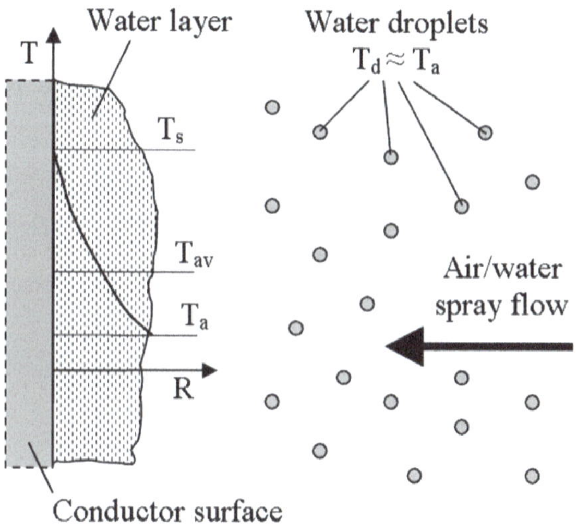

Fig. 4.1 Schematic temperature distribution through water layer (Péter, 2006)

4.3 Mechanisms of Ice Adhesion

Ice adheres to virtually any solid material. It is necessary to understand, from a theoretical point of view, what are the physico-chemical mechanisms responsible for its adhesion on a given substrate to study and find the proper strategies to remove or melt ice and snow from a given substrate, to prevent its formation, or to reduce its adhesion strength.

Adhesion is an important and complex physical concept (Darque-Ceretti & Felder, 2003; Myers, 1999). In fact, three different types of adhesion can be identified depending on the problem scale:

1. "Thermodynamic adhesion" or "ideal adhesion" defined by the reversible work needed to separate two surfaces bonded by molecular interaction forces across the interface (van der Waals and hydrogen type bonding), which are long-range interactions (>0.3 nm).
2. "Chemical adhesion" which involves the covalent, electrostatic, or metallic atomic bonding across a given interface, and which are short-range interactions (0.15–0.3 nm).
3. "Mechanical adhesion" which involves the mechanical interlocking of microscopic asperities at the interface between the two materials.

In the case of ice-solid interactions, the various factors involved are as follows:

1. Thermodynamics related to surface tension
2. Influence of intermolecular forces: electrostatic, van der Waals, and hydrogen bonding

3. Topographic influence, also referred to as roughness on several scales:

    a. Atomic roughness (vacancies, terrace, steps, or ad-atom)
    b. Nanoroughness
    c. Macroscopic roughness (several μm)

4. Influence of the water quasi-liquid layer (QLL) at the interface
5. Influence of the chemical heterogeneity of the solid surface
6. Other factors such as type and conditions of precipitation.

4.3.1 Surface Tension

Adhesion strength between two materials is referred to as adhesion work, W_a, which corresponds to the change in free energy when two separated surfaces are created for a given system. The liquid–solid–vapour system is schematically represented in Fig. 4.2 where γ_{sv}, γ_{lv}, and γ_{sl} are the surface energies of the solid, the liquid, and the solid/liquid interface, respectively. The angle Θ is the contact angle between the liquid and vapour phases.

Adhesion work W_a is given by Eqs. 4.5 and 4.6. It can be seen from Eq. 4.6 that the adhesion work can be determined from contact angle measurements. The binding energies of H_2O molecules to different solids are expected to be similar for water and ice (Petrenko & Whitworth, 1999). It can also be assumed that the W_a values on different ice-solid interfaces are correlated with the contact angle of liquid water. In fact, some authors, Bascom et al. (1969), Crouch and Hartley (1992), and Landy and Freiberger (1967), have measured the shear strength of ice at the contact interface on different materials and compared their results with the corresponding water contact angles. Figure 4.3 shows results compiled by Petrenko and Whitworth (1999). The correlation between shear adhesion strength and contact angle is somewhat weak, especially considering the log scaling on the vertical axis. This may be attributed to the fact the surfaces are not equivalently smooth and that the mechanism of failure is not considered in the measurement of W_a. Rearranging Eq. 4.6 yields Eq. 4.7, named the Young Equation.

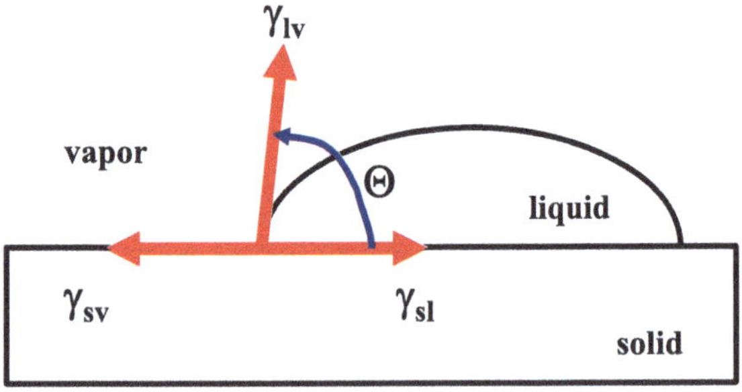

Fig. 4.2 Schematic representation of free surface energies at a triple phase system.

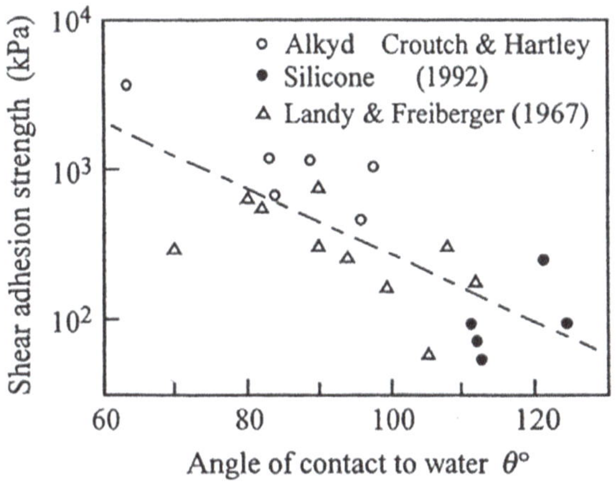

Fig. 4.3 Correlation between shear strength of ice adhesion and contact angle for various plastics (Petrenko & Whitworth, 1999)

$$W_A = \gamma_{sv} + \gamma_{lv} - \gamma_{sl} \tag{4.5}$$

$$W_A = \gamma_{lv}(1 + \cos\Theta) \tag{4.6}$$

$$\gamma_{lv}\cos\Theta = \gamma_{sv} - \gamma_{sl} \tag{4.7}$$

In the literature, many calculations of the surface free energies can be found (Table 4.1). Considering the values displayed in this table, materials of surface energy lower than liquid water (72 mN m^{-1}) should exhibit low ice adhesion strengths. For instance, PTFE (or Teflon®) and PDMS would appear to be the best candidates to counter ice adhesion. However, it must be emphasized that surface energy measurements must be performed on ultra clean and very flat surfaces. While the first requirement can be achieved relatively easily, a given material such as industrial surfaces or coatings such as those of OHL conductors may have a degree of roughness large enough to affect contact angle measurements. Also, very rough surfaces may exhibit superhydrophobicity or ultrahydrophobicity with contact angle ($\Theta > 150°$).

4.3.2 Intermolecular Forces

Adhesion of ice on a solid substrate is governed by a combination of a number of forces and factors that occurs at the ice/solid interface, including electrostatic forces, hydrogen bonding, van der Waals forces, mechanical interlocking, and liquid-like layer (CIGRE WG B2.44, 2015; Farzaneh, 2008; Petrenko & Peng, 2003). The mechanisms of ice adhesion are reported in CIGRE TB 631 (CIGRE WG B2.44, 2015). The knowledge of the following intimate factors that determine ice adhesion on surfaces is the starting point in designing coatings and treatments to reducing ice adhesion and facilitate ice detachment.

Table 4.1 Surface free energies for different substrates (Boluk, 1996)

Material	Surface free surface energy(mN m^{-1})
Poly(tetrafluoroethylene) or PTFE	20
Poly(dimethylsiloxane) or PDMS	22
PVDF	25
Polystyrene	33
Nylon 6,6	46
Copper	60
Liquid water	**72**
Silica (dehydrated)	78
Anatase (TiO_2)	92
Iron oxide (Fe_2O_3)	107
Aluminium	>100
Steel	>100
Rutile (TiO_2)	143

4.3.2.1 Electrostatic Forces

Electrostatic interactions occur at material/substrate interfaces when they have different electronic bond structures (Kasaai & Farzaneh, 2004), and both materials gain charge through an in-balance of charges. Electrostatic forces at the ice-substrate interface are related to the presence of ionic and Bjerrum defects in the crystalline ice structure. Electrostatic attraction theory is based on Coulomb's law and receptor–donor interactions. Petrenko and Ryzhkin (1997) studied the theoretical electrostatic interaction taking place at ice/metal or ice/dielectric interfaces. Their theory, summarized next, is based on the Jaccard theory stating that the electrical charge in ice is transferred by protonic point defects, L, D, H_3O^+, and OH^-, as shown in Fig. 4.4.

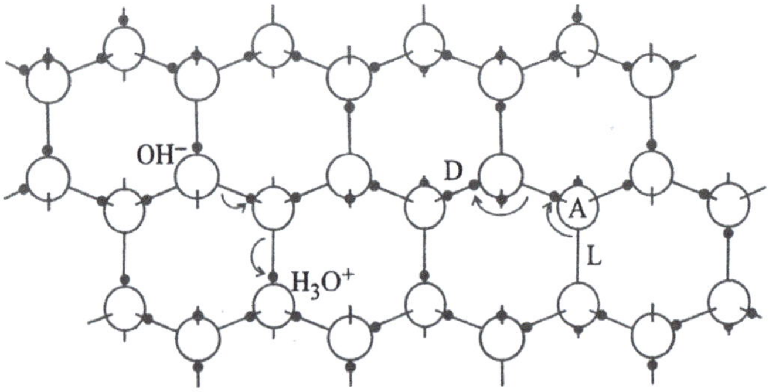

Fig. 4.4 Ionic and Bjerrum defects in ice structure (Petrenko & Whitworth, 1999)

Empty bonds (L-defect), bonds with two protons (D-defect), and ionic defects resulting from the water ionization reaction are electrically charged and distributed in the solid (Petrenko & Ryzhkin, 1997-08). These defects play a role like electron and hole conduction in electronic semiconductors. The empty bond is an L-defect, and the bond with two protons is a doubly occupied D-defect. The other two defects correspond to ionic defects resulting from the water ionization reaction Eq. 4.8:

$$2H_2O \leftrightarrow H_3O^+ + OH^- \tag{4.8}$$

At an ice crystal surface, some of the protonic defects may be captured in the surface states, which have energies lower than those in the bulk of the ice. The capture of charged protonic defects in the surface states will result in a surface charge build-up and therefore in the creation of a surface electric field. Additionally, at the metal or dielectric surface, a surface charge is created. The force of this charge is described in Eq. 4.9, where the solid has an effective dielectric constant (ε) larger than that of ice (ε_{ice}).

$$F = \frac{q^2}{16\pi\varepsilon_0\varepsilon r^2}\,\frac{\varepsilon_{ice} - \varepsilon}{\varepsilon_{ice} + \varepsilon} \tag{4.9}$$

The image force F in Eq. 4.9 is inversely proportional to the square of the distance to the interface r, q is the charge, and ε_0 the absolute permittivity in vacuum.

Using the energies of these different defects, the theoretical adhesive energies for an ice/metal interface were between 0.08 and 1.3 J m^{-2}. These values were comparable or even higher than those found experimentally at –20 °C. However, these authors did not consider the surface oxide or hydroxide layer which, in the case of non-noble metals or industrial metals or alloys, is always present at the surface. In fact, the thickness of such a layer can have either a very low conductivity (e.g. TiO_2), or be very thick in the case of anodized aluminium (Al_2O_3). They assumed that the very same mechanism is applicable to the ice/insulator interface, but no evaluation of the corresponding energy of adhesion was performed (Petrenko & Ryzhkin, 1997). A charge q_{ice} on the ice surface induces the "image charge" in a metal, while the very same charge q_{ice} will induce a smaller "image charge", q_{diel}, in the insulator, as shown in Equation (4.10), where ε_{diel} is the dielectric permittivity of the insulator. In most solid dielectrics, ε_{diel} is much larger than 1, and the induced charges are comparable with ones induced in metals. The lower ε is, the lower is the electrostatic-related adhesion. For instance, Teflon®, which has a very low ε ($\sim$2.1), exhibits icephobic properties.

$$q_{diel} = q_{ice}\left(\frac{\varepsilon_{diel} - 1}{\varepsilon_{diel} + 1}\right) \tag{4.10}$$

4.3.2.2 Surface States

Some of the charged defects may be captured in the surface states, which have energies lower than those in the bulk of the ice, resulting in a surface charge build-up and a consequent surface electric field. This in turn induces the generation of imaginary charged states in the underlying substrate, proportionally to the dielectric permittivity. The attractive electrostatic force is dependent on the dielectric constant of the ice and of the substrate, and it is a short-range interaction (0.15–0.3 nm) (CIGRE WG B2.29, 2010; CIGRE WG B2.44, 2015). Hydrogen bonding is the interaction between hydrogen atoms and electronegative atoms. Hydrogen atoms in the ice tend to form a bridge between ice and the surface, thus contributing to ice adhesion strength. This is a long-range interaction (>0.3 nm).

4.3.2.3 *Acid–Base Hydrogen Bonding Forces*

Hydrogen bonding, which can also be regarded as an electrostatic interaction, results from the distribution of a proton (hydrogen atom) between two electronegative atoms such as oxygen, nitrogen, or fluorine. In fact, these attractions are responsible for the cohesion of solid ice and are also present in liquid water (H_2O). Therefore, this type of bonding plays an important role because oxides, $-CH_n$ or $-CF_n$ (n = 1–3) types of radicals, are often present on the solid side. Chemical bonding between ice and other solids is largely hydrogen bonding, but it has not been studied and quantified in detail yet. Van Oss et al. (1992) measured the contact angles of a flat, polycrystalline ice surface with several liquids. They concluded that the contribution of the Lifshitz–van der Waals non-polar component to the ice surface tension (26.9 mJ m^{-2}) was less than that of the polar Lewis acid–base component (39.6 mJ m^{-2}), which in the case of ice is due to hydrogen bonding.

4.3.2.4 *Lifshitz–van der Waals Forces*

The Lifshitz–van der Waals interaction forces are always present. It is the most common interfacial force, resulting from a temporary dipole–dipole interaction. The Lifshitz–van der Waals interaction between ice and several metals and insulators was calculated by Wilen et al. (1995), and they concluded that this mechanism is not a dominant factor for ice adhesion. This confirmed results of Petrenko and Ryzhkin (1997-08), Sojoudi et al. (2015), Van Oss et al. (1992).

4.3.2.5 Mechanical Interlocking

Mechanical interlocking is related to the water penetration, expansion, and solidification in the three-dimensional structures of the practical surfaces. This mechanism can lead to strong interlocking between ice and the substrate (CIGRE WG B2.29, 2010; Farzaneh, 2008).

4.3.2.6 Quasi-Liquid Layers

Liquid-like layers existing on the ice surface play a significant role in ice adhesion mechanism and strength (CIGRE WG B2.29, 2010; CIGRE WG B2.44, 2015). The presence of quasi-liquid layer (QLL) may increase the contact surface area between the ice and the solid since it wets the surface of the solid. The thickness of QLL varies with the temperature, ranging from 10 to 100 nm at −4.5 °C, which corresponds to a 30–300 water molecule depth layer. Many studies revealed that the QLL familiar near 0 °C is also present at temperatures as low as −30 °C (Dash et al., 2006; Jellinek, 1962).

4.3.2.7 Capillary Force

Capillary force is a function of the liquid water content of the snow and of the air quantity that occupies the micropores and channels. The overall effect is that capillarity forces reach a maximum at LWC levels around 20–25%, also depending on temperatures, crystals dimensions, and shape, and decreases at lower and higher LWC (Shore, 2000).

4.3.3 Influence of Surface Roughness on Ice Adhesion

Industrial or functionalized surfaces are never perfectly plane. Aluminium and stainless steel-based alloys always exhibit a certain surface roughness or even porosity. Water can penetrate their three-dimensional surface structures and subsequently solidify. Strong joints are created through enhanced mechanical interlocking resulting from the unique property of solid water to expand during the freezing process as shown in Fig. 4.5a. A comprehensive review of the ice adhesion due to mechanical interlocking can be found in (Kasaai & Farzaneh, 2004). The degree of surface roughness therefore affects the strength of the joint. The expansion coefficient of water around 0 °C is greater than those of metals and oxides: water expands upon freezing, whereas metals and the oxides contract. On the other hand, air may be entrapped in some pores, and the resulting pressure build-up can be significant and may lead to crack initiation and propagation as well as ice de-bonding, as shown in Fig. 4.5b (Penn & Meyerson, 1992). Therefore, internal pressure build-up or residual stress can result in ice de-bonding.

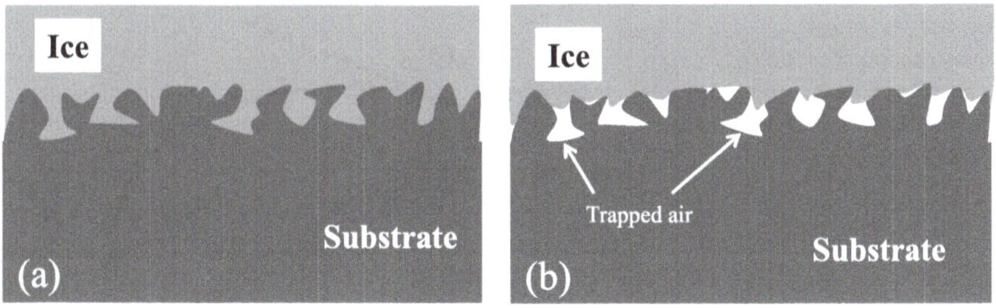

Fig. 4.5 Schematic representation of ice adhesion on a rough substrate. **a** Mechanical interlocking and **b** low adhesion strength due to air entrapment

The CIGELE team at University of Quebec at Chicoutimi discovered that certain superhydrophobic surfaces, with contact angle $\Theta > 150°$, can drastically reduce ice adhesion strength (Noormohammed, 2009; Safaee, 2008). A condensed description is presented here. More details on the theoretical approach of superhydrophobicity can be found in Bico et al. (2002), Quéré (2002, 2005).

Surface nano and micro-asperities are usually responsible for the superhydrophobic phenomenon. The value of the contact angle measured on a rough surface corresponds to an apparent contact angle (Θ_r), which depends on both the surface roughness and the behaviour of the liquid drop at the interface. Two different models were proposed to describe this phenomenon, known as the Wenzel (1936) and the Cassie and Baxter (1944) models, represented schematically in Fig. 4.6.

These models suggest that contact angle measurements alone (obtained by the sessile drop method) are not sufficient to characterize properly superhydrophobic surfaces, and that the evaluation of contact angle hysteresis, as shown in Fig. 4.7 and Eq. 4.11, is necessary. The contact angle hysteresis can be evaluated when a liquid droplet is allowed to slide on a given surface. Advancing and receding angles, Θ_{adv} and Θ_{rec}, are measured on the distorted liquid droplet, and $\Delta\Theta$ is computed. If $\Delta\Theta \to 0$, the surface exhibits high superhydrophobicity. It has been demonstrated that superhydrophobic surfaces with $\Delta\Theta < 5°$ are icephobic (Kulinich & Farzaneh, 2009).

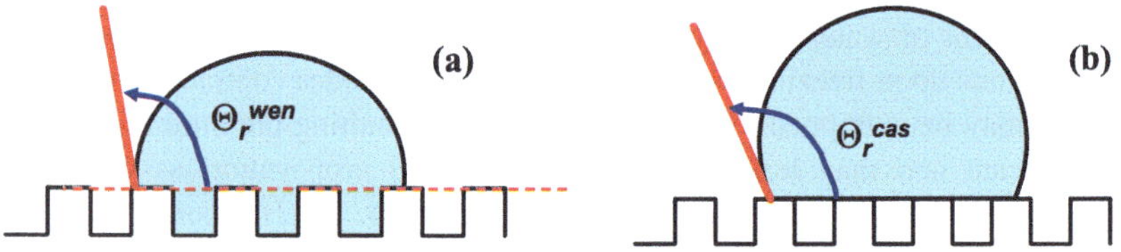

Fig. 4.6 Wenzel (**a**) and Cassie (**b**) models for a water droplet deposited on a rough surface

Fig. 4.7 Schematic representation of a sliding water droplet for contact angle hysteresis ($\Delta\Theta$) evaluation

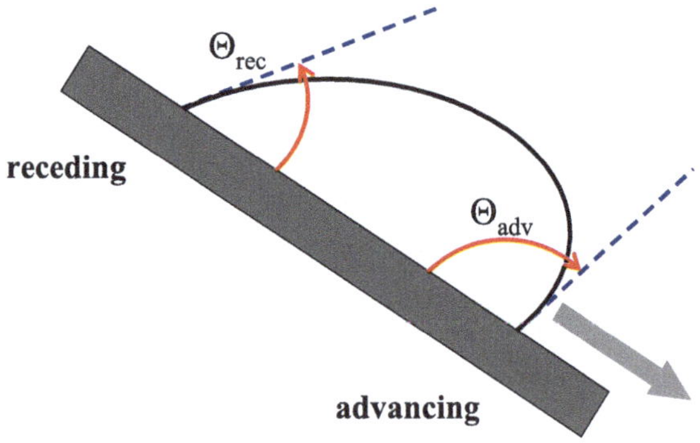

$$\Delta\Theta = \Theta_{adv} - \Theta_{rec} \quad (4.11)$$

Additional mathematical treatment of Wenzel and Cassie models is found in Sect. 4.4.2.

4.3.4 Influence of the Quasi-Liquid Layer

Since ice frequently exists at ambient temperatures very close to the melting point, a quasi-liquid layer (QLL) is present on the ice surface as well as at the ice-solid interface, which constitutes an influential parameter on ice adhesion strength. Early studies by Jellinek (1962) have been devoted to the behaviour of an ice surface and its interaction with solid surfaces, accounting for the existence of a QLL relative to temperature. It was estimated that the thickness of the QLL can extend from 100 to 1000 Å at –4.5 °C, which corresponds to a 30–300 water molecule depth layer. A recent review by Dash et al. (2006) cited many laboratory studies that have explored the interfacial fusion of ice with various substrates such as quartz, metals, graphite, glass, silica, polystyrene, and PVC. A variety of techniques have been used for that purpose, namely atomic force microscopy, mercury porosimetry, quartz microbalance, and X-ray scattering. Many of these cited studies noted that interfacial fusion starts at temperatures much lower than that of surface fusion. For example, in the case of graphite, interfacial fusion is detected as low as –30 °C. The presence of QLL is clearly a factor to be considered, since it wets the surface of the solid and therefore increases the contact surface area between the ice and the solid, which promotes ice adherence.

4.3.5 Heterogeneous Surface

In 1994 and 1997, Murase et al. (1994),Murase and Fujibayash (1997) published their research results on heterogeneous polymer coatings aimed at decreasing ice adhesion. These studies involved coatings with organopolysiloxane chains (presence of $-CH_3$ radicals) grafted with fluoro-polymer chains (presence of $-CF_2$): such coatings were found to drastically reduced ice and snow accretion. To explain their findings, Murase performed molecular orbital energy calculations for three molecules: ethane (C_2H_6), dimethylesiloxane (DMS) and hexafluoroethane (C_2F_6). The hydrogen bond length (O–H and F–H) as well as the various interaction energies was evaluated, and the former were found to differ widely depending on the molecular group examined. In fact, there was a slight repulsion between a water molecule and a siloxane group, while a strong attraction was observed for the fluorocarbon group and H_2O molecules. It is worth mentioning also that the water molecule orientations at the surface of the fluorocarbon group and the polysiloxane group were completely different. This work shows that by creating at the molecular level various disparities in terms of energy bonding and water molecule orientation, the ice-material interface can be weakened with the probable creation of a wide range of dislocations and slips in the ice structure immediately adjacent to the solid.

4.3.6 Influence of Ice Structure on Its Adhesion

The ice structure at the ice-metal or ice-metal oxides interface usually evolves from the presence of large polygonal crystals or grains to the formation of smaller ones with similar shape due to recrystallization. The overall process occurs within a few hours after initial freezing (Bascom et al., 1969; Crouch & Hartley, 1992; Jellinek, 1962). This process leads to the accumulation of dislocations at ice surfaces and occurs in any polycrystalline material subject to a steady stress. This stress is induced by differences in thermal coefficients of expansion and contraction of the ice and the solid. There are also volume changes as the ice forms and cools (Crouch & Hartley, 1992).

As a rule of thumb, the greater the grain size, the lower the ice adhesion. Druez et al. (1978) performed wind tunnel studies on ice microstructure for water droplets having variable speed and temperature during their impacts on aluminium conductors. It was found that the adhesion strength decreased with increasing ice grain size, which was in turn dependent on temperature and air velocity. In fact, ice grain size decreased as wind velocity increased and decreased as temperature decreased.

Another study (Penn & Meyerson, 1992) has shown that the ice crystals (or grains) not directly attached onto steel, cement concrete, asphalt concrete, and glass substrates were randomly oriented polygons, while the crystals directly attached to the substrate surfaces were air bubble-free, with a much smaller polygon structure (or needles, in the case of steel). The layer formed by these crystals can be assumed to be stronger than bulk ice, this added strength being derived from the small size of the crystals in the absence of visible flaws. The presence of such a strong layer would explain the high adhesion strength of ice on these materials. It would also suggest that mechanical ice removal procedures are likely to leave a thin layer of adhered ice behind. The study also addressed the appearance of ice crystals on polystyrene and found that the ice interference fringes on this material were significant. Interference fringes are an indication of mechanical stress within the ice crystals themselves. This stress was attributed to the low thermal conductivity and the high coefficient of thermal expansion of polystyrene.

4.3.7 Concluding Remarks

The rigorous scientific process of linking experimental results to the corresponding theoretical developments is a difficult task in the case of ice-solid interfaces. Firstly, on the ice side, many types of ice are encountered in nature: wet and dry snow, rime, and glaze ice. Moreover, contaminants and air bubbles may be present at the ice-material interface, and this is almost the case in OHL conductor icing. Secondly, on the materials side and from an engineering point of view, the presence of atomically clean and flat surfaces is never achieved, which makes experimental work on adhesion very difficult to compare with theory. Nevertheless, it can be clearly stated that the two main influences on ice adhesion strength are:

i. the electrostatic forces present at the ice-solid interface
ii. the surface roughness of the substrate.

Concerning the electrostatic forces, materials, or coatings with very low dielectric permittivity such as fluoropolymers, generally exhibit low ice adhesion strengths. As for substrate roughness, this undoubtedly plays a crucial and complex role in ice adhesion strength. In the presence of strong mechanical interlocking, ice cracking and even ice de-bonding can be observed due to excessive pressure build-up resulting from air trapped in closed pores. Therefore, the engineering of an optimum surface roughness, having superhydrophobic properties for instance, may create internal stresses large enough to de-bond ice, particularly when the substrate is composed of a low dielectric material.

Other parameters should also be considered to fully understand the ice adhesion mechanism: temperature (as it affects the QLL) and the type of precipitation (e.g. impact velocities of supercooled water droplets) that influence water penetration within a three-dimensional surface structure, as well as the shape and size of the ice crystal grains affecting ice adhesion strength. Furthermore, factors usually leading to poor ice adhesion strength include substrate low thermal conductivity, relatively large differences between ice and substrate thermal expansion coefficients, as well as some heterogeneous materials.

In addition to the above concerns, the use of hydrophobic coatings to reduce ice adhesion on OHL conductors has raised some other operational concerns such as the audible noise from the protected conductors. Furthermore, the influence of such coatings on corona noise levels and related interferences is not well understood and should be investigated.

4.4 Principles of Hydrophobicity and Icephobicity

Hydrophobicity and icephobicity are two essential functional characteristics necessary for elaboration of coatings for protecting power network equipment from icing and pollution. These characteristics are discussed to acquire a better understanding on how these coatings function.

Wetting is the result of molecular interactions at the liquid/solid interface which determines how well a liquid adheres to a surface. The extent to which a solid surface can be wetted with any liquid is commonly described in terms of wettability. The wettability of a surface depends mainly on surface energy and surface roughness (CIGRE WG B2.44, 2015).

Surface energy is the energy needed to generate a surface from a bulk material. The value of surface energy is one of the most important factors determining the wetting behaviour of a surface. A surface with higher energy, such as most metals, glasses, or some ceramics, leads to higher wettability and therefore moves towards hydrophilicity. On the other hand, low surface energy materials, such as some polymers or polymeric functions, tend to move towards hydrophobicity. Theoretically, the contact angle on a smooth surface cannot exceed 120°.

Surface roughness is another important parameter which can further increase water contact angle, theoretically up to 180°. The effect of roughness on the wetting behaviour along with wetting models for rough surfaces is important in both wetting and icing conditions.

4.4.1 Characterization of Wettability/Hydrophobicity

The shape of a water droplet on a solid and rigid horizontal surface is roughly a truncated sphere. The angle at which the sphere is truncated, called the contact angle (CA), provides a quantification of wettability. The definition of the contact angle (θ_c) is shown in Fig. 4.8.

$$\cos\theta_c = \frac{\gamma_{SV} - \gamma_{SL}}{\gamma_{LV}} \tag{4.12}$$

The contact angle (CA), called sometimes static contact angle, is defined as the angle at which the liquid-air interface meets the solid–liquid interface. For an ideally flat and homogeneous solid horizontal surface, contact angle, θ_c, can be alternatively described as the angle between the solid surface and the tangent of the droplets as shown in Fig. 4.8. The static contact angle is determined from Young equation as shown in Fig. 4.8. In this equation, θc is the equilibrium contact angle and γ is the surface energy for which the subscript denotes the relative interface. Here, the subscript in γ_{SL} refers to the solid–liquid interface, the subscript SV refers to solid–vapour interface and LV to the liquid–vapour interfaces shown.

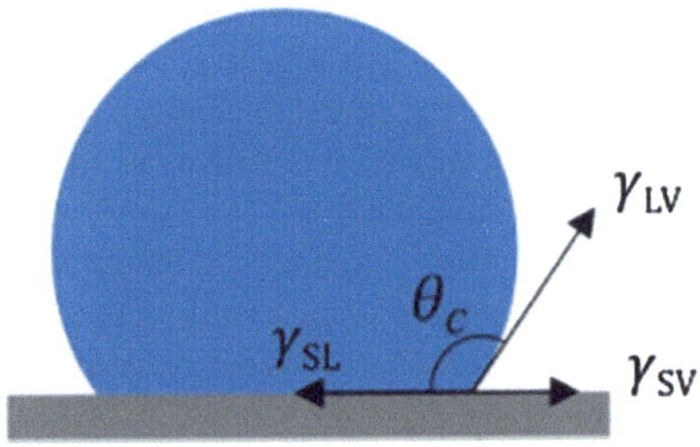

Fig. 4.8 Contact angle between a liquid and a solid surface (CIGRE WG B2.44, 2015)

Visual interpretation of contact angle on wetted insulator surfaces is provided by a seven-level guide, originally from STRI and now adopted as IEC TS 62073 (IEC, 2003).

4.4.2 Process of Wetting and Hydrophobicity

Unlike ideal surfaces, real surfaces do not have perfect smoothness, rigidity, or chemical homogeneity. To describe the wetting behaviour on real surfaces, other models leading to other equations should be described.

A mathematical model was developed by Wenzel to formulate Young concept of wetting behaviour on practical surfaces. On such a rough surface, the water can penetrate features of the surface as shown in Fig. 4.6a. This model is now known as the Wenzel wetting regime (Wenzel, 1936), governed by Eq. 4.13.

$$\cos\theta_w = r_s \cdot \cos\theta_c \tag{4.13}$$

In this equation, r_s is the roughness ratio and is defined as the ratio of true area of the solid surface to the apparent area, while θ_w and θ_c refer to the apparent Wenzel contact angle and the equilibrium Young contact angle, respectively.

The Cassie and Baxter (1944) assumes that the surface features are small enough that the capillary effect will prevent the water penetration Fig. 4.6b. Equation 4.14 governs this model.

$$\cos\theta_{CB} = f_s(\cos\theta_c + 1) - 1 \tag{4.14}$$

In the above equation, f_s is the fraction of solid/liquid interface where the water droplets in contact with surface, while θ_{CB} and θ_c refer to the apparent Cassie–Baxter contact angle and the equilibrium Young contact angle, respectively.

All the above concepts including Young, Wenzel, and Cassie–Baxter models consider the contact angle of a water droplet resting on the surface (static contact angle). To fully characterize the wetting behaviour on a surface besides static contact angle considerations, dynamic contact angles and contact angle hysteresis (CAH) should be also considered. Dynamic contact angles are defined as the contact angles related to moving liquid fronts (interfaces). For instance, when a droplet is moving on a surface, the contact angle in the movement direction is referred to as the advancing contact angle and the contact angle on the opposite side is referred to as the receding contact angle, as shown in Fig. 4.9.

Advancing contact angle (θ_a) corresponds to the maximum stable angle, whereas the receding contact angle (θ_r) is the minimum stable angle. Figure 4.9 also shows that the difference between θ_a and θ_r is defined as contact angle hysteresis (CAH).

Another important concept in wetting process of a surface is sliding angle, called also roll-off angle (α). It corresponds to the angle between the sample surface and the horizontal plane, α in Fig. 4.9, at which the liquid drop starts to slide off the sample surface under gravity influence.

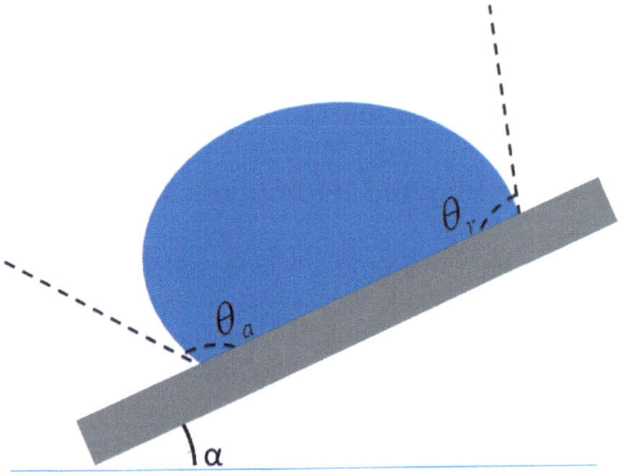

$$CAH = \theta_a - \theta_r$$

Fig. 4.9 Contact angle hysteresis on inclined surface (CIGRE WG B2.44, 2015)

Contact angle hysteresis and sliding contact angle are important factors in roll-off behaviour and consequently lower adhesion strength between the droplet and the surface, and easier sliding of the droplet on the surface (Shirtcliffe et al., 2001). Such surface properties have been proven to be an essential factor in self-cleaning surfaces and anti-icing applications (Guo et al., 2011). As concerns self-cleaning of such surfaces, water droplets will roll off the surface easily, absorbing dirt particles (Shulz & Sinapius, 2015).

Unlike ideal surfaces, real surfaces do not have perfect smoothness, rigidity, or chemical homogeneity. In order to describe the wetting behaviour on real surfaces, other models leading to other equations should be described. Wenzel (Fig. 4.10a) and Cassie–Baxter (Fig. 4.10b) models are two of the most common models defined for rough surfaces (Alizadeh et al., 2012).

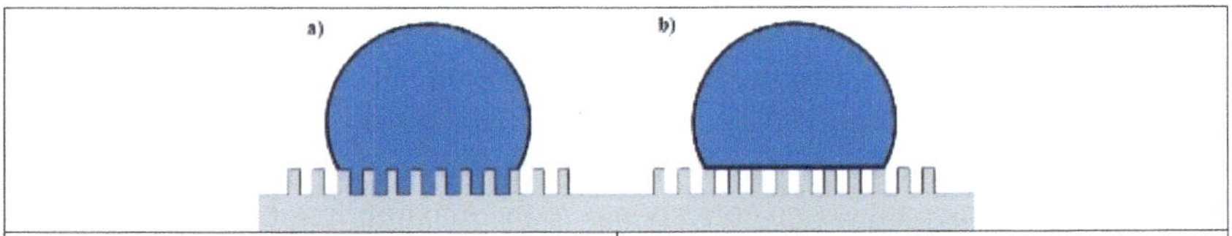

$\cos\theta' = r\cos\theta$	$\cos\theta' = -1 + f_s(\cos\theta + 1)$
a) Wenzel wetting model	b) Cassie-Baxter wetting model

Fig. 4.10 Hydrophobicity models

In the Wenzel model, it is assumed that water can penetrate the surface features, thus completely wetting the surface. In this case, the apparent contact angle of the rough surface, θ', correlates with the surface roughness and the contact angle of the smooth surface, θ, following the Wenzel equation where r, the roughness factor, is the ratio between the actual to projected solid–liquid contact area.

In Cassie–Baxter model, water cannot penetrate the surface features, so that air is trapped between the water droplet and the substrate, forming an air/solid composite interface. Using the Cassie–Baxter equation, the apparent contact angle on this surface, θ', can be calculated from the contact angle of the smooth surface, θ. In Cassie–Baxter equation, f_1 represents the fractions of solid–liquid contact area.

Under either of these regimes, a high contact angle does not necessarily translate to droplet mobility and self-cleaning. The ability to shed water droplets depends on the sliding angle, which defines the surface tilt required for a droplet to mobilize on the surface. The sliding angle is dependent on the interaction strength between water and solid, which can be further described by the Furmidge equation:

$$mg\sin\alpha = \sigma d(\cos\theta_r - \cos\theta_a) \tag{4.15}$$

where m is the mass of the water droplet, g is the gravitational acceleration, α is the sliding angle, σ is the surface tension of water, d is the droplet contact area diameter, θ_r is the receding angle, and θ_a is the advancing angle. In this equation, the left side represents the gravitational force, while the right side represents the capillary force. The term in the brackets is a measure of the interaction strength between water and the solid on the interface. The physical consequences of this equation can be further illustrated by characterizing two extremes:

- In the Casssie–Baxter regime, the actual water/solid contact area is very small which results in poor attachment of the water on the surface. Such surface has a small sliding angle and hence good self-cleaning properties.
- In the Wenzel regime, the actual contact area is much larger than the apparent area resulting in a much higher attachment strength. Under this regime, a surface with a very high contact angle may have a large sliding angle and no self-cleaning behaviour.

Considering the above quantitative concepts, a **superhydrophobic** surface is defined in Table 4.2 as a surface on which the water contact angle θ_C is greater than 150° and which has low contact angle hysteresis (CAH < 10°).

Table 4.2 Definition of hydrophilicity, hydrophobicity and superhydrophobicity (CIGRE WG B2.44, 2015)

	Hydrophilic	Hydrophobic	Superhydrophobic
Contact angle θ	$\theta < 90°$	$150° > \theta \geq 90°$	$180° > \theta \geq 150°$

4.4.3 Superhydrophobicity Links to Icephobicity

In analogy with the definition of hydrophobicity, the term icephobicity has been invented to indicate those surfaces which behave "virtuously" with respect to ice and snow, for example, by

- Preventing the accumulation of ice or snow on the surface
- Reducing the adhesion of ice and snow, thus facilitating the shedding of the accreted mass through action of gravity or external force.

Some correlation has been reported between superhydrophobicity and icephobicity (Alizadeh et al., 2012; Antonini et al., 2011; Bahadur et al., 2011; Tourkine et al., 2009; Zhang et al., 2012). It has been shown that a low contact angle hysteresis is a defining criterion in icephobicity (Kulinich & Farzaneh, 2009). Surfaces with low contact angle hysteresis showed lower ice adhesion strength along with longer delays in freezing times. This suggests that icephobicity is more likely to be achieved in Cassie–Baxter wetting regime. Recently, a few papers investigated this correlation and suggested that a superhydrophobic surface can reduce or prevent the ice accretion by one or more of the following mechanisms:

1. The heat transfer between the droplet and the surface is hindered by the insulation effect of the nanostructured roughness on the surface. In other words, the micro/nanostructure on a superhydrophobic surface acts as an insulating layer in the interface (Alizadeh et al., 2012; Antonini et al., 2011).
2. For a superhydrophobic surface, contact area between the surface and the water droplet is significantly smaller, and therefore, less potential nucleation points are present on the surface. In other words, lower contact area results in higher activation energy for nucleation and growth of ice crystals. It has been shown that for a superhydrophobic surface, the contributions of solid/liquid interface to the solidification process are negligible (Bahadur et al., 2011). Moreover, smaller contact area can lead to lower heat transfer area between the droplet and the surface, which subsequently delays the freezing point (Tourkine et al., 2009).
3. Unlike hydrophilic surfaces, water droplets are extremely mobile on a superhydrophobic surface and will rebound upon impact as shown in Fig. 4.11.

The time for which the water droplet is in contact with the surface before rebounding on a superhydrophobic surface is called *the drop rebound time* or *shedding time*. On a superhydrophobic surface, this period is usually a few tens of milliseconds. If the drop rebound time is smaller than the time needed for ice nucleation, then it can be expected that the ice formation is reduced or prevented (Antonini et al., 2011; Bahadur et al., 2011; Kulinich & Farzaneh, 2009; Mishchenko et al., 2010).

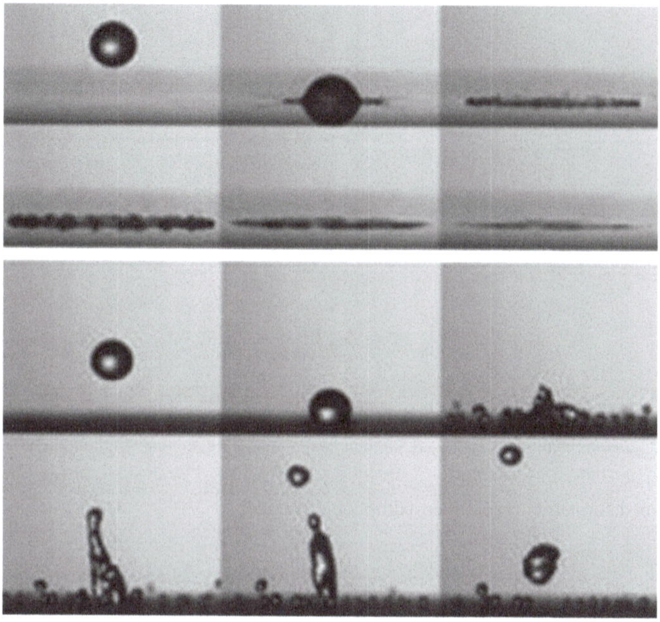

Fig. 4.11 Impact of a water droplet on a hydrophilic surface (top) and on a hydrophobic surface (bottom). The rebound phenomena on a superhydrophobic surface may be responsible for the delayed ice formation on superhydrophobic surface (Antonini et al., 2011)

Mishchenko et al. (2010) showed that in icephobic applications, frost formation in a supersaturated environment prior to freezing is significant. Before water solidification, frost can cover the micro/nanostructure on the surface, thus reducing the hydrophobicity of the surface (Fig. 4.12) and increasing the ice adhesion strength. In this case, ice adhesion strength was shown to increase linearly with the surface area, suggesting that frost can cover the whole structure, including post tops, side walls, and valleys (Varanasi et al., 2010). Therefore, extra care should be taken while dealing with the environments where frost formation is probable.

It is clear from Fig. 4.12 that frost can cover the whole structure and subsequently increase ice adhesion strength (Varanasi et al., 2010).

Generally, icephobicity is related to surface roughness, i.e. micro/nanostructured coatings. However, some studies suggest that it is possible to develop a surface which is smooth at molecular level and yet extremely icephobic (Menini & Farzaneh, 2009; Jafari et al., 2010). These surfaces are mainly characterized by their low contact angle hysteresis. The absence of micro/nanoroughness on the surface prevents the formation of frost or mechanical interlocking of ice. Moreover, low contact angle hysteresis should increase the mobility of supercooled droplets on the surface and thus further increase icing delay.

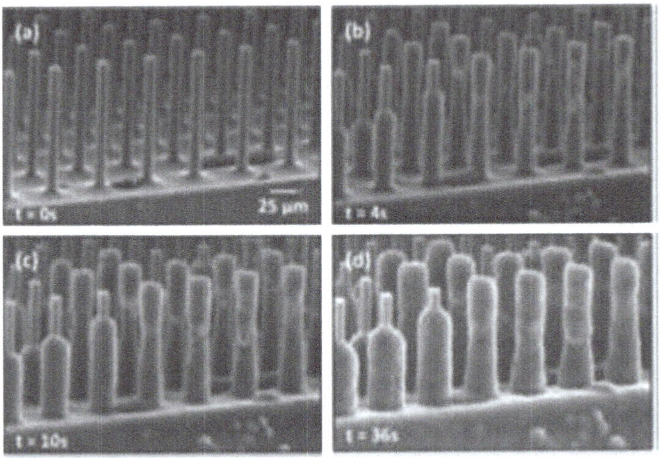

Fig. 4.12 Frost formation on a superhydrophobic surface

Few studies have investigated the icing mechanism on porous pavements (Antonini et al., 2011; Bahadur et al., 2011). Research showed that the pressure build-up due to water expansion upon freezing was significant. If air is entrapped in the surface structure, air pressure increases rapidly. For example, it was found that when 2% of the ice was frozen, air pressure was 36 atm. This pressure increased to 420 atm when 20% of the water was frozen (Varanasi et al., 2010). This pressure increase can lead to initiation and propagation of cracks in the ice structure, resulting in ice de-bonding. These results suggest that although surface roughness can lead to potential interlocking and therefore to high adhesion strength, carefully optimized roughness can facilitate the ice shedding process.

4.4.4 Principles of Icephobicity

Icephobic properties should be measured by determining the ice accumulation prevention effectiveness and/or the ice adhesion reduction effectiveness. Unfortunately, at present, a standard procedure for the determination of the above cited properties has not yet been defined. Each laboratory applies its own testing procedure, which complicates the comparison of results.

Many variables in study of icephobicity depend on

- The requirement of the anti-icing coating related to the specific application
- The measuring apparatus and methods adopted and
- The procedure used to produce the ice or snow amount on the test piece.

An extensive review of the literature on the concept and techniques for characterizing icephobic coatings was reported in CIGRE TB 631 (CIGRE WG B2.44, 2015). A review and update of these concepts and techniques are presented in the following sections.

The adhesion and the detachment of ice from a surface are ascribable to two different forces: adhesive and cohesive forces. The separation of ice from a surface can occur on two ways: adhesive failure at the interface between ice and substrate or cohesive failure in the bulk of ice layer (CIGRE WG B2.29, 2010; CIGRE WG B2.44, 2015). The adhesive failure depends on the surface energy of the underlying substrate (and therefore upon the coating characteristics), while the cohesive failure depends on the internal structure of the ice accretion and in principle is not modified by a coating.

For what concerns adhesion, coatings and engineered surfaces may present different topographic, chemical, and physical properties, who can be combined to obtain more effective anti-icing behaviour. The scientific literature reports a wide variety of materials and structure combinations, giving rise to widespread results, whose interpretation is not always clear.

On the other hand, ice and snow obtained by natural phenomena, in laboratory or in climatic chambers along different procedures and in different environmental conditions, show different cohesion properties, depending on density, liquid water content, crystalline structure, and voids. All these factors influence the shedding or detachment behaviour of the ice and can lead to different experimental results.

The behaviour of complex coating strategies is built on an understanding the individual factors that influence ice adhesion and shedding. The following sections discuss the most relevant of the surfaces and ice coating properties.

4.4.4.1 *Surface Characteristics Relevant to Icephobicity*

Surface Roughness

In thermal rating, the "rugosity" or roughness of a round-strand conductor R_f is defined by the strand diameter d (mm) and external diameter D (mm) as

$$R_f = \frac{d}{2(D-d)} \tag{4.16}$$

Rugosity plays a role in convection heat transfer (CIGRE WG 22.12, 2002) at high wind speeds and is a standard measure of surface roughness that can also be used in discussions of icephobicity.

A rough surface offers a larger contact area to the adherence of the ice and favours mechanical interlocking. Moreover, surface asperities impede the sliding of the ice during shedding. Many experimental studies show a clear correlation of shear stress versus surface roughness parameters (R_a, R_z, S_a) (CIGRE WG B2.44, 2015; Farzaneh, 2019; Sojoudi et al., 2015; Zou et al., 2011).

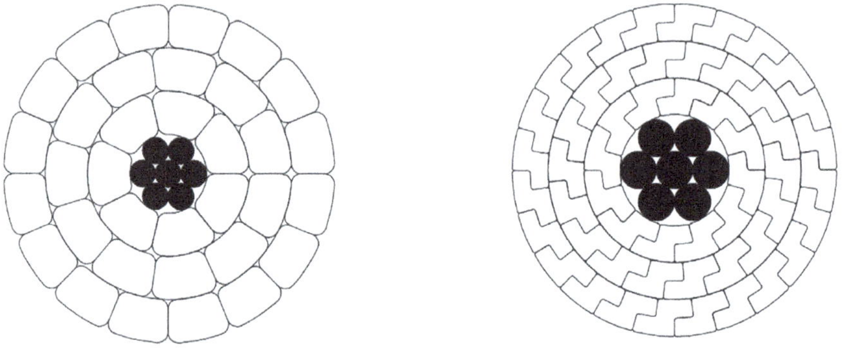

Fig. 4.13 Compact and "Z"-shaped conductor cross sections with reduced ice accretion characteristics (CIGRE TF B2.11.06, 2007-06)

Roughness plays a fundamental role in increasing hydrophobicity, as discussed below. Finally, rough surfaces promote the durability of anti-icing coatings because they are less prone to abrasion (Balordi et al., 2019; Kako et al., 2004).

Depending on the type of ice, the deposit may be altered by the rugosity of the conductor. This is sometimes a function more of the torsional stiffness of the conductor than its value of R_f from Eq. 4.16. For example, Z-shaped and compact trapezoidal strand conductors (Fig. 4.13) with diameter larger than 25 mm have torsional stiffness that is twice that of round-wire conductors. Additional torsional stiffness reduces the ability of the conductor to rotate, leading to asymmetrical ice deposits

Twisted-pair conductors are designed to limit aeolian vibration by changing the profile along the line. These pairs of wires may also affect the aerodynamics of the conductor with ice and affect stability conditions in the presence of ice and wind. The twisted pair create a kind of permanent air flow spoiler (drag damper) (CIGRE TF B2.11.06, 2007-06).

Surface Energy

Wettability is one of the most important characteristics of solid surface, and it can affect many physical and chemical processes, such as adhesion, adsorption, and lubrication. Surfaces with low affinity to liquid water (i.e. low wettability) are defined as hydrophobic.

Hydrophobicity is mainly determined by surface energy, and it is measured by the contact angle (CA) of a water drop deposited on the surface. A more complete measure includes advancing contact angle (ACA), receding contact angle (RCA), and their difference, named contact angle hysteresis (CAH).

Decreasing surface energy is a strategy to make anti-icing and anti-snow coating (Balordi et al., 2019-06). It is widely demonstrated that high values of CA, RCA, and low CAH are related to lower ice adhesion data, when no other factor is considered (i.e. roughness) (Cassie & Baxter, 1944; Zou et al., 2011).

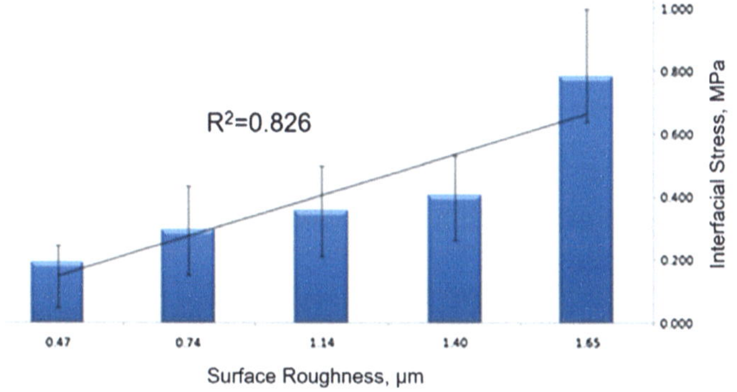

Fig. 4.14 Variation of interfacial stress with surface roughness

Surface energy influences the viscosity value at the water–solid interface of wet snow. The low surface energy leads to low viscosity which makes the snow slip easily and as a result leads to a low adhesion of snow

Surface wettability is also affected by microscopic roughness. A combination of low surface energy and proper roughness may lead to higher level of hydrophobicity.

Superhydrophobicity

When the contact angle is higher than 150°, Table 4.2 defines the surface as superhydrophobic. This condition is obtained combining roughness (at micrometric scale and at nanometric scale) with a low surface energy thin film, mimicking the surface microstructure of water-repellent plants or leaves (e.g. lotus leaf). Although superhydrophobic surfaces show very marked water repulsion (allowing a water drop to jump away), their behaviour with respect to ice and snow adhesion (Andersson et al., 2017) is quite variable and complex to describe. The main reason is that roughness and surface energy play contrasting roles (CIGRE WG B2.44, 2015; Farzaneh, 2019; Zou et al., 2011). In fact, on one hand, roughness increases contact area at the interface, favouring mechanical interlocking of ice and substrate. On the other hand, roughness, and specifically nanoscale and microscale roughness, is crucial in obtaining superhydrophobic surfaces.

Johnson and Dettre (1964) simulated the microstructure surface based on Wenzel equation and Cassie–Baxter equation and found the existence of a critical value for the roughness factor of solid surfaces. Beyond this threshold, the wettability model changes from Wenzel wetting model to Cassie–Baxter wetting model. With the greater roughness factor, a more stable Cassie–Baxter wetting state with enhanced non-wettability is obtained. It is well known that on a rough surface containing microscale structure, the main contact interface between droplet and solid surface is in infiltration state, and it is not easy to have a very large apparent contact angle. Since the micro–nanoscale composite structure on the surface of the lotus leaf causes a high apparent contact angle and a very low contact angle hysteresis, the structural effects of the lotus leaf surface and the bionic preparation techniques have attracted significant attention. The micro–nanoscale composite structure can easily eliminate the infiltration contact state caused by the primary microscale structure. It can increase the contact angle up to 150° (superhydrophobic) and decrease the contact angle hysteresis to less than 10°. This makes the research of the intrinsic relationship between special non-wettability and microstructure to have a leap-forward development.

According to the definition of wetting regimes known as Wenzel state and Cassie–Baxter states, a similar definition has been given to ice accreted materials whose contact area to the underlying substrate is fully infiltrated (Wenzel model) or partially suspended on air pockets situated in the roughness valleys (Cassie–Baxter model). A great difference in adhesion forces can be expected in the two cases (CIGRE WG B2.44, 2015).

The applicability of Wenzel or Cassie–Baxter ice adhesion model is also related to the solidification process and on its velocity, temperature, and humidity. In fact, it has been demonstrated that if the surface is already covered by a thin layer of frost due to vapour condensation in humid environment, the hydrophobic properties of the surface are compromised, and the subsequent precipitation (water or snow) will be deposited on an ice film and not on the hydrophobic coating. On the contrary, small grains of dry snow falling in very cold environments will probably sit on a rough and dry surface presenting many air pockets, leading to a Cassie–Baxter ice, easier to remove. This aspect may give a reason to the different results obtained by experimental laboratories where ice samples are prepared along very different procedures.

Many researchers found a correlation between superhydrophobicity and icephobicity (Bormashenko & Grynyov, 2012; Kako et al., 2004; Marmur, 2003; Wolansky & Marmur, 1999-10). However, many factors come into play, experimental results of ice adhesion on superhydrophobic surfaces are rather disperse and subject to different interpretations (Yeong et al., 2015).

Thermal Conductivity

The solidification heat of about 80 cal/g released in the ice accretion process can be dissipated through the underlying solid substrate or in the air, possibly favoured by the wind. A low thermal conductivity towards the substrate decreases the heat transfer and therefore delays the freezing process (Tourkine et al., 2009).

When considering metallic components (conductors, ground wires, and towers), the thermal conductivity of the ice/substrate interface can be limited by the conductivity of the coating and by the contact area. Therefore, the thermal flux can be lowered by acting on the coating material (choosing a bad thermal conductor, for example, Teflon or silicon) or by reducing the contact area (e.g.in Cassie–Baxter regime). On insulators, thermal conductivity in the substrate is already at a minimum level.

The contact area between the surface and a water droplet is relatively small for a superhydrophobic surface in Cassie–Baxter regime (Cassie & Baxter, 1944). In this case, the micro/nanoroughness on a surface acts as an insulating layer on the interface to reduce the heat flux between the liquid and the substrate (Alizadeh et al., 2012; Asadollahi, et al., 2019; Guo et al., 2012; Zhang et al., 2012).

Significant delay in freezing time can imply lower ice accreted volumes, especially when it combines with reduced time of contact of supercooled droplets with the surface, as it occurs for example on a superhydrophobic surface where water droplets have great mobility and rebound phenomena upon impact. On the contrary, when snow (already crystalline) falls on the surface, low thermal conductivity has no effect, since solidification is not the main physical process in the snow accretion and adhesion.

Time of Contact

As mentioned above, water droplets have high mobility on some hydrophobic or superhydrophobic surfaces, often characterized by high receding contact angle (RCA) and very low hysteresis. On these surfaces, water droplets run quite freely or suffer strong rebound when they impact at a certain velocity. This phenomenon allows too short a time to nucleate a solid ice particle, and, in practice, the droplets run away before starting solidification. This aspect is particularly relevant when the droplets impact with some velocity, due to strong winds, but loses its relevance in low velocity regimes.

Dielectric Constant

The dielectric constant of the coating material affects the electrostatic attractive force. Surface materials with lower dielectric constant can exhibit lower levels of snow adhesion (Ryzhkin & Petrenko, 1997). Since vander der Waals forces decay much more rapidly with distance than electrostatic forces, a low electrostatic force plays a significant role in ice adhesion. As a demonstration of this principle, electrically insulating materials such as Teflon(R), Poly-perfluoroalkylacrylate (PFAA), polypropylene, and polydimethylsiloxane (PDMS) had, in order, large to small effects on the sliding angle of supercooled water drops in ice adhesion studies (Murase & Fujibayashi, 1997).

Surface Heterogeneity

The research published in Murase and Fujibayashi (1997) and Murase et al. (1994-12) showed that a coating presenting both –CH_3 and –CF_2 radicals at the surface was able to drastically reduce ice and snow accretion. Calculation demonstrated that the hydrogen bond length (O–H and F–H) as well as the interaction energies differs widely depending on the molecular group, resulting in slight repulsion (between a water molecule and a siloxane group) or a strong attraction (between fluorocarbon group and H_2O molecules). Moreover, the water molecule orientations at the surface of the fluorocarbon group and the polysiloxane group were completely different. This in turn can lead to the creation of a wide range of dislocations and slips in the ice structure immediately adjacent to the solid that weaken the interface adhesion.

4.5 Anti-icing Coating Concepts for Power Network Equipment

A coating is a material applied to the surface of an object to impart surficial properties different from those of the underlying substrate. The coating can be in the form of thin or thick film, varnish, tape, generally covering all the surface. Different types of materials and technologies can be used to form a coating. Some technology does not apply a different material on the surface but generates a modification of the surface by using mechanical, chemical, or more complex physical–chemical treatments, and these are called “engineered surfaces”. In a wider sense, devices that can be applied to a surface could be also called coatings.

As a general distinction, coatings can be classified as passive or active (CIGRE WG B2.29, 2010; CIGRE WG B2.44, 2015; Farzaneh, 2019; Farzaneh & Volat, 2008). Passive coatings require no external energy to be effective. They are based on the exploitation of one or more of the factors that influence ice accretion, adhesion and shedding described above. While a few active coating techniques were described in TB 438 (CIGRE WG B2.29, 2010), none has developed further in 2021. Active coating methods for de-icing are applicable to conductors. However, some semi-conducting coatings applicable to insulator surfaces can also be categorized as active coatings and provide significant benefits in winter fog and flashover conditions.

Much effort has been devoted to the development of coatings for power network equipment to reduce the amount of ice accretion for preventing potential damage to overhead lines during icing periods. A review of some of the developed coatings with potential application to conductors, ground wires and insulators is reported in CIGRE TB 631 (CIGRE WG B2.44, 2015) and more recently in Farzaneh (2019). Based on these reviews, several conceptual coating strategies have been developed and tested. In the following, an updated overview of these concepts with examples and experimental results will be presented. Conceptual application examples and results verified on a pilot scale or in field tests are discussed in the sections that follow.

4.5.1 Concept 1—Decrease Mechanical Friction by Reducing Surface Roughness

A simple way to increase the anti-icing properties of aluminium surface is to reduce the mean surface roughness, to damp mechanical friction on crevices, as illustrated in Fig. 4.14 (Hassan et al., 2010)

4.5.2 Concept 2—Decrease Mechanical Interlocking by Reducing Contact Ice/Surface Area

Polydimethylsiloxane (PDMS) microrod film has effective anti-icing properties due to the hydrophobic surface. Hydrophobic surface will increase the contact angle and

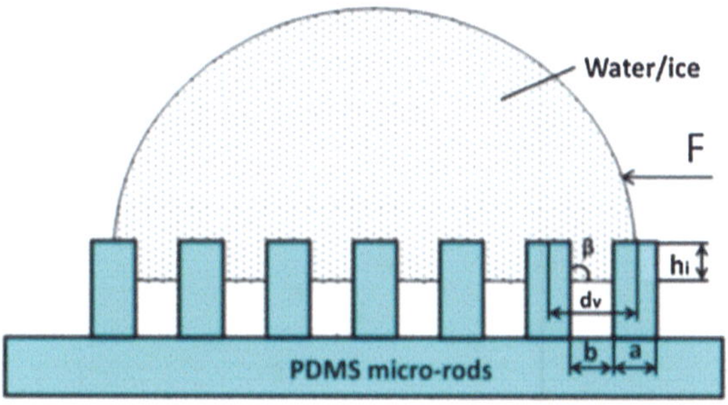

Fig. 4.15 Intruding depth model

decrease contact angle hysteresis, reducing the water/surface contact area. This results in a corresponding decrease in ice-solid contact area. Moreover, the intruding depth decreases when contact angle increases and contact angle hysteresis decreases, as shown in Fig. 4.15 (He et al., 2015). From the reference (He et al., 2015), the rod height, width, and space were set at $h = 20$ μm, $a = 10$ μm, and $20 \leq b \leq 50$ μm. A decreased intruding height denotes that less ice penetrated into the rod, thus resulting in decreased mechanical interlock.

4.5.3 Concept 3—Delay Freezing by Reducing Contact Area and Thermal Conductivity

One of the effective ways to reduce the accumulation of ice on the surface of a substrate is to delay the freezing of incident supercooled water droplets. In the following, some different methods are presented.

Example 1—Copper nanostructured hydrophobic surfaces show anti-icing properties. Compared with the smooth surface, a distinct delay of freezing was observed for nanostructured surfaces which have static contact angle greater than 150°. Moreover, the freezing delay was negligible among different droplet adhesion which indicates the actual reduction of the ice surface contact area increases the freezing time (Beshkar et al., 2017-06; Safaee et al., 2008).

Example 2—A uniformly textured hydrophobic surface with nanopillars of small top diameter can delay the freezing time. A smaller areal fraction corresponds to a smaller total contact area that reduces the total heat transfer rate from cold substrate to droplet. This delays the icing process, as shown in Fig. 4.16 (Nguyen et al., 2018). Figure 4.17 shows that ice adhesion force and delay of freezing time are not sensitive to pillar height of the texture.

Example 3—The rougher hydrophobic surface leads to a higher contact angle, therefore the longer freezing delay time. By mixing acrylonitrile-styrene-acrylate (ASA) resin with fluorine-modified titanium dioxide (TiO_2), an anti-ice coating was realized. As displayed in Fig. 4.18 (Qi et al., 2019), the freezing delay time exhibits a positive correlation with the value of contact angle.

Example 4—The nucleation rate during drop impact is significantly reduced by the mix of hydrophobic surface and hydrophobic particles, in Figs. 4.19 and 4.20

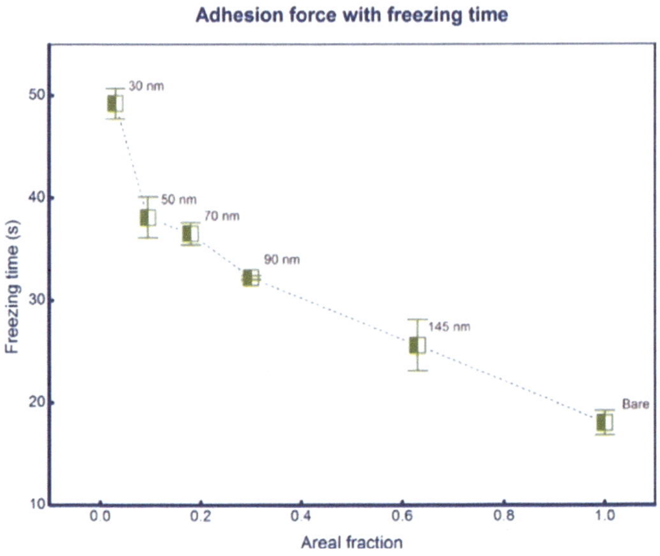

Fig. 4.16 Correlation between freezing time and areal fraction

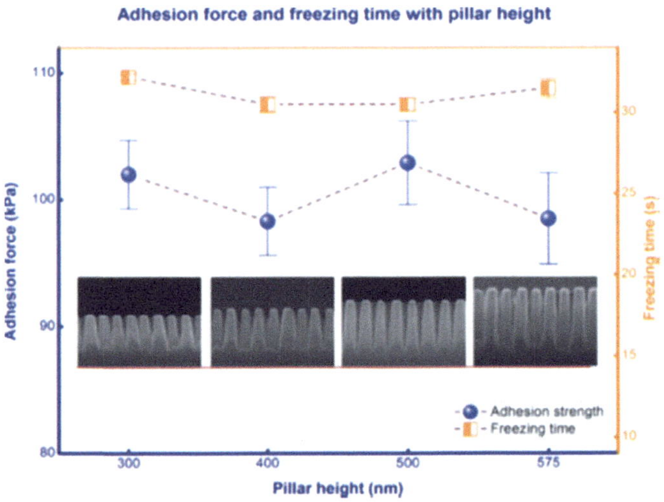

Fig. 4.17 Adhesion force and freezing time value for same pillar diameters of 90 nm with different pillar heights (300, 400, 500, and 575 nm).

(Wang et al., 2016). Experimentally, the freezing is completely prevented on the surface made of amphiphilic Janus particles even after a repletion of 100 drop impact experiments.

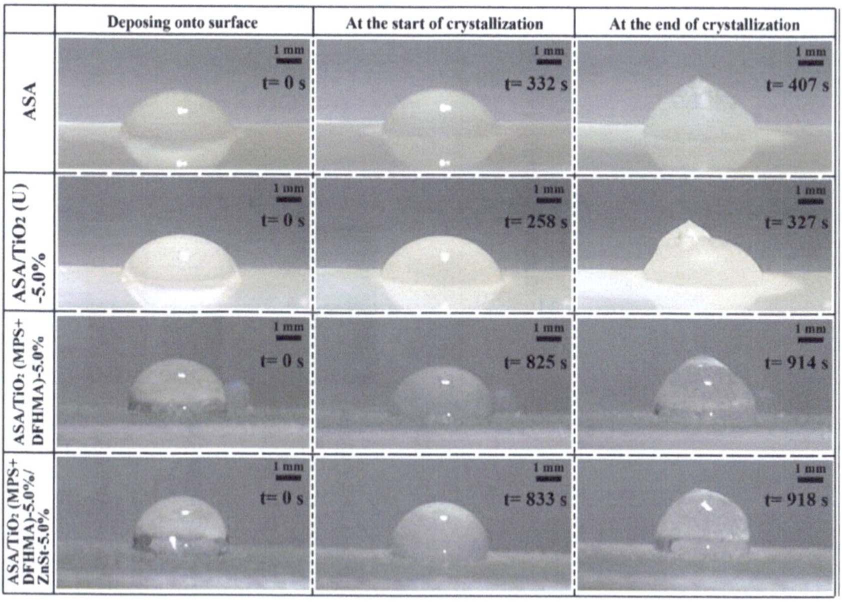

Fig. 4.18 Freezing process of a water droplet on sample surface

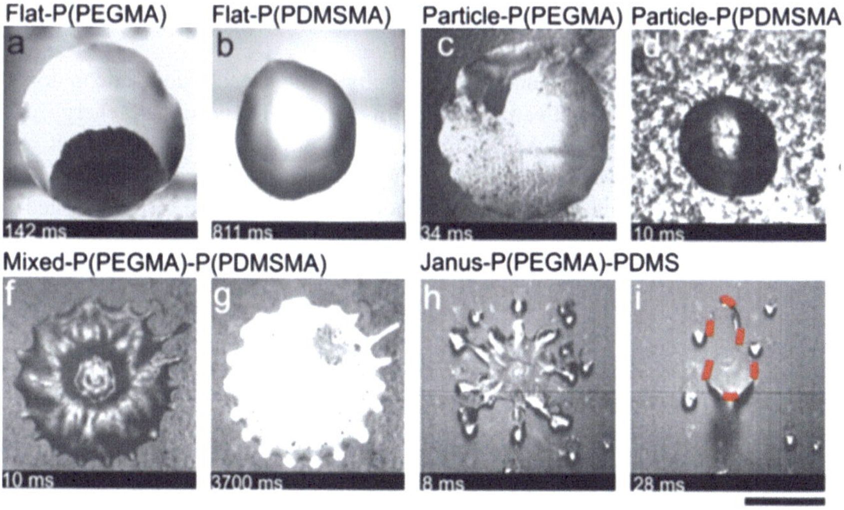

Fig. 4.19 Kinematics and freezing of a drop impacting onto flat surfaces

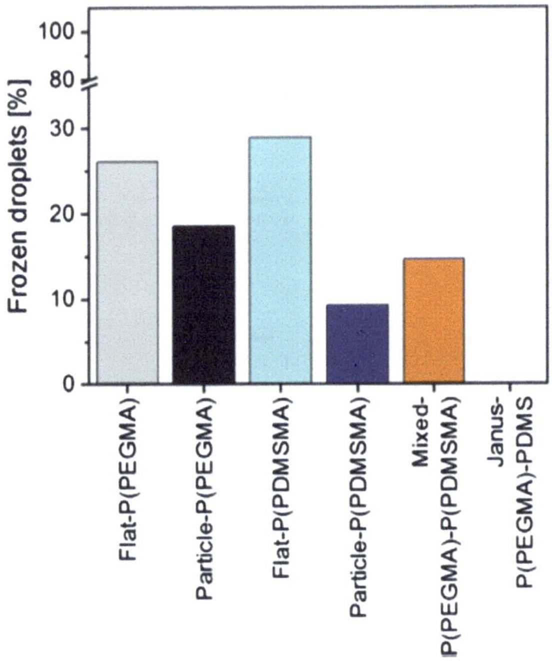

Fig. 4.20 Relative number of frozen drops out of all drops

4.5.4 Concept 4—Reduce Adhesion Forces by Reducing Surface Energy

To show the effects of surface energy and roughness on ice adhesion, silicon doped hydrocarbon and fluorinated-carbon thin films with different surface energy and roughness were investigated. As illustrated in Fig. 4.21 (Zou et al., 2011), it was found that the ice adhesion force decreases with the increasing water contact angle but only for similar roughness surfaces. In addition, ice adhesion on sandblasted is much larger than the smoother as-received surface, which confirms that roughness and surface energy play contrasting roles.

4.5.5 Concept 5—Reduce Ice Adhesion and Facilitate Its Shedding by Applying SLIPS Coatings

Biomimetic slippery lubricant-infused porous surface (SLIPS) can be achieved by combining a porous structure with a low-surface-tension lubricant which is both immiscible in water and can be easily deposited on a rough substrate (Liu et al., 2015). These structures gave such a surface reliable anti-ice properties. Due to the immiscibility of water and lubricants of low surface tension, the lubricant layer will

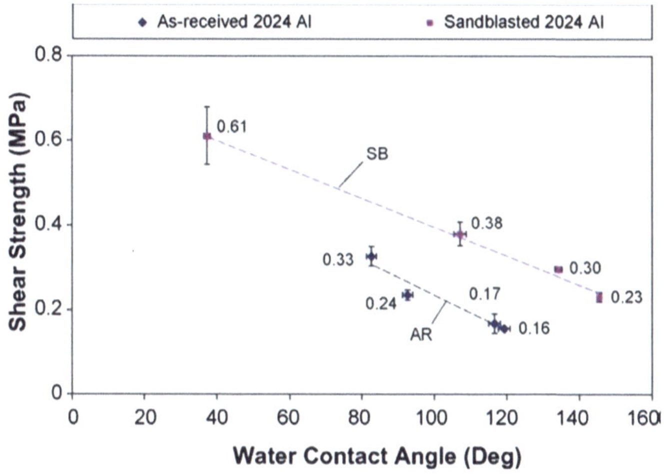

Fig. 4.21 Relationship between water contact angle and ice adhesion strength of as-received and sandblasted samples

elevate the condensed droplets to the outer surface and rapidly replace the water–solid contact into water-lubricant contact. The water and ice droplet is removed with the removal of the lubricant, which leads to a minimized ice adhesion which is 1-2 orders of magnitude lower than the conventional surface. Moreover, due to the capillary force, the lubricant stored in the porous was driven to the surface to supplement the lubricants, so SLIPS has self-healing properties. An effective approach to optimize the SLIPS anti-icing performance is to minimize the spreading coefficient, which is a measure of the tendency of a liquid phase to spread (complete wetting) on a second, liquid or solid phase, as shown in Fig. 4.22 (Nguyen et al., 2019).

The most challenging point of SLIPS is its durability. According to the recent research, SLIPS with a fluorinated hierarchical micro/nanoscale substrate maintains the best anti-icing capability in terms of efficiency. As can be seen from Fig. 4.23 (Nguyen et al., 2019), lubricant in the microscale pores is drawn to the solid-ice interface by strong capillary force only when the contact angle between lubricant and surface is zero.

4.5.6 Concept 6—Use of Freezing Point Depressant Fluids

This technique makes it possible to avoid ice adhesion and formation on an exposed object by applying a fluid. These fluids are generally prepared by mixing some commercial liquid products with water. They are commonly used on roads,

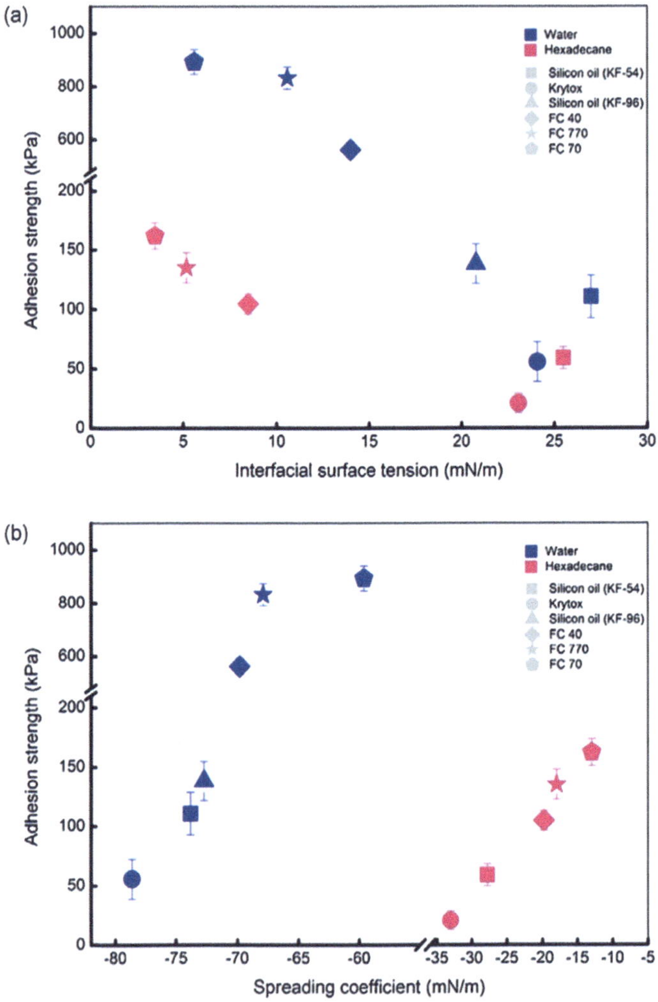

Fig. 4.22 Relationships between adhesion strength and (a) interfacial surface tension and (b) spreading coefficient

runways, aircraft, and other structures to prevent icing. The durability of these coatings on aircraft wings, for example, is limited to about one hour (CEATI, 2002-04). The application of these fluids to power network equipment does not seem practical considering their limited durability, cost effectiveness, and their harm to the environment (CIGRE WG B2.29, 2010; CEATI, 2002-04; Farzaneh, 2019).

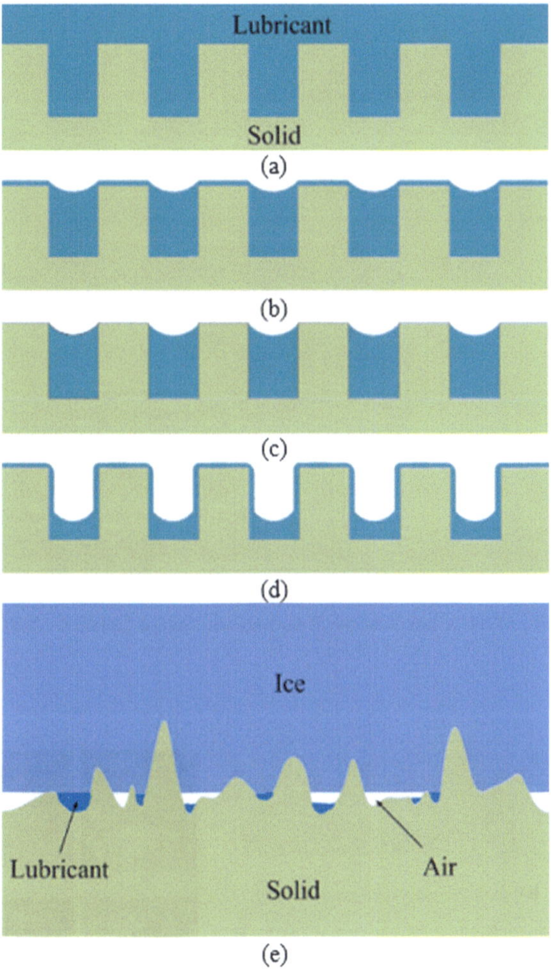

Fig. 4.23 Shear force on the surfaces after icing/de-icing

4.5.7 Concept 7—Use of Combined Passive and Active Coatings

By combining closely spaced micropores on polymer-based electric heating coating with SLIPS, the authors get an anti-icing coating (SEHC) which possesses superior oil film isolating effect for ultralow ice adhesion (Fig. 4.24). Up to 53% electric heating energy was saved in the de-icing progress with remarkable shorter de-icing time. Also, 23% power density and 11% energy consumption were saved in laboratory tests, as shown in Fig. 4.25 (Liu et al., 2019). In contrast to Concepts 2 and 3 (water drops that freeze), Concept 7 Fig. 2.13 experiment shows spray in chilled chamber to supercool the drops.

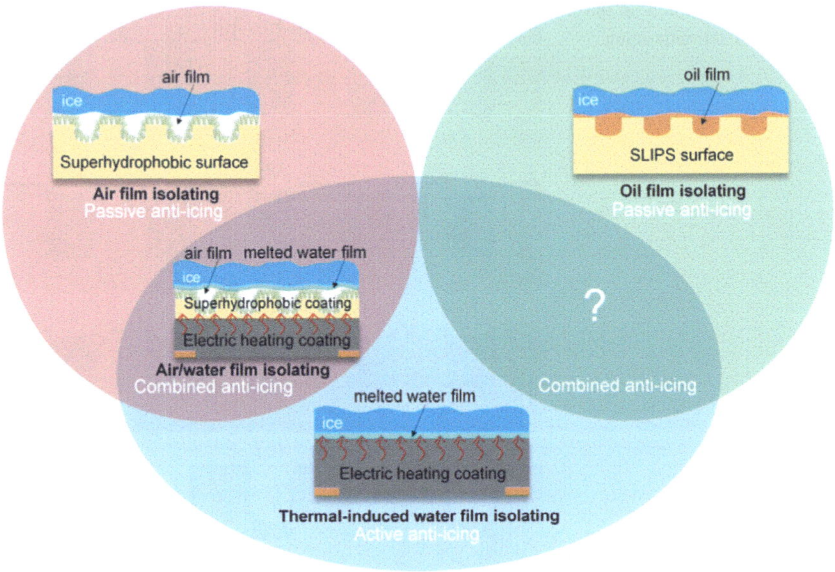

Fig. 4.24 Classification of various ice-surface isolating effects

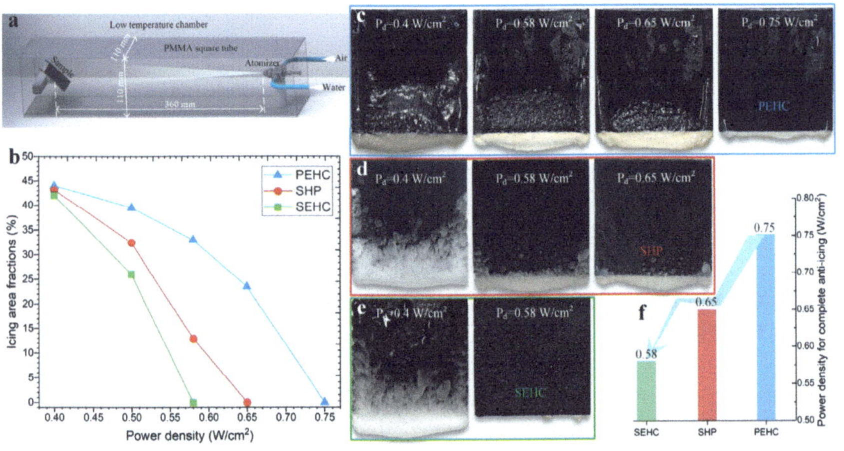

Fig. 4.25 Dynamic anti-icing tests of PEHC/SHP/SEHC samples with different heating power densities

4.5.8 Concept 8—Active Ice Electrolysis system

One of these methods against icing is based on ice electrolysis (Petrenko, 2000) by applying a DC voltage between a grid electrode and a conductive surface to be protected when ice forms on the surface and bridges the circuit (see Fig. 4.26) (Farzaneh, 2019; Farzaneh & Volat, 2008; Petrenko, 2000). The grid electrode is insulated from the conductive surface (Fig. 4.26a) and can have different configurations, as shown in Fig. 4.26b, c. Based on this concept, the configuration illustrated in Fig. 4.27 was proposed for protection of energized conductors against

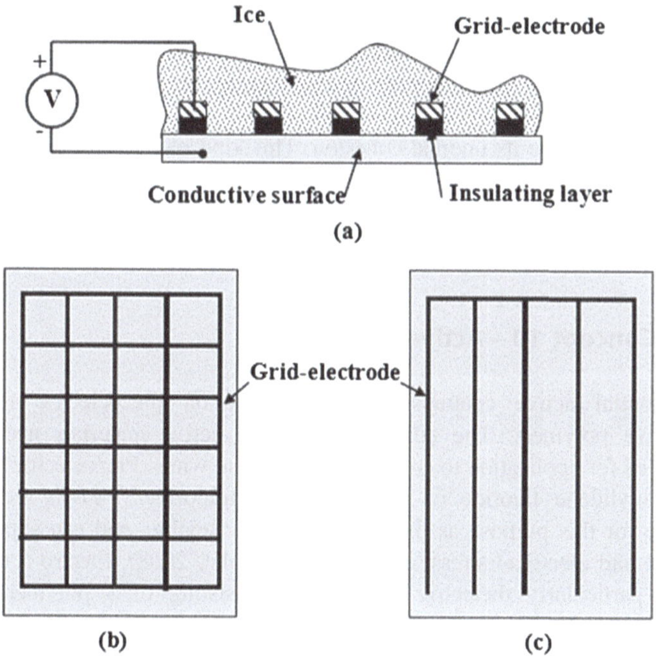

Fig. 4.26 Principle of generation of ice electrolysis by application of a DC voltage

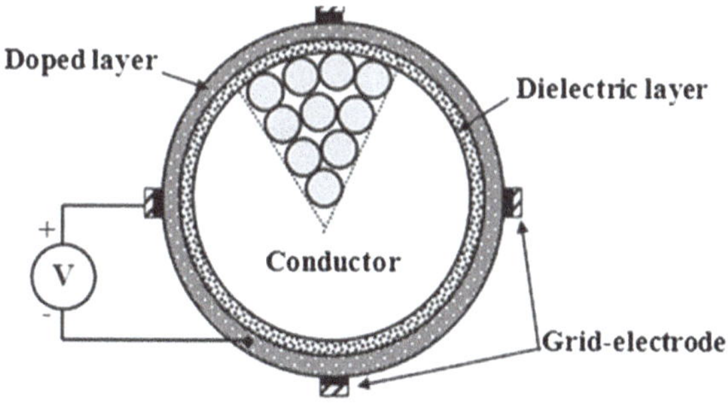

Fig. 4.27 Ice electrolysis method for conductor application

icing despite the fact that this technology seems to be more convenient for application to the ground wires (Petrenko, 2000). One of the difficulties with this method is to set up the axial grid electrode surrounding the wire which must be electrically insulated from the wire (Farzaneh & Volat, 2008).

4.5.9 Concept 9—Induce Ice Internal Cracking by Using Inhomogeneous Surfaces

When the ice block adheres to a surface presenting different domains of adhesion strength (as for example a surface only partially coated with a PTFE tape (Nakagami, 2007-10), the gravitational forces will cause an uneven slipping of the block, causing its fragmentation and facilitating the shedding. Similarly, a mechanical obstacle posed on the slipping path of the accreted ice (as for example snow-resistant rings (Mizoe et al., 2007-10) on conductors) can cause sleeve deformation or fracture, weakening its internal cohesion. This kind of concepts are extensively experimented and adopted, particularly in Japan, where they are effective under wet snow precipitation (Nakagami, 2007-10; Marcacci et al., 2019; Mizoe et al., 2007-10).

4.5.10 Concept 10—Active Materials

Other potential active coatings may be based on piezoelectric films and electro-active polymers. The efficiency of these active materials needs to be demonstrated for application to conductors or ground wires. Piezoelectric polymers like polyvinylidene fluoride or polyvinylidene difluoride (PVDF) seem to be appropriate for this purpose as they are very thin, flexible, and can support high mechanical and electrical stresses (Farzaneh & Volat, 2008). Electro-active polymers and particularly dielectric elastomers, consisting of a polymer material sandwiched between two compliant electrodes, can also provide thin-film active coatings. By applying a high electric field to the electrodes, it is possible to stretch the polymer under Maxwell stress. 7 MPa at the ice/substrate interface seems to be sufficient to separate the ice layer, as reported in Farzaneh and Volat (2008). Further research is necessary to verify the efficiency and applicability of PVD and electro-active polymers to overhead power line conductors and wires.

4.5.11 Concluding Remarks

Much effort has been devoted to the development of coatings for conductors and ground wires to reduce the amount of ice accretion for preventing potential damage to the lines during icing periods. As passive methods are more attractive than active ones, most of this work has been recently concentrated in that direction.

Table 4.3 Icephobic coatings developed for conductor materials (aluminium or stainless steel substrate)

Deposition technique	Main chemical component	Properties and applications	Durability	Comments	Ref.
Chemical etching + dip coating	Stearic acid	Lower ice accumulation compared to RTV	Not tested	No quantitative measurement on ice adhesion	Wang et al. (2010)
Al substrate anodization + dip coating or plasma sputtering	PTFE	Ice adhesion strength is less than half of that of bare Al	Not tested		Jafari et al. (2010), Menini and Farzaneh (2009), Menini et al. (2011)
Al substrate anodization + Low-pressure plasma polymerization	HMDSO	ARF 3.5 for the first icing/de-icing cycle	ARF reduced to 1.4 after 15 icing/de-icing cycles		Mobarakeh et al. (2013)
Slippery liquid infused porous surface (SLIPS)	Polypyrrole	ARF 85	Not tested	Usable in humid conditions	Kim et al. (2012)
Various tools to be directly used onto coating on Al samples (foam roller, float, foam brush, bristle brush, and plastic putty knife)	Different ratios of MP-55 (PTFE-based powder) and Rain-X commercial products	ARF 17	Three cycles tested on all coatings	Study conducted by NASA to evaluate the efficiency of the coatings marketed as icephobic and to determine an optimal application method	Ferrick et al. (2012)
Attaching a PDMS layer to the substrate using an adhesion promoter	PDMS	ARF of around 100 Low mechanical properties	Not tested	PDMS-based elastomer coating was proven to be superior to several commercial and under development coatings	Susoff et al. (2013–10)
Spin coating	Alumina nanoparticles + RTV	ARF 4.5 good corrosion resistance 4A (very good) coating/substrate adhesion	Not tested		Momen and Farzaneh (2014)

(continued)

Table 4.3 (continued)

Deposition technique	Main chemical component	Properties and applications	Durability	Comments	Ref.
Dip coating	Stearic acid	ARF 8 Significant freezing delay compared to control sample	Stable superhydrophobic behaviour after 10 icing/de-icing cycles	No icephobicity durability experiment performed	Li et al. (2013)
Chemical etching + dip coating	PTES	ARF 6	After 20 icing/ice-breaking cycles, ARF is reduced to 4 After 40 icing/ice-melting cycles, ARF is reduced to 4.5	Droplet rebound was observed at low temperatures and high humidity when the samples were tilted at least 30 degrees	Wang et al. (2013)
Chemical etching + dip coating	Silica nanoparticles + fluorosiloxanes	Decreased ice accumulation	Stable icephobic and superhydrophobic behaviour after 100 in-cloud icing and subsequent ice-melting	Ice adhesion strength is not measured Stainless steel substrate Highly adhesive coating	Boinovich et al. (2013)

There are many alternatives to consider when modifying the surface properties of the overhead line components to prevent ice accretion. These range from thin layers of chemicals with desirable properties, such as superhydrophobicity, through to treatments to form engineered surfaces with regular features on nano or micrometre scales all the way to features that increase the rugosity of the conductor itself.

A review of some of the developed coatings with potential application to conductors and ground wires is presented in Table 4.3.

5 Systems for De-Icing Overhead Power Line Conductors and Ground Wires

Ice deposits on overhead power networks, and particularly on ground wires and phase conductors have always been a major challenge in cold climate regions (Farzaneh, 2008). This chapter describes the process of reduction of the ice and snow accretion on ground wires and phase conductors that has accumulated naturally. Understanding the natural ice and snow shedding processes may help improve performance when using or designing de-icing methods.

Considerable research and development have been carried out, and successful large-scale technologies can address the challenges of anti-icing and de-icing. A review devoted to anti-icing and de-icing methods for overhead power lines was presented by Polhman and Landers (1982). A few years later, Hesse (1988) presented a comprehensive description of two methods used by Manitoba Hydro, namely ice breaking by rolling, a "mechanical method" and ice melting by short-circuit current, a "thermal method". However, these reviews presented only *ad-hoc* techniques and, consequently, were not exhaustive or complete enough to apply on large scale. A first classification of these methods in four categories was proposed: passive, thermal, mechanical, and "miscellaneous", reflecting the physical principles used for ice removal (Laforte et al., 1998). A more recent report proposed a six-category classification: passive techniques, active coatings and sheathings, active methods on bare conductors, thermal methods, mechanical methods, and miscellaneous methods (CEATI, 2002–04).

The anti-icing and de-icing methods are classified into the following four following categories:

1. passive methods
2. active coatings and devices
3. mechanical methods

M. Farzaneh and W. A. Chisholm, *Techniques for Protecting Overhead Lines in Winter Conditions*, Compact Studies, https://doi.org/10.1007/978-3-030-87455-1_5

4. thermal methods.

Active coatings and methods are defined as anti-icing strategies, described in Sect. 4.5.

Another classification can be used which takes into consideration the permanent or temporary character of the method, the need for line modification, and whether it is automated (Volat et al., 2005). With this classification system, the existing potential methods can be categorized into online, limited use and permanent methods.

Online methods are those using Joule effect, with energy from or external to the line, with no extra device or coating attached to the energized conductor or ground wire. Limited-use methods are those which are used only at specific locations and are not permanently installed on the line. Finally, permanent methods are those which are permanently installed on the conductors or ground wires.

No permanent method for de-icing overhead conductors has been deployed widely. Many promising approaches remain as work in progress. Developing and validating such methods as coatings, is a complex task, as they must perform successfully and have a relatively long service life. In the meantime, to respond to regulatory requirements focused on network resilience, traditional de-icing methods retain a role in operation of modern electric power systems in many countries.

5.1 Mechanisms of Ice Shedding

Ice shedding from ground wires and phase conductors, and any phenomenon resulting in ice mass reduction, may occur through any combination of melting, sublimation, and mechanical breaking processes. These mechanisms are characterized by the prevailing atmospheric conditions, shedding rates, duration of ice shedding, variations of ice load, and ice strength.

5.1.1 Melting

Ice mass reduction by melting consists of two phases under the condition that air temperature remains above 0 °C. In the first phase, characterized by low shedding rates, melting is limited to the external part of the ice deposit, as it is affected by air temperature, solar radiation and wind velocity. When the air temperature is above 0 °C, this phase usually does not last long before being replaced by the second phase. The short duration of this initial phase partly explains its lack of field observation. In the second phase, melting also occurs at the ground wire/conductor-ice interface resulting in ice chunks dropping off under the effect of wind and gravity. Shedding occurs when adhesive forces on fractured ice deposits are overcome by the aerodynamic and gravitational forces involved. This process is hastened by the fact that ice-free sections of the ground wire/conductor will heat up faster. The second phase may last a few hours and is associated with a significantly higher shedding rate than the first, typically about 0.3 kg/m/h, but it may exceed 1 kg/m/h in some cases (Druez et al., 1990, 1995).

5.1.2 Sublimation

During sublimation, vapour molecules are released to the ambient air from the ice surface. The process depends on the vapour concentration gradient between the water vapour concentration for saturated conditions and that of moist air. The most important atmospheric parameters influencing sublimation are the relative humidity of air, air temperature, and wind velocity. Since sublimation occurs at the ice surface, its rate increases with the external surface area of the accretion, and consequently, with the ice mass and ice porosity. The shedding rate is always low during sublimation, in the order of 0.01 kg/m/h only (Druez et al., 1995). The sublimation process, however, may last up to several days, in which case the mass reduction may become significant. Therefore, ice sublimation does not generally cause ice chunks falling from the ground wires and phase conductors.

Lambrinos et al. (1987) examined the effect of atmospheric parameters on the ice mass loss during sublimation and on the sublimation flux, defined as the mass flow rate divided by the corresponding area of the ice sample. They undertook wind tunnel experiments on a cylindrical ice sample by varying air temperature, T_a, air velocity V_a, and relative humidity of air, φ_a, in the following ranges: 243 K < T_a < 270 K, 0.5 m/s < V_a < 4 m/s, and 40% < φ_a < 90%. Experimental observations are summarized as follows.

- The mass loss increased linearly with time in the range of the conditions examined.
- The sublimation flux increased with air temperature. The dependence was found linear for low temperatures and became exponential as the temperature approached the freezing point of water. This result is explained by the significant change in the partial vapour pressure gradient and by the dependence of diffusion on temperature.
- The sublimation flux increased linearly with air velocity due to the linear dependence of the convective mass transfer coefficient on air velocity. The slope of this linear relationship was steeper for higher air temperatures and for lower relative humidity values.
- The sublimation flux decreased with relative humidity of air. This dependence may be realized by nonlinear steeply decreasing curves for low relative humidity values, while these curves tended to become linear for higher values of humidity.

5.1.3 Mechanical Ice Breaking

Ice shedding by mechanical breaking is a consequence of adhesive or cohesive failure of the accretion. Sudden ice shedding can cause the conductors to jump sometimes leading to flashovers, especially when they vertically arranged. In turn, conductor jumps and flashover-induced mechanical shocks may cause more subsequent shedding. Since both static and dynamic loads may induce mechanical ice breaking, the following factors exert an influence on this mass reduction mechanism: wind velocity, air temperature, ice mass, and strength. Ice shedding by mechanical breaking usually occurs at temperatures below 0 °C. The most important factor in the breaking of the ice at such temperatures is the large displacements of the ground wire/conductor due to aerodynamic forces. Galloping conductors undergo very high-amplitude vibrations, usually resulting in fast ice breaking. Low-amplitude, high-frequency wind-induced vibrations may also remove ice from the ground wire/conductor due to fatigue type failure. Thus, the meteorological factor which has the greatest influence on natural mechanical ice breaking is wind velocity; this has been confirmed by observations on test lines (McComber et al., 1990; Druez et al., 1995) showing that the decrease in ice mass was proportional to the square of the wind velocity. It was also observed that the slope of this relationship increased with the elevation of ground wire/conductor above ground level, which correlates with the increase of wind velocity with elevation as predicted by the classical boundary layer wind model.

Druez et al. (1990) examined the effects of some parameters on the rate of ice shedding from mechanical ice breaking, or more precisely, on shedding rate. These are organized in Table 5.1.

Druez found that $V_{a,s}$ and P_0 were the most important factors, related to aerodynamic forces and thus to ground wire/conductor vibration. The other important parameters in all affect the mechanical properties of ice.

Table 5.1 Ice shedding parameters and range studied by Druez et al. (1990)

Variable	Wind velocity, m/s		Mass of accreted ice at the start of sheddingP_o (kg/m)	Air temperature (°C)	
	$V_{a,a}$ during accretion	$V_{a,s}$ during shedding		$T_{a,a}$ during accretion	$T_{a,s}$ during shedding
Minimum	1.9	1.0	0.5	–20.9	–21.3
Maximum	4.4	8.2	5.5	–4.5	–6.0

Table 5.2 Mechanisms of ice shedding (T_a—air temperature, V_a—air velocity, ϕ_a—relative humidity of air, P_0—ice load)

Mechanism	Process	Main influencing parameters	Place of appearance	Duration	Shedding rate
Melting, 1st phase	Simple melting due to $T_a > 0°C$	T_a, V_a, solar radiation	External part of ice	Short (minutes), depending on T_a	Low (lack of data from field observation)
Melting, 2nd phase	Ice drop-off under wind and gravity	V_a, P_0	Conductor-ice interface	Few hours	High (0.1–1 kg/m/h)
Sublimation	Vapour molecules released to ambient air	ϕ_a, T_a, V_a, ice surface area	ice-air interface	Several days	Low (about 0.01 kg/m/h)
Mechanical ice breaking	Adhesive or cohesive failure of accretion	V_a, T_a, P_0, ice strength	ice deposit structure	Variable (from few hours to few days)	Variable (below 0.1 to ca. 0.5 kg/m/h)

The authors cautioned that their assessment of the relative importance of the aerodynamic forces and mechanical properties of rime ice required further investigation for other types of ice.

The shedding rate during mechanical ice breaking may vary within a wide range, from below 0.1 kg/m/h to around 0.5–0.6 kg/m/h. The duration of such shedding events may also vary from a few hours to a few days (Druez et al., 1990, 1995).

Of the three mass reduction processes, mechanical breaking is the most potentially damaging for transmission lines because it is generally associated with a high shedding rate, and it involves the sudden fall of large ice chunks. The main characteristics of the different ice shedding mechanisms are summarized in Table 5.2.

5.2 Mechanisms of Snow Shedding

The shedding of snow accretion from ground wires and phase conductors differs from ice shedding mainly because the process and structure of ice and snow accretion are different. The principal condition of snow growth is that adhesive forces between the snow and ground wire/conductor surface as well as cohesive forces between the snowflakes are high enough to keep the snowflakes together on the ground wire/conductor. Snow sheds from the ground wire/conductor when aerodynamic and gravitational forces exceed these adhesion forces.

5.2.1 Adhesion Forces in Snow Accretion

The most important parameters governing the processes of snow accretion and snow shedding are air temperature, wind velocity and precipitation intensity. Dry and wet snow are the main types. Adhesive forces during snow accretion have different sources depending on if the accretion process is dry (at low air temperature) or wet (when air temperature is higher) (Sakamoto, 2000).

5.2.1.1 *Adhesion Forces for Dry Snow*

Dry snow accretion occurs below freezing temperature and involves low adhesive forces; dry snow will be blown off the conductor if the wind velocity reaches about 2 m/s. Thus, dry snow may only accumulate under calm conditions and usually does not give rise to significant overload since it is likely to fall due to gravity, especially if the conductor is small in diameter and has low torsional rigidity.

5.2.1.2 *Adhesion Forces for Wet Snow*

The adhesive forces are significantly larger between the snowflakes in wet snow accretion above 0.5 °C. Wet snow may not shed even if exposed to high wind speed, thus promoting the growth of heavy loads. If ambient conditions change during snow accretion such that the air temperature drops below freezing point, the wet snow will freeze in place, thus causing the adhesive forces to increase and leading to further snow growth.

Overload due to wet snow accretions is more frequently observed. The factors associated with wet snow accretion are of greater interest and are discussed next. The liquid water content (LWC) of snow is a common factor in the relevant research studies as it determines whether wet snow accretes on conductors and whether it sheds.

5.2.2 Cohesive Limit

The LWC of snow controls the mechanical cohesion in the snow matrix formed during wet snow accretion. Increasing the LWC of the snow matrix weakens the internal cohesive forces which keep snow particles together. Above a maximum diameter of snow deposit, aerodynamic and gravitational forces overcome the cohesive forces, the snow sleeve breaks up and the snow deposit drops off. This maximum diameter decreases rapidly with LWC as illustrated in Fig. 5.1. Wind–tunnel experiments were carried out by Sakamoto et al. (1988) to simulate wet snow accretion and to study the conditions of wet snow accretion and shedding. Although the authors proposed an empirical relationship between initial snow water content and air temperature, they concluded that it was not possible to obtain a clear relationship between the snow sleeve water content and the cohesion of the sleeve. However, they proposed the critical LWC value of 40% when the maximum snow sleeve diameter is around 5 cm. Above this LWC level, the snow sleeve breaks up before its diameter reaches 5 cm, meaning it will fall off the transmission line. At

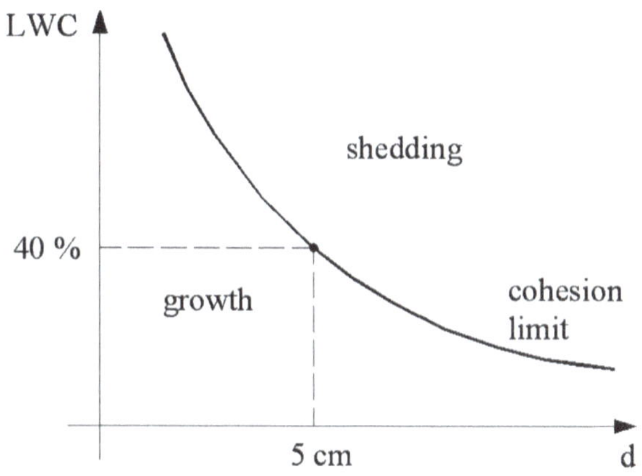

Fig. 5.1 Qualitative relationship between LWC of snow sleeve and its maximum diameter

this critical LWC level, wet snow accretion continues if the ambient conditions remain favourable, but wet snow always sheds, at least partially, before the sleeve diameter reaches 5 cm (Admirat et al., 1988). The existence of this critical LWC threshold was also verified experimentally by Roberge (2006), Roberge et al. (2007, 2007–10).

The LWC of snow sleeves was reported to be in the range of 20–30% during natural events in France, and between 10 and 50% in Japan (Admirat et al., 1986). In both countries, snow sleeves of large diameter may grow when LWC approaches the lower limit of the ranges observed, so that the resulting overload may be significant and cause damage to transmission lines.

5.2.3 Factors Influencing Liquid Water Content (LWC)

The LWC of snow accretion is the result of thermodynamic exchanges between air and the snow sleeve. The most important parameters influencing LWC are air temperature, relative humidity of air, wind velocity and precipitation intensity. Admirat et al. (1988) presented a thermodynamic model to calculate the mean LWC of a snow sleeve as a function of sleeve growth, air temperature, differential vapour pressure (defined from the relative humidity of air), wind velocity, vertical snowflake velocity, precipitation intensity, and initial snow water content which is proportional to the square of the air temperature. An alternate thermal model was developed by Poots and Skelton (1995) to calculate LWC during axial growth of wet snow on a fixed cylinder. This model considered more parameters than those listed above, and the authors provided a critical precipitation rate which corresponded to the cohesive limit. The critical value depends on air temperature and wind velocity, and if the precipitation rate is less than this critical value, then LWC exceeds 40% and the snow deposit will shed.

5.3 Mechanical Methods for De-Icing

In most cases, mechanical methods can be considered as de-icing methods as they are used to speed the shedding process after packed snow or ice have formed on conductors and ground wires. Laforte et al. (1994–12) demonstrated that mechanical methods require around 10^5 times less energy than thermal methods to force ice shedding. Generally, most of the active mechanical methods are based on two strategies. One strategy consists in breaking the ice by scraping it and the second in releasing energy from shock waves, vibrations or ground wire/conductor twisting to break and pull off the ice. One of the main advantages of mechanical methods is their relative ease of application compared to thermal methods. In fact, mechanical methods are those preferred for timely and fast intervention to de-ice short critical sections of a power network. However, in the absence of any precise operational instructions, methods involving significant bending of ground wires/ conductors should be avoided for OPGW to prevent damage to the optical fibres.

5.3.1 Passive Methods

Passive methods do not require an external source of energy, but use only natural forces such as wind, gravity, or solar radiation. Consequently, they can function on both energized and non-energized phase conductors as well as ground wires. This group includes most of the anti-icing methods used to prevent or reduce the accretion of wet snow and ice on conductors. To achieve this, different strategies are used:

- Weakening ice adhesion strength.
- Preventing freezing of supercooled water droplets on impact.
- Using a combination of specific devices for limiting the impact of ice overload on conductors

The third strategy is based on passive methods, which exploit natural forces such as wind, gravity or solar radiation in order to limit the adverse effects of ice loads on overhead lines. Some of these methods are already effective for wet snow but their efficacity for ice needs to be studied. A well-known method among these consists in using counterweights Fig. 5.2 to increase the torsional stiffness of conductor spans.

Field observations in Japan, Iceland and France on wet snow have shown that this device can limit the formation of cylindrical deposits of wet snow by limiting the rotation of a conductor resulting from eccentric snow loading on its windward side. The counterweight also decreases the torsional oscillation natural frequency and can de-tune the galloping activity that occurs when this is synchronized with a natural vertical frequency of the span (CIGRE TF B2.11.06, 2007–06). With highly eccentric snow loadings, shedding caused by gravity and wind forces is facilitated. Based on these observations on wet snow, numerical and experimental studies were

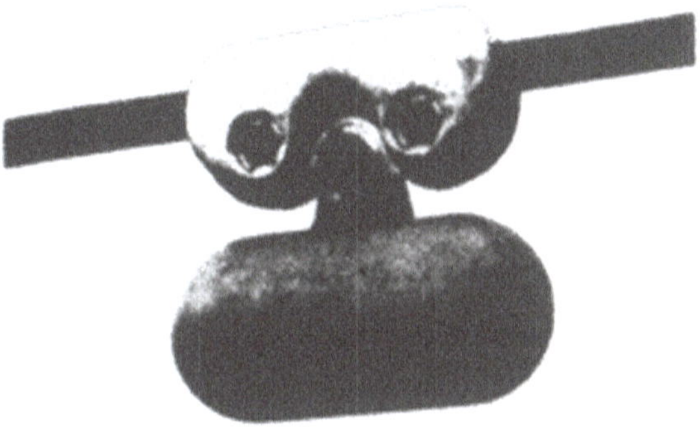

Fig. 5.2 Illustration of a counterweight

conducted on atmospheric ice (McComber, 2000; Blackburn et al., 2002). The results showed that limiting the rotation of a ground wire under icing conditions can effectively reduce ice load as well as shedding time under natural warming conditions.

Bundled conductors have higher rotational stiffness than single conductors which leads to differences in ice accumulation and shedding. Field measurements on two parallel 400 kV lines in heavy in-cloud icing area have shown that a subconductor in a duplex bundle accumulates more elongated ice shapes that tend to fall off before the semi-circular accretions on a single conductor (Elíasson et al., 2009). Other field measurements of ice shedding on a 80-m long test span exposed to heavy in-cloud icing have revealed that ice shedding tends to occur first on conductors of larger diameter (Elíasson et al., 2009) and those with smoother surface (trapezoidal strands vs. circular) at the same diameter (Thorsteins & Eliasson, 1998). This last observation was confirmed on a natural site on a 200-m segment of a conductor line with trapezoidal strands compared to ACSS and ACSR round-stranded conductors (Laforte et al., 2005).

Another interesting passive method to reduce wet snow accumulation is the use of snow rings around the conductors (Higuchi, 1972; Saotome et al., 1988). The snow tends to accumulate on the top of the conductor and slide down along the direction of the strands, as shown in Fig. 5.3 (top). Adding rings (Fig. 5.3, middle) or wires in opposite direction to the strands (Fig. 5.3, bottom) causes the sliding wet snow to shed by breaking the surface tension as it impacts the obstacle at the bottom of the conductor.

All overhead lines in Japan, except for the Okinawa region where there is no snow in winter, are equipped with snow rings. The overhead lines with single conductors are also fitted with counterweights, as shown in Fig. 5.4.

The effectiveness of these prevention systems has been determined by sophisticated automatic monitoring equipment which confirmed a significant reduction in accreted snow after the installation of these devices (Saotome et al., 1988).

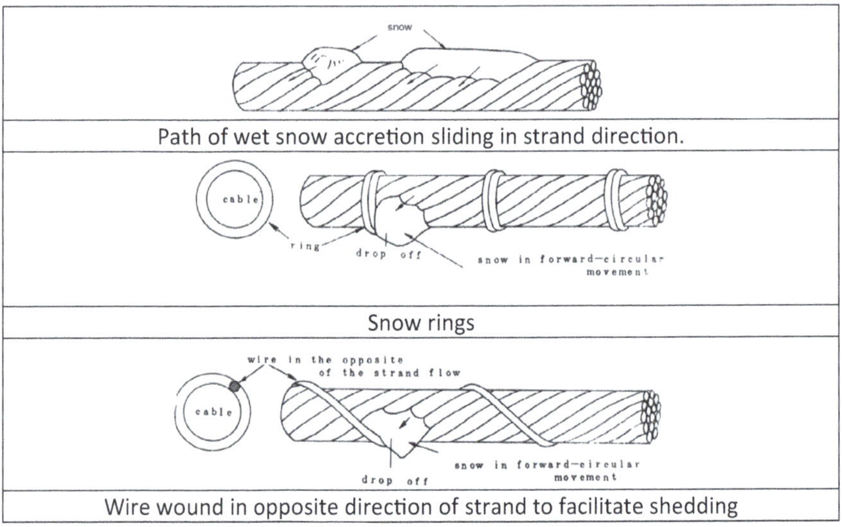

Fig. 5.3 Accessories to reduce snow accretion on stranded conductors (Higuchi, 1972)

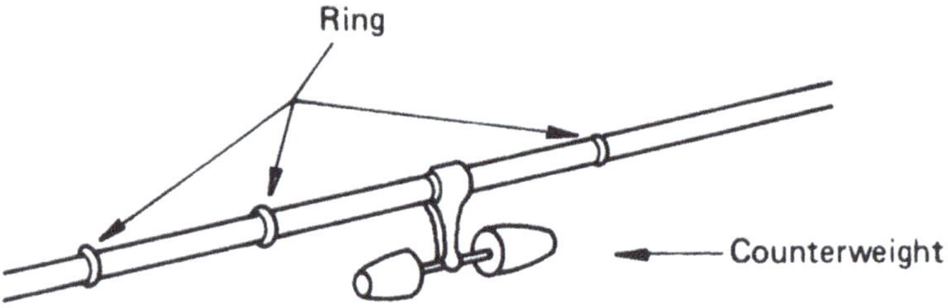

Fig. 5.4 Combination of rings and counterweights for wet snow removal from conductors (Saotome et al., 1988)

5.3.2 Scraping Methods

The simplest scraping methods are manual and use scrapers, rollers, or cutters attached to a rope which is pulled by line crews to release the ice. These methods can only be used when lines are accessible from the ground (CEATI, 2002–04; Farias, 1999). For example, Manitoba Hydro uses an aluminium roller wheel to de-ice energized distribution lines up to 12 and 25 kV (Farias, 1999). More recently, this method has been upgraded by using automatic robots. One such robot, the remotely operated vehicle (ROV), has been developed at Hydro-Quebec's Research Institute (IREQ) (Leblond & Côté, 2004-05-03; Montambault et al., 2000). Robust, lightweight, and compact, the ROV device has high traction force, which allows it

Fig. 5.5 Prototype of the ROV de-icer (Leblond & Côté, 2004-05-03)

to perform demanding tasks. It has been successfully tested on 315 kV energized line conductors. Its electronic circuitry is protected against electromagnetic interference, and it has an operational range of 1 km. The ice-scraping tool in Fig. 5.5 comprises a set of steel blades mounted on the ROV.

The third-generation automated prototype was tested in field conditions in 2010. In service, the ROV is anticipated to be installed from a helicopter or an insulated boom truck, since the icy tower structures prevent line crews from climbing them safely to reach the wires.

5.3.3 Shock Wave Methods

Energy releasing methods use conductor or ground wires for transmitting mechanical energy to induce ice shedding. As ice is a very brittle material at high strain rates ($>10^{-3}$ s^{-1}) (Petrovic, 2003), relatively little mechanical shock energy is required for breaking the ice because energy is not wasted in plastic deformation.

5.3.3.1 Mechanical Shocks with Live-Line Tool or Rope from Ground

One way to mechanically induce ice shedding is to create a shock wave which is propagated along the conductor. A method that has been used for more than a century creates this shock wave manually by hitting the conductor with an insulated pole, as illustrated in Fig. 5.6a. An alternative way is to pull a rope looped over a de-energized conductor, as shown in Fig. 5.6b. This method is effective for de-icing

a) Striking with an insulated live-line tool

b) Pulling down with a rope and releasing

Fig. 5.6 De-icing an overhead conductor line with live-line tool or rope (CIGRE WG B2.29, 2010)

one span at a time, but it requires that the ground below the conductor be easily accessible.

An improvement of the rope method was proposed by Laforte and Allaire (1998). As illustrated in Fig. 5.7, the rope is equipped with a hammer/hook head activated by a pneumatic piston charged with compressed air. This rope can be used from ground by line crews or mounted on a telescopic lift. These methods require accessibility of the iced lines by line crews and, consequently, are not efficient to de-ice lines over rough terrain or crossing rivers. In such special cases, the rope may be activated directly from the tower or from a helicopter (CEATI, 2002–04).

Hydro-Québec then improved the pneumatic hammer approach into a new device to de-ice ground wires by shockwaves. This is a portable cylinder–piston system called De-icer Actuated by Cartridge (DAC) (Gagnon, 2012; Leblond & Côté, 2004-05-03; Leblond et al., 2005–06). The concept in Fig. 5.8 is that a shock wave is established from high velocity piston impact, activated by firing blank 7-mm cartridges in a six-chamber magazine.

Explosive charges of different strengths are used, depending on conductor size. The de-icing operation with the DAC is carried out entirely from the ground, which represents a major advantage. First, a commercially available line-thrower is used to throw a projectile which tows a rope that passes over the ground wire/conductor to be de-iced. Next, the DAC is pulled up to the ground wire/conductor and held in place by a taut rope. The DAC is equipped with a revolver barrel that stocks six blank cartridges that can be fired remotely from the ground. Numerous tests have been carried out to assess the efficacity of the method and to optimize its physical parameters (Leblond et al., 2005–06). For example, a series of trials on a ground wire 100-m test span indicated that one firing only could de-ice the full span with

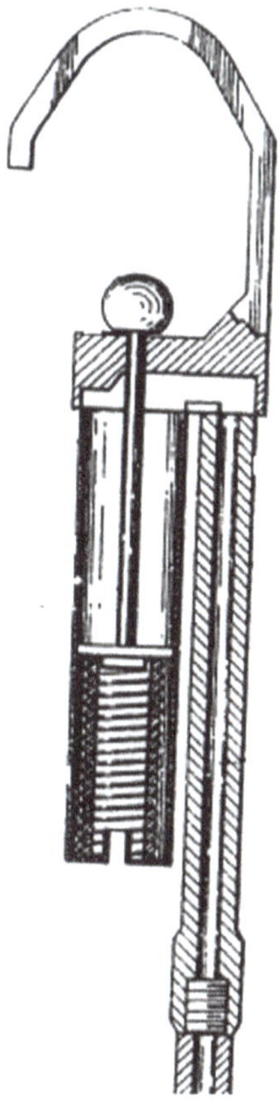

Fig. 5.7 Pneumatic hammer/hook for de-icing overhead conductors (Laforte & Allaire, 1998)

eccentric accretions, but multiple firings were necessary for an equivalent concentric ice accretion (Fig. 5.9).

Since 2011, five sets of DAC have been deployed in Québec and have successfully de-iced ground wires in the Saguenay and Matapédia regions (Gagnon, 2012).

5.3.3.2 Mechanical Shocks with Live-Line Tool or Rope from Helicopter

In 2010, Hydro-Québec and helicopter company Héli-Boreal developed an alternative method for removing ice accumulation from transmission line conductors and groundwires. The method consists in applying a mechanical impact to a conductor/bundle and groundwire to induce ice shedding. A wood distribution pole is attached to a helicopter belly hook via an insulated rope and a quick-release fitting. As shown in Fig. 5.10, two methods can then be used to de-ice a 56-mm phase conductor and an optical fibre ground wire (OPGW): (1) impact, or hitting; and (2) sliding, or rolling.

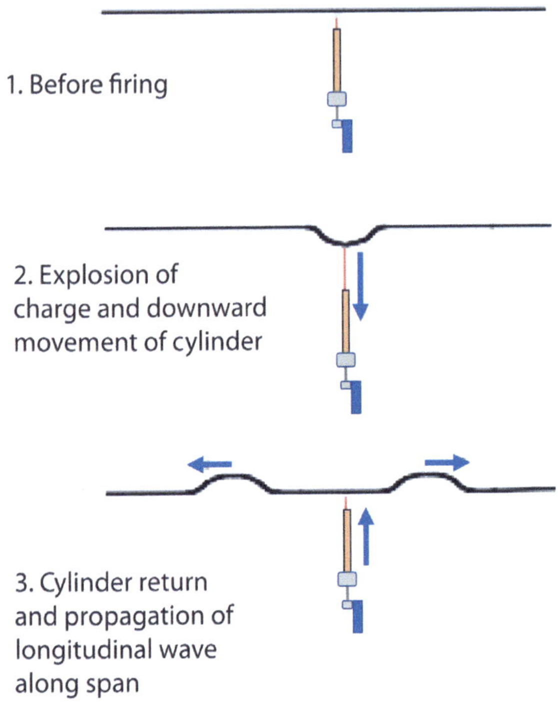

Fig. 5.8 Concept of mechanical shock wave de-icing using "DAC" De-Icer Actuated by Cartridge (Gagnon, 2012)

The line between the helicopter and the pole must be long enough to prevent elastic recoil of the conductor when ice is shed and short enough for the pilot to maintain control of the motion of the pole. The de-icing process goes faster with the mechanical impact method, but the slide-and-roll method, which strictly speaking is a scraping technique, does not damage the sensitive OPGW communication fibres. Two helicopters had to be deployed over four days to de-ice 400 spans, mostly using the impact method, the temperature a persistent—20 °C. This part of Labrador had been subjected to a prolonged ice storm and the accretion was hard ice with an outer layer of rime ice.

In higher altitude regions only accessible by helicopter, British Columbia Hydro crews have used a 90 kg mass attached to a 30-m rope with a series of large knots at about 0.5 m spacing as shown in Fig. 5.11.

The impact of the rope on the conductor forces conductor motion and sheds accreted wet snow. Initially, the bottom part of the knotted rope is set vertically below the conductor and the top part of the rope is pulled up at a 30° angle against the conductor causing the knots to catch (impact, lift, and drop) the conductor as they are pulled across. However, the operator must make sure that the suspended mass does not touch the ground or foul on a structure. Live-line rope is used to guard against the possibility of the knotted rope falling across two conductors at different potentials. Typically, half of the sticky wet snow of a 500-m span can be removed in less than half an hour resulting from about 30 multiple-knots pulls per

Fig. 5.9 DAC prototype held in place by a taut rope and ready to be fired (Leblond et al., 2005–06)

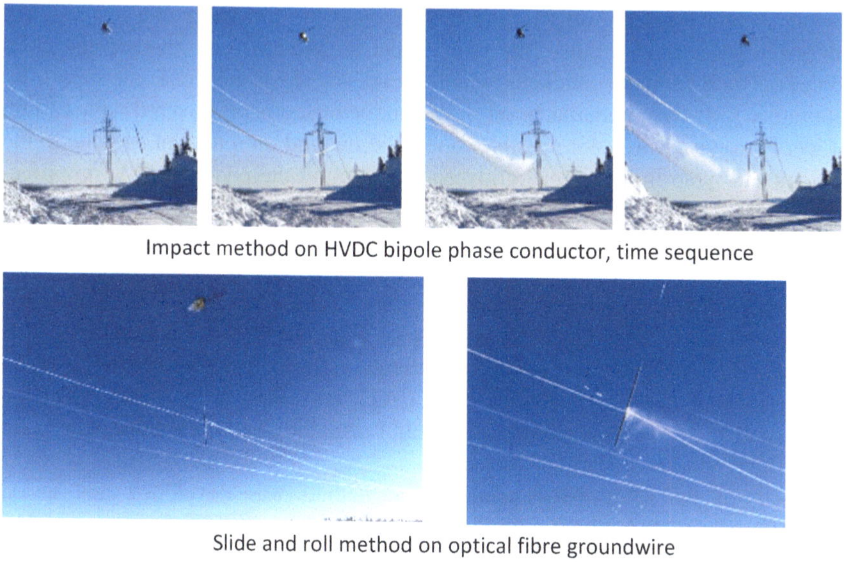

Fig. 5.10 Helicopter de-icing with telephone pole (courtesy Hydro-Québec)

Fig. 5.11 Helicopter use of 90 kg mass with large, knotted rope

half hour along the length of the span. This is considered by BC Hydro to be much more effective than simply bumping the conductors with insulated poles from a helicopter as described in Farzaneh (2008).

5.3.3.3 Electromechanical Shocks

Electro-impulse methods have been proposed for de-icing conductors and ground wires by shock waves. One of these methods initially developed for de-icing aircraft wings (EIDI) was tested on electric lines (Egbert et al., 1989–07). The EIDI technique adapted to power line de-icing consists of stacking two insulated strips of copper ribbon together and winding them around the full conductor span. When energized by a current pulse, the two copper strips repel one another and exert forces outward from the conductor which breaks and sheds the ice. Some successful laboratory tests were undertaken on short conductor segments covered with 12 mm of ice. However, it was not possible to test the method on a natural site due to the difficulty to wrap the actuator tightly enough around the conductor. An improvement of the EIDI method was proposed by Laforte et al. (1998), where the shock wave actuator is formed by a pair of wire strands of the external layer of the conductor which are isolated and connected at one extremity, while the other ends are connected to the impulse current generator. Tests conducted in situ with this improved method have permitted de-icing of a 260-m OPGW span. Even if this technology appears effective, further investigations are necessary before going onto prototype installations on ground wires (CEATI, 2002–04). A practical operational issue is the insulation degradation of the EIDI actuator when the ground wire is hit by lightning strokes.

Based on the same EIDI principle, a method was developed and tested Hydro-Québec to de-ice lines with twin or quad conductors at rated voltages of 315 and 735 kV, respectively (Landry et al., 2001). In this case, the EIDI actuator is

Fig. 5.12 Electro-impulse De-icing (EIDI) of twin bundle using 10 kA and appropriate reclosing sequence (Landry et al., 2001)

formed by the bundled conductors. The high current impulse is generated by a short-circuit current (I_{SC}) at the rated voltage of the line and by the subsequent action of the electromagnetic forces forcing the conductors to knock against each other and de-ice the span, as shown in Fig. 5.12. Asymmetrical I_{SC} and reclosing sequences are necessary to reduce the amplitude and duration of the short-circuit currents as much as possible. The conductors must be excited at a frequency close to their fundamental subspan frequency to get a maximum dynamic motion which is synchronized with the reclosing sequences.

Studies on the Hydro-Québec power system indicated that this method with multiple reclosing sequences could be applied on 315 kV lines, but only for emergencies during severe ice storms. For 735 kV lines; however, the required short-circuit currents and reclosing sequences posed unacceptable threats to network stability, and therefore, the method was not suitable for the EHV system.

5.3.4 Vibrating Devices

5.3.4.1 Vertical Oscillation at Span Resonant Frequency

The "skipping rope" resonant frequency of a wire, evident as galloping motion, depends on the span length and mode of oscillation. For a wire with fixed end-points, the frequency is approximately:

$$f = \frac{n}{2l}\sqrt{\frac{T}{m}} \tag{5.1}$$

where f is in Hz, n is the number of galloping loops, l is the span length (m), T is the conductor tension (N), and m is the conductor mass per unit length (CIGRE TF B2.11.06, 2007–06).

Mechanical de-icing methods can use devices attached to the conductors or ground wires to induce sustained vibrations to shed ice, exciting one of the galloping frequencies. Two devices, similar in principle, have been proposed and tested.

The first device, Automatic Ice Control (AIC), has been reported in Hansen (2004). It comprises a current transformer for power supply, a camera, different sensors for ice detection, a control box with a HF emitter/receiver, and a commercial electromagnetic vibrator. All these elements are encapsulated in a rigid protective housing directly mounted on the conductor as shown in Fig. 5.13.

The AIC is permanently installed at the midspan and is completely autonomous. Due to its ice detector and HF communication capabilities, de-icing sequences can be fully automated or controlled by signal order transmitted through HF. The AIC was installed on a 230 kV line located near St. John's, Newfoundland in 2004–2006, and the test was not successful. The system experienced major problems ranging from repeated sensor malfunction to communication failure and was never operational during the two winter seasons (CEATI, 2008).

Fig. 5.13 Automatic ice control (Hansen, 2004)

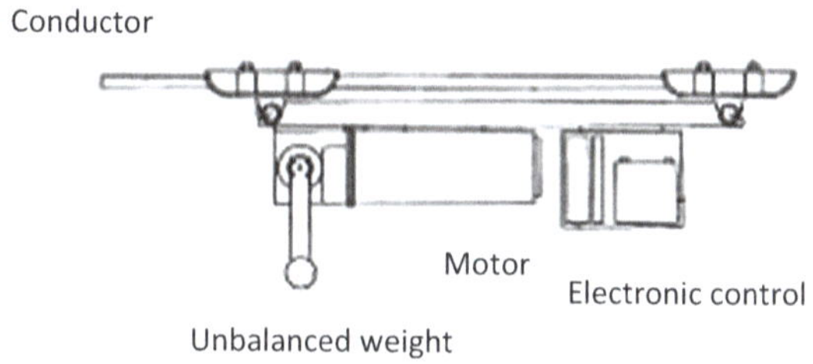

Fig. 5.14 Ice-shedder apparatus (Nourai & Hayes, 2003)

The second vibrating apparatus in Fig. 5.14 is called iceshedder. It is mounted on the conductor and includes a motor activating an eccentric weight whose unbalanced motion is tuned to a natural frequency of the span, estimated initially using (5.1). Preliminary de-icing tests were conducted on power lines with a span $l = 152$ m and 3 cm diameter conductor (Nourai & Hayes, 2003).

By operating the ice-shedder device within a frequency range of approximately 1.5–8.0 Hz, conductor displacements of 10–33 cm (4–13 inches) were observed, with accelerations between 0.5 and 14 g. Accumulated ice on power lines can be adequately removed within these ranges of forced motion. The ice-shedder can be installed easily on bare conductors and ground wires. Moreover, it can easily be automated and driven by power from the bare conductor line or from an external power supply. It could also be used with bundled conductors, with minor modifications, but this had not been tested in 2010. In fact, the de-icing capacity of this method has not been quantified yet, in terms of maximum radial ice thickness and maximum span length that can be excited with the limited power of the motor. It is also possible that these large oscillations of conductors or ground wires could eventually cause long-term mechanical damage to the power lines or support insulators.

Thermal rating considerations may conflict with the most advantageous location of a vibration system. Adding 50-kg mass at midspan increases the sag considerably as shown in Fig. 5.15 and thus reduces the thermal rating of the line. Adding the

Fig. 5.15 Effect of 50-kg mass on sag of 300-m span of 795 kcmil (403 mm^2) Drake 26/7 ACSR conductor

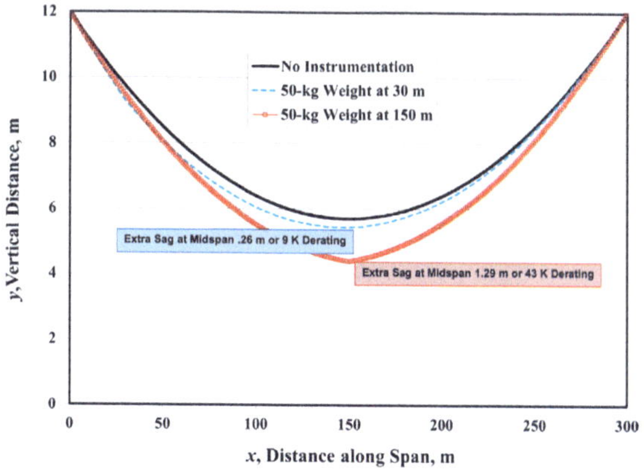

same 50-kg mass near one end of the span, in Fig. 5.15 at 10% of the span length, has less effect on the thermal rating, with a permanent change in midspan sag of 0.26 m corresponding to derating of 9 K.

In thermal rating, CIGRE practice now reports temperature differences in K (Kelvin) rather than as C°, with the same value and this practice is used in Fig. 5.15.

In this regard, the system in Fig. 5.14 may be preferred to the midspan-mounted equipment in Fig. 5.13 if the ice shredder can be mounted close to an insulator and still excite a galloping frequency.

5.3.4.2 Torsional Oscillation at Span Resonant Frequency

Galloping occurs mainly when ice accretion reduces the rotational resonant frequency of a conductor to match one of the frequencies in Eq. 5.1. When this match occurs, wind input to eccentric ice can couple energy from rotational to vertical oscillation modes, leading to large conductor displacement (CIGRE TF B2.11.06, 2007–06).

A de-icing method proposed by Laforte et al. (2005–06) exploits this understanding of galloping detuning methods based on torsional oscillation of the conductor and its torsional stiffness. An apparatus in Fig. 5.16 slowly twists the ground wire or the conductor around its longitudinal axis and then suddenly releases the elastic strain energy accumulated by torsion (Allaire & Laforte, 2003). The torque on the conductor can be applied by hand, using a live-line tool, or by a motor depending on the torsional stiffness of the conductor span. The efficacy of this method was demonstrated on a 15-m-long ground wire span which was entirely

Fig. 5.16 Twisting device installed at midspan of a ground wire (Laforte et al., 2005–06)

de-iced using the manual twisting device installed at midspan as shown in Fig. 5.16 (Laforte et al., 2005–06).

The main advantages of this torsional de-icing system are its simplicity, efficacity for all types of ice accretion and conductors and reduction of conductor movements like those in Fig. 5.6 that can lead to faults. The torsional method requires very low mechanical energy and can be operated safely and quickly by line crews to de-ice strategic ground wire spans or non-energized conductors. The main disadvantages are that the automation of this de-icing process may be quite complex as it requires an electric motor mounted on a rigid attachment at the height of the conductor, a reduction gear box and a magnetic clutch coupled at the motor, a control module, and an ice detection unit. Finally, this technique cannot be applied to bundled conductors.

5.4 Thermal Methods for De-Icing

In cases where anti-icing with Joule effect has not been effective, utilities may consider the use of higher energy (higher current or longer application) to de-ice the line. More energy is required for de-icing than anti-icing (Prud'Homme et al., 2005a). Joule or "thermal" methods for de-icing can be divided in two categories: (i) methods based on pure Joule effect, I^2R where R is the conductor resistance and (ii) methods based on dielectric losses, radiative waves and external heat sources.

5.4.1 Joule-Effect Methods: Historical Experiences

The oldest method for de-icing transmission lines uses Joule effect heating. A search of the IEEE Xplore database for the term <sleet> , describing ice pellets (IP) mixed with snow (S) or rain (R) in present-day WMO terminology, yields references dating back to 1912. Early references show that many original transmission lines had poor operating experience in icing conditions. It was appreciated that lines designed for icing levels in northeast USA were twice as costly as those in the southern USA with milder climate.

For years, in USA, Canada, and the former USSR, thermal de-icing and anti-icing methods based on Joule effect have been refined and implemented in service. In 1952, a special focus of the AIEE Transactions consolidated sleet-melting practices at four utilities in northeast USA (Bartlett et al., 1952–08; Corey et al., 1952–08; Shealy et al., 1952–08; Smith & Wilder, 1952–08). These review papers summarize experiences dating back to original de-icing trials.

5.4.1.1 *Joule De-Icing, New England Electric, 1920*

Joule heating was successfully used by the New England Company during the storm of 1920 (Oliver, 1925–11). A current of 300 A was applied to a 1/0 copper conductor, giving a current density of about 2.8 A/kcmil. This is equivalent to a current density of about 1.7 A/kcmil in an aluminium conductor with higher resistance per km than copper of the same size.

5.4.1.2 *Joule De-Icing, New England Electric, 1939*

Early developments in ice melting and ice detection methods on the American Gas and Electric System were described in 1939 papers (Bartlett et al., 1952–08). These methods basically consisted in detecting ice formation by its effect on power-line carrier current, and then isolating and connecting sufficient lengths of a 132-kV circuit in series to obtain the required ice melting current when the circuit was short-circuited at one end and connected directly to a 132 kV source at the other end. In planning ice melting procedures, some of the considered factors were:

- The selection of a line combination and grounding and switching arrangement which will give as high a current as possible without endangering equipment and service to customers.
- The importance of melting as quickly as possible to minimize the circuit outage duration. Early detection and the use of as high a de-icing current as possible can help to accomplish this.
- Any necessary adjustments in system load distribution and the need, if required, for help from interconnected companies to diversify supply via imports.
- The short-time current carrying capacity of buses, by-pass facilities, current transformers, and so forth. In some instances, it was necessary to rebuild or replace equipment to support de-icing requirements.
- The adequacy of conductor joints to withstand ice melting currents.
- The need for sending staff to unattended stations and keeping them there throughout the switching period.

5.4.1.3 *Joule De-Icing, American Gas and Electric, 1954*

Using 132-kV power sources, necessarily of very substantial capacity, the ice-melting current was applied to short-circuited line sections within the range of 60–80 miles (96–128 km) in length as compared with 120- to 150-mile (192–240 km) sections for existing 132-kV lines (St.Clair & Imburgia, 1954–01).

For the lines with marginal mechanical design margins, systems were engaged in icing conditions to energize loops of conductor with summer-limit currents on the order of 1 A per kcmil of aluminium cross section (2 A per mm^2). However, with an increasing demand for energy, transmission lines became longer and more efficient with the use of higher voltage, larger conductors, and conductor bundles. As a result, using the Joule effect for de-icing has become a challenging task considering the minimization of electric losses in the process.

The American Gas and Electric Company adopted ice melting as a part of the project of designing a 330-kV system, in 1954. Ice melting was a factor in both the 330 kV line design and the conductor selection.

5.4.1.4 *Joule De-Icing, Manitoba Hydro, 1970s*

Manitoba Hydro, Canada, began using high currents to melt ice in the early 1970s as an experimental procedure (Adolphe, 1992–11). To date, they have the capability to melt ice off several thousand kilometres of lines with conductors ranging in size from 2/0 to 336.4 kcmil (11–18 mm diameter) ACSR.

The following is the sequence of events adopted by this utility:

- Prior to the ice storm season, all required engineering calculations are made to determine the currents that can be obtained. Station equipment ratings are checked to see that all equipment is satisfactory for the expected current levels. Provisions are made to allow the connection of required lines and sources. This may require the upgrade of existing or installation of new equipment. Switching procedures are written up to accomplish these connections and emergency manuals produced and kept up to date.
- Once icing is detected, the required melting current for the specific conductor size, weather, wind speed and ice thickness is determined. The location of jumpers for the desired length of line to be cleared is calculated.
- The line to be treated is isolated. The shorting jumpers are connected at the calculated distance down the line. Any required protection settings are applied to the equipment. Finally, the line is connected to a station transformer of appropriate voltage and size.
- The line to be melted is energized. Staff at the station monitors the current and staff at one or more points along the line watch the ice removal. When the ice has dropped, the melting process is terminated, and the system is returned to normal.
- The details of ice accumulation, weather conditions, melting times, and success rate are recorded for further analysis.
- If icing continues, a line may have to be treated more than once during a storm and the process may be repeated.

Ice melting procedures are routinely carried out by Manitoba Hydro during severe widespread ice storms and for anti-icing during less severe weather conditions.

5.4.1.5 *Joule De-Icing, Present Day*

Both AC and DC have been used in different countries to melt ice (CEATI, 2002–04), and effective technologies for both types of current are now available. Generally, using AC does not involve costly equipment, since the required current is supplied from the power network through a transformer of modest cost. However, under certain atmospheric conditions, the melting power must be sufficiently high, especially with long transmission lines, to obtain the right ice-melting current. A variety of electrical system reconfiguration schemes have been considered for ice melting technologies, as described in Chap. 7.

5.4.2 Equations Governing De-Icing with Joule Effect

The heating energy required by thermal methods depends principally on whether the application is for anti-icing or de-icing. In both cases, the nominal current in the overhead conductor must be increased for the conductor surface temperature to exceed 0 °C. The required current depends on many parameters, and particularly upon ambient temperature, wind velocity, heat exchange to droplets, and in the case of de-icing, the thickness of accretion.

The current required to prevent ice formation on overhead line conductors was modelled by Clem (1930-12-06). Considering the heat lost by convection, Clem proposed Eq. 5.2 for calculating the temperature rise of a bare conductor surface above ambient air for wind speeds greater than 1 m/s to prevent ice accumulation.

$$\Delta T = 4.43 \cdot 10^{-4} \frac{I^2 R_{AC}}{\sqrt{dv}} \tag{5.2}$$

where

ΔT is the temperature rise (K).

R_{AC} is the conductor AC resistance (Ω/km).

I is the conductor current (A).

d is the conductor diameter (mm).

v is the wind speed (m/s).

Clem also proposed Eq. 5.3 to determine the current required to melt the ice already accumulated on the conductor.

$$I = 1772.5\sqrt{\frac{(w_i + w_c)d}{R_{AC}}} \tag{5.3}$$

where

R_{AC} is the conductor AC resistance (Ω/km).

w_i is the melt-through power (W/mm^2).

w_c is the power required to maintain temperature rise (W/mm^2).

d is the conductor diameter (mm).

The model proposed by Clem has served as the basis for different improved models used by various utilities to determine the current required to improve de-icing performance. One of these models is used by Manitoba Hydro (Hesse, 1988). In the Hesse model, heat loss due to radiation and heat gain due to solar radiation, which were neglected by Clem, are considered, as presented by Eq. 5.4. This equation provides the current required to melt the volume of ice V_{MELT} within a time Δt:

$$I = \sqrt{\frac{1}{R_{AC}}\left(P_c + P_S - P_{SOL} + \frac{\rho_I\left(L_F + C_{pI}(T_F - T_A)\right)}{\Delta t}\right)V_{MELT}} \tag{5.4}$$

where

R_{AC} is the conductor AC resistance (Ω/km).
P_C is the convective heat transfer (W).
P_S is the radiative heat transfer (W).
P_{SOL} is the solar heat transfer (W).
ρ_I is the ice density (kg/m^3).
L_F is the latent heat of fusion of ice (334 kJ/kg).
C_{pI} is the specific heat of ice (2.05 kJ/kg/K at 0 °C, corrected for temperature).
T_F is the fusion temperature of ice (0 °C).
T_A is the ambient temperature (°C).
Δt is the required time for a melt (s).
V_{MELT} is the volume of ice sector to be melted above the conductor (m^3).

The general validity of Eq. 5.3 was verified with accumulated considerable experience on conductor de-icing for twenty-five years (Farias, 1999). However, the Manitoba Hydro model in Eq. 5.4 makes some limiting assumptions. It works well for dry-grown ice because all impinging precipitation is assumed to be captured and frozen and because heat transfers associated with the impinging precipitation are ignored. However, it is not adapted for mild temperatures when ice accumulates in a wet regime and when some impinging precipitation drips off the ice surface and contributes or extracts heat energy to or from the iced conductor (Huneault et al., 2005-04a).

In order to take into account the wet-grown ice regime as well as the heat transfer associated with impinging supercooled droplets, new melting models were developed and tested by Hydro-Quebec to evaluate and manage the required current during ice storms (Huneault et al., 2005-04a, 2005-04b). One model has been tested at length and provides good results under typical ice storm conditions but overestimates the melt energy in colder conditions. This model was then improved with the consideration of the trapped water and ice, elements that are also missing in the Manitoba Hydro model (Huneault et al., 2005-04a). This new model has been formulated but has not been validated yet. The general current equation is given by Eq. 5.5.

$$I = \sqrt{\frac{r_C}{r_I}\frac{\rho_I\left(L_F + C_{pI}(T_I - T_A)\right)}{R_{AC}\Delta t}V_{MELT}} \tag{5.5}$$

where

R_{AC} is the conductor AC resistance (Ω/km).
L_F is the latent heat of fusion (334 kJ/kg).
ρ_I is the ice density (kg/m^3).

Table 5.3 De-icing current required for various types of conductors

Conductor Description	Diameter (mm)	Ice thickness (mm)	Wind speed (km/h)	Ambient Temperature (°C)	De-Icing Current (range) (A_{rms})
Beretsford 1354 kcmil ACSR	36	20	60	−2, −5	1900
Bersimis 1360 kcmil ACSR	35	10, 20, 50	10, 30	−1, −5	1320–1850
Condor 795 kcmil ACSR	28	10, 20	10, 30	−1, −5, −10	970–1350
OPGW	23	20	30, 60	−2, −5	920
OPGW	17	20	30, 60	−2, −5	600
Ground wire ½ in	13	10, 20, 50	10, 30	−1, −5	120–170
Ground wire 7/16 in	11	20	60	−2, −5	170

r_C is the radius of the conductor (m).
r_I is the outer radius of the ice sleeve (m).
C_{pI} is the specific heat of ice (2,05 kJ/kg/K at 0 °C, corrected for temperature)).
T_I is the temperature of the ice (°C).
T_A is the ambient temperature (°C).
Δt is the required time for a melt (s).
V_{MELT} is the volume of ice sector to be melted above the conductor (m^3).

Table 5.3 provides some values of de-icing current obtained for different types of conductors and different environmental conditions (Prud'Homme et al., 2005a).

5.4.3 Ice Melting Power Requirements for Conductors

For an ice melting duration of 0.5–1 h, a current density of about 2.5–3.0 A/mm^2 is required in aluminium conductors. The lower value of current density corresponds to ambient conditions with low wind speed and ambient temperature about 0 °C, minimizing the convective heat loss term P_c in Eq. (5.4). The higher value of 3 A/mm^2 (1.5 A/kcmil) is needed for rapid de-icing with wind speeds of 5–10 m/s and temperature of about −10 °C. Typical heating currents required for single conductors are illustrated in Fig. 5.17.

The efficiency of ice melting greatly depends on the procedure being carried out in due time. It is also especially important that the lines to be heated are taken out of service for the shortest time possible, which requires the application of well-rehearsed or automatic switching operations on the lines before and after the ice melting procedure.

The active power needed for ice melting may vary from hundreds of kW for short lines with small conductors, up to tens and hundreds of MW for long lines with large conductors. Active power does not depend on the current type, AC or

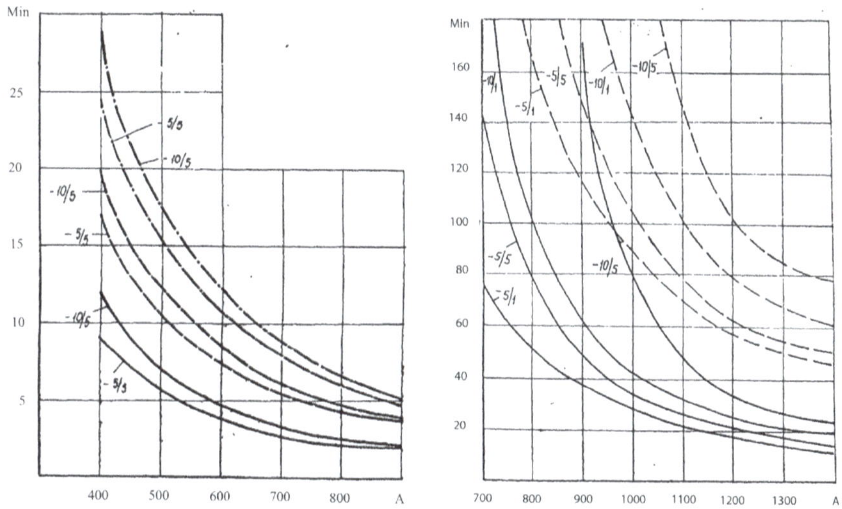

(a) ACSR 150/24 mm² (d=17.1 mm) conductor and 0.2 g/cm³ density hoar frost, for ice sleeve diameters of:

80 mm (— · — ·)

60 mm (– – –)

40 mm (solid lines)

(b) ACSR 400/51 mm² (d=27.5 mm) conductor and 0.9 g/cm³ density glaze ice, for ice sleeve diameters of:

90 mm (– – –)

50 mm (solid lines)

Ice melting duration (minutes) as a function of AC rms melting current (A)

Fraction numerators indicate air temperature (°C) and denominators indicated wind velocity (m/s)

Fig. 5.17 Examples of ice melting currents for single conductors (Timoshova et al., 2003)

DC, but an AC source must also supply reactive power and a DC source may have significant loss in its rectifiers. Thus, the full power needed for ice melting is much larger than the active power.

All ice melting schemes can be divided in two groups: ice melting without outage and ice melting with outage. Ice melting without outage is feasible for the following conditions:

- shifting or re-distributing the load between other lines of the same voltage or parallel lines of different voltages
- phase-by-phase melting using AC or DC.

Ice melting with outages applies to lines where short-circuit methods are used on radial lines. Thus, depending on the conditions, the three phases are connected in series or in parallel, sometimes using the earth as a return conductor.

The power system reconfiguration options needed to develop summer-limit current flow, exceeding 1 A_{rms} per kcmil of conductor cross section (2 A/mm^2), are discussed in Chap. 7.

5.4.4 De-icing Conductors Using Skin Effect

Heating the surface of the conductor, rather than the entire conductor, reduces the electrical energy needed to melt ice by a large factor. Even at power frequency, skin effect in larger ACSR conductors causes higher current density in the outer wire layers. Systems using high-frequency energy have been considered and tested for improved ice melting efficiency in the laboratory but have not proved practical, compared to the tendency to use DC to mitigate the inductive reactance of the conductors.

The first skin-effect de-icing method uses a superimposed high-frequency electric field of around 100 kHz to induce dielectric losses in ice combined with the skin effect in the conductor to melt the ice sleeve (Sullivan et al., 2003). Theoretically, this method can induce ice melting on line sections as long as 50 km. However, this solution, which was tested in a laboratory on 1-m sections of bare conductors, could generate electromagnetic perturbations likely to interfere with telecommunications. Also, to speed ice melting, an increase in high-frequency voltage is required as well as the development of specific inductors to support this high voltage.

5.4.4.1 *Dielectric Coating and 60 kHz*

Joule-effect methods based on dielectric coatings require some electrical energy to be effective. One of the proposed methods makes use of losses in a specific dielectric coating covering the entire surface of the conductor (Petrenko & Sullivan, 2003). By choosing an adequate dielectric coating among ferroelectric materials, it would be possible to maintain the surface conductor temperature above the freezing point to melt the ice/substrate interface. However, these coatings would need to be applied in a factory, and it is unlikely they would survive the environment of an overhead line for more than 5 years. Moreover, this method would require the use of a higher AC frequency of 60 kHz, instead of the 50 or 60 Hz service voltage frequency during the ice-melting process. The use of such high frequencies can lead to electromagnetic disturbances and interference. Also, high-frequency generation would require addition of an external source and coupler to superimpose a 60 kHz electric field onto normal 50/60 Hz line voltage. Suitable wave traps, like those used for older power line carrier systems, would also be needed to divert the 60 kHz energy away from station equipment. The use of high frequency could also affect ACSR conductors differently than AAAC ones as most energy loss (and so heat gain) would be in the zinc layer on the galvanized steel core strands. Active coating methods have not been offered commercially or used widely.

One advantage in automating this approach is that the attenuation of the 60 kHz signal increases rapidly during ice accretion, a characteristic noted in operating experience of power line carrier communication systems in the past as described in Sect. 3.5.

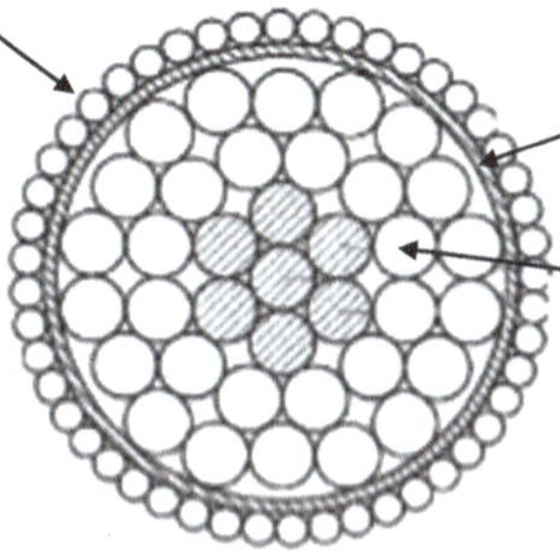

Fig. 5.18 TB483 Figure 7.15: Principle of pulse electrothermal de-icer applied to phase conductors (Petrenko & Sullivan, 2005)

5.4.4.2 Electrothermal Pulse

In the same way as AC and DC Joule-effect methods, the pulse electrothermal de-icer uses the current pulse to heat an external conductive coating surrounding the conductor (Petrenko & Sullivan, 2005). This conductive coating can be made of a layer of conductor strands insulated from the main conductor body by a dielectric layer, as illustrated in the Fig. 5.18.

The main advantage of using current pulses is that the required average power can be reduced by a factor of 100 as compared to AC or DC current. Due to the short duration of the current pulse, the thermal energy from the Joule heat of the external layer is released adiabatically at the ice/external layer interface, with a minimum thermal diffusion into the dielectric layer. Both anti-icing and de-icing can be obtained with this method. However, the approach requires some significant modifications of the bare conductor by the addition of an insulated and a conductive layer sheath over the length of the conductor. This can cause problems during the installation, but also in the summertime when the thermal limitation of the conductor can be reached rapidly, leading to a reduction in ampacity during warmer days. Finally, the efficiency of this method must be demonstrated as no laboratory or field tests on real stranded conductors have been done yet.

5.4.4.3 Magnetic Melting Methods

Other such methods are based on the use of a ferromagnetic coating for the purpose of sustaining a positive temperature of the energized conductor surface (CEATI, 2002–04). Instead of absorbing energy from the electric field, the ferromagnetic coating absorbs energy from the magnetic field, which is at a maximum intensity at the conductor surface. Ferromagnetic coating heating is based on hysteresis and the induced eddy current loss generated by the AC magnetic field. Thus, such methods do not require any external source of energy. This method, known as LC-Spiral Rod method, has been implemented successfully in Japan. Figure 5.19 shows a typical installation of LC-Spiral Rods on a conductor and their relative performance to prevent snow accretion.

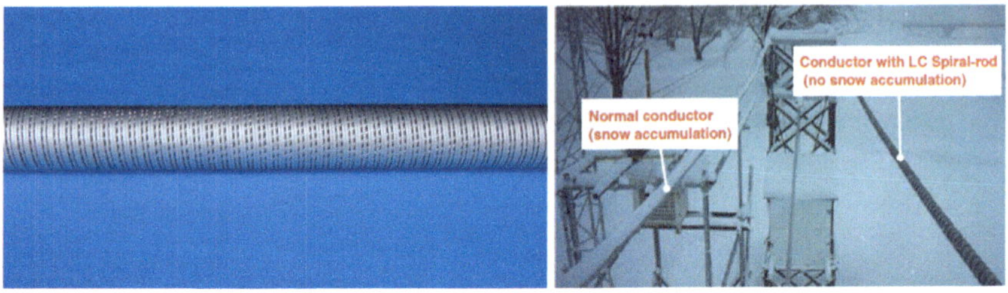

Fig. 5.19 Snow-melting magnetic wire used in Japan (Kitamura et al., 2003)

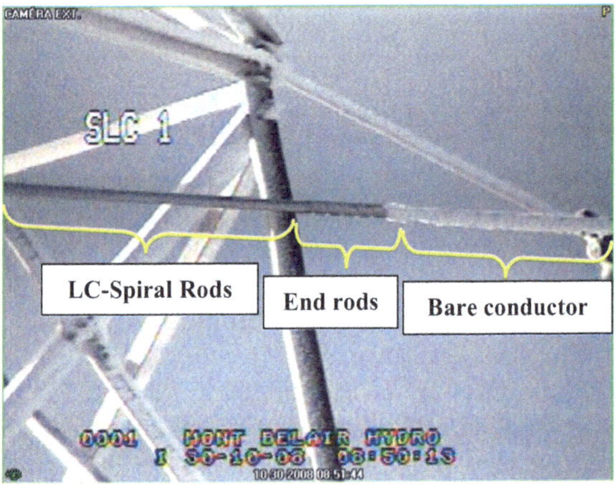

Fig. 5.20 Anti-icing performance of LC-Spiral Rods at Mont Bélair icing test site (Leblond et al., 2009)

These spiral rods have been manufactured and installed for more than 20 years to prevent accidents caused by the sudden fall of large chunks of snow. In Japan, LC-Spiral Rods have been satisfactorily used under wet snow conditions and are installed on more than 100 spans (50 transmission lines). According to the observation results at the Mont Bélair icing test site (Canada) by Hydro-Québec, LC-Spiral Rods are also effective under natural icing conditions (freezing rain and in-cloud icing) under favourable weather and conductor current conditions (Leblond et al., 2009). Figure 5.20 shows satisfactory anti-icing performance of LC-Spiral Rods at a Mont Bélair field trial.

Magnetic melting techniques were used successfully on high-altitude line sites at Tomintoul in Scotland over 50 years ago (Wareing, 1992–09) but were discontinued due to metallurgical limitations in shaping the material.

5.4.5 De-Icing Ground Wires using Heat Tracing or Ice Electrolysis

Another proposed Joule-effect solution resides in using electrical tracers used in chemical plant for heating pipes (CEATI, 2002–04). Electrical tracers are electrically insulated resistive heating wires which can be wound around towers or ground wires to heat their surface. They provide long service life in protecting pipes from freezing, even at temperature of –40 °C in sheltered conditions. Self-regulating heat tracing cables are more suitable because they conduct current when the temperature falls below 0 °C and do not require an independent controller. Although heat tracing is a mature technology, so far it has mainly been used to keep anemometers free of ice in power system weather observation systems.

More recently, thanks to new developments in ice adhesion, a method based on ice electrolysis was proposed (Petrenko, 2000). Ice electrolysis is produced via a small DC voltage applied between a grid electrode and the surface to be protected. This method requires low external energy compared to conventional Joule (thermal) methods. An electrically insulating layer separates the grid electrode from the conductive surface. When ice forms on the surface, it bridges the circuit, and the DC voltage is applied to the interface between the conductive surface and the ice. This leads to the accumulation of gas bubbles between the ice and the solid surface which loosen the ice layer's bond onto the surface. As presented in Fig. 5.21, a configuration was proposed for energized conductors (Petrenko, 2000). However, it seems more convenient to use it on ground wires, as the conductive surface could be the wire itself and the energy required is very small. More studies and experiments need to be carried out under different icing conditions and using different electrode sizes and materials to optimize the efficiency of this method for ground wires. In its present state, however, this method seems to be impractical as it has not been manufactured and tested on a large scale.

Ground wire de-icing can be useful as part of a global approach to avoid major breakdown of transmission networks during severe ice storms. Ground wire de-icing by Joule effect requires a current source as well as an electrical insulation of the ground wires at towers, as shown in Fig. 5.22 (Bourdages, 2000–05).

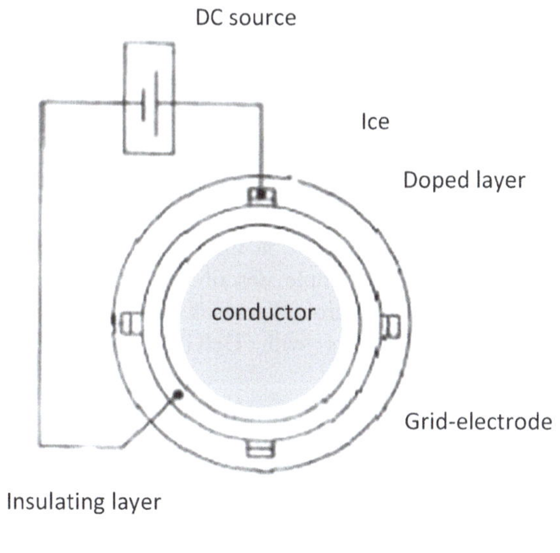

Fig. 5.21 Ice electrolysis method applied to a bare conductor (Petrenko, 2000)

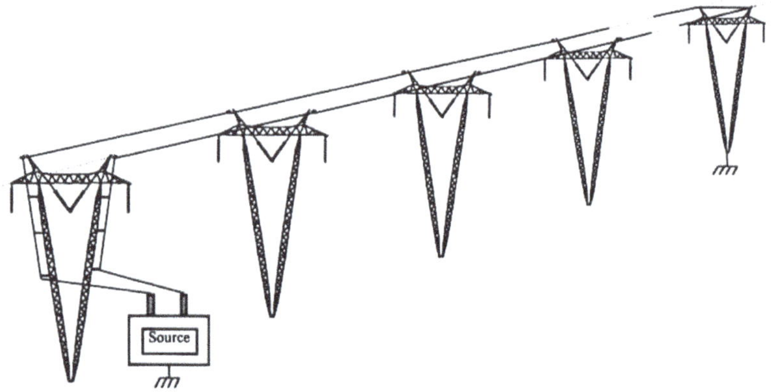

Fig. 5.22 Simultaneous de-icing of two ground wires (loop configuration) (Bourdages, 2000–05)

A medium voltage AC transformer (25 kV), to which current can be supplied by the main AC circuit, may be used as the current source. This method can de-ice several kilometres of ground wires. The range of de-icing is limited only by the withstand voltage of the insulators and arcing horns when covered with ice. In remote areas, it is possible to use an auxiliary diesel generator to de-ice the ground wires on strategic line sections, such as river crossings. However, opinions are still divided on the final approach to use. One philosophy recommends reinforcing the ground wires and the tower top members to withstand the amount of ice accumulation forecast associated with the geographical area. Another approach recommends changing the ground wires only once broken, as the cost of this operation is little more than insulating the existing ground wire and replacement often affords an opportunity to fit an optical fibre ground wire (OPGW). Finally, the last solution recommends ground wire insulation for Joule-effect de-icing as the associated costs are partly compensated by the elimination of induction losses. The most important induction loss in overhead ground wires is at peak load when each kW of generation capacity is presently valued at approximately 2000 €. Peak generation needs to make up an additional 2 kW/km for a 735 kV line and 1 kA phase current, and up to 26 kW/km for double circuit lines with super-bundle phasing.

There are other factors in the decision to insulate OHGW. CIGRE Technical Brochure 694, for example, describes how to manage ac fault current step and touch potential exposures around transmission structures. The division of fault current among many towers with OHGW connections reduces the touch voltage significantly.

Fig. 5.23 Remote steam de-icing vehicle (Lanoie et al., 2005)

5.4.6 Steam De-Icing

Another method for de-icing uses steam as a heat source. This method is used by Hydro-Québec to de-ice electrical equipment of substations like disconnect switches and post insulators (Lanoie et al., 2005). For locally limited interventions, a vehicle has been adapted with an insulated telescopic arm for remote steam de-icing operation in high-voltage environment, as illustrated in Fig. 5.23; it could easily be used to de-ice strategic or critical spans of power lines.

5.4.7 Radio frequency and Radiant Energy De-Icing

Other proposed methods use external heat sources to shed the ice sleeve. One such method consists in using radiation and specifically radio-waves to heat the conductor/ice interface (Berry et al., 1993–04). Though efficient for de-icing railway

Table 5.4 Main features of anti-icing and de-icing techniques

De-icing type	Technique	Description	Characteristics (status, disadvantages, line length, installation)	Country of Use	Cost level
Passive	Counterweights	Used for reducing ice or wet snow accretion	• Operational • Any line length • Permanent installation	Japan, Iceland France	Low
	Snow rings (SR), wires, combination of SR and counterweight	Used for reducing wet snow accretion	• Operational • Any line length • Permanent installation	Japan	Low*
	Modifying crossarms, using separate structures for each phase, rearranging phasing configuration	Used for minimizing the impact of icing on electrical clearances in double circuit lines	• Operational • Any line length • Permanent installation	New Zealand and others	Low-Moderate
	Installing inter-phase spacers (insulators)	Reduces probability of phase clashing	• Operational • Possible conductor fretting • Any line length • Permanent installation	New Zealand and others	Moderate
Active coatings and devices	LC spiral rods	Used for melting and shedding snow at milder temperatures	• Operational and promising short-term for anti-icing applications • Induces extra losses • Any line length • Permanent installation	Japan Scotland	Low*
	Heated tracers	Used for melting and shedding ice and snow at milder temperatures	• Operational but used only for water pipes • Needs external supply • Short line length due to need for supply • Permanent installation	Japan	Moderate-High
	Ice electrolysis	Applicable to wires equipped with electrodes energized with DC	• Not tested for real applications • Needs external supply. May degrade in UV light • Short line length due to need for supply • Permanent installation	N/A	High

(continued)

Table 5.4 (continued)

De-icing type	Technique	Description	Characteristics (status, disadvantages, line length, installation)	Country of Use	Cost level
Mechanical	De-icing using ropes or roller wheels	Used for manual de-icing of distribution lines up to 25 kV	• Operational • Labour intensive • Short line length • Temporary installation	Canada	Moderate
	Remotely Operated Vehicle (ROV)	Used for de-icing of ground wires and line conductors up to 315 kV	• Promising in short-term • Installation onto line may be difficult • Up to 1 km range • Temporary installation	Canada	Moderate
	Pneumatic hammer	Used for de-icing line conductors one span at a time	• Not tested for real applications • Labour intensive • Short line length • Temporary installation	Canada	N/A
	De-icer Actuated by Cartridges (DAC)	Used for de-icing conventional ground wires	•• Promising in short-term • Labour intensive • Short line length • Temporary fitting	Canada	Low
	Electro-impulse methods	Used for de-icing of ground wires and line conductors	• Not tested for real applications • Risk of damage by lightning. May cause telephone interference • Short line length due to need for supply • Permanent installation	N/A	High
	Ice-shedder devices	Used for de-icing of line conductors and ground wires by inducing certain vibration ranges	• Not tested for real applications • May damage line if used for long periods • Any line length • Permanent installation	Canada	N/A
	Twisting devices	Used for de-icing of single conductor lines one span at a time by mechanically twisting the conductors	• Not tested for real applications • Any line length • Permanent installation	Canada	Low
	Weight attached to rope with large knots	Used for partial snow shedding in mountainous areas using helicopters	• Operational • Needs helicopter • Short line length	Canada	Moderate-High

(continued)

Table 5.4 (continued)

De-icing type	Technique	Description	Characteristics (status, disadvantages, line length, installation)	Country of Use	Cost level
			• Temporary use		
Thermal	Dielectric coating	Application of dielectric coating onto the line and injection of high frequency supply	• Theoretical • May cause telephone interference • Short-medium line length • Permanent installation	N/A	High
	Load shifting method	Used for anti- and de-icing of line conductors by transferring or shifting loads from other circuits	• Operational • Only works on single conductor lines • Medium-long line length • Permanent installation	Many	Low
	Reduced voltage short-circuit method	Used for anti- and de-icing of line conductors by using 3-phase short-circuits	• Operational • Only practical on MV lines due to high power demand • Medium line length • Permanent installation	Many	Moderate
	DC current method	Used for de-icing of long sections of high current lines by running DC current at stages through the phase conductors. Can also be applied to ground wires	• Operational • Large installation • Medium-long line length • Permanent installation	Former USSR / Canada	High
	On-load Network De-Icer method (ONDI)	Used for de-icing of lines by using a phase-shifting transformer to vary the current from one conductor to another, without requiring disconnecting the section to be de-iced from the network	• Promising • Only works on single conductor lines • Medium-long line length • Permanent installation	USA/Canada	High

(continued)

Table 5.4 (continued)

De-icing type	Technique	Description	Characteristics (status, disadvantages, line length, installation)	Country of Use	Cost level
	Contactor load transfer (bundle shifting) method	Used for de-icing of lines by allowing the current flowing in all conductors of a bundle into one	• Promising but not tested • May prove difficult to instal into a real transmission line • Any line length • Permanent installation	N/A	N/A
	Pulse electrothermal de-icer method	Proposed for de-icing of line conductors by allowing the current pulse to heat an external conducting coating	• Not tested for real applications • Reduces line rating • Medium line length Permanent installation	N/A	N/A
	Ground wire de-icing method	Used for de-icing many km of ground wires using a medium AC voltage transformer as current source. This requires the GW to be insulated from the towers	• Operational • Short-medium line length • Permanent installation	N/A	High
	High-frequency electric field method	Proposed for de-icing of line conductors by using a high-frequency electric field to induce dielectric losses	• Not tested for real applications • May cause telephone interference • Medium line length • Permanent installation	N/A	N/A
	Steam generating device	Used for de-icing equipment like disconnect switches and post insulators	• Operational • Labour intensive • Short line length • Temporary use	Canada	Moderate

* Cost is relatively low if installed at the time of line installation, otherwise conductors may need to be replaced

lines, this method seems not so well adapted to power lines as it is limited to de-icing one span at a time and requires a mobile and powerful radio wave source.

Radiant heat sources such as an array of incandescent heat lamps have been assembled and used for anti-icing and removing ice from substation apparatus. These need to be within 10 m, often mounted at the base of a vertical post insulator as at the Kinzua Dam in Warren, PA (Farzaneh & Chisholm, 2009). Focused arrays of heat lamps are not so practical for overhead line de-icing.

5.5 Concluding Remarks

Table 5.4 summarizes the main anti-icing (AI) and de-icing (DI) techniques. Each method is presented according to efficacity, operational status, disadvantages, country of use, and cost level.

The strategies adopted by the utilities converge towards the thermal methods based on the Joule effect for de-icing of overhead power lines on a large scale. Many thermal methods are well known, and some have been used for more than 100 years in several countries. However, to be effective, these methods must be selected and used adequately within the framework of a well-established procedure which must consider both the availability of the power on the network at the desired time and the weather data regularly updated on the icing sites. This suggests that the efficacity of a thermal de-icing strategy requires an adequate detection system of ice accretion on a line which could provide a real-time monitoring of the evolution of the ice build-up. Developing an efficient ice prevention strategy also requires timely cooperation among the various maintenance services and stakeholders to take full advantage of the time windows available to set up effective sequences of anti-icing and, should that be ineffective, de-icing.

Mechanical methods seem to be preferred for specific localized interventions aimed at protecting short sections of strategic lines. Indeed, shorter de-icing times can be achieved with mechanical methods compared to thermal methods, and they are well suited to extreme emergency situations, to avoid the collapse of overloaded towers, for example.

6 Protective Coatings for Overhead Lines in Winter Conditions

In power networks, many economic losses are linked to the corrosion of line sections located in coastal and/or industrial areas. In fact, marine and industrial pollutants cause corrosion on metal components, such as conductors, fittings, insulator electrodes, and support structures. Corrosion does not stop in winter weather; instead, the use of road salt and accumulation of pollution during long periods without rain may lead to rapid degradation of power system components in unfavourable locations.

Field testing of a coating, including exposure in normal and extremely cold winter conditions, is the most appropriate method for evaluating its performance even though such evaluation usually takes several years. Furthermore, some products may be useful in some environs and conditions but not in others. Factors such as rain, sand, or salinity are decisive parameters in this regard.

6.1 Anti-corrosion Coatings and Materials

The most economical way to prolong the lifetime of transmission towers installed in areas that prove to have higher-than-anticipated level of corrosivity, for example winter exposure to road salt, is to apply protective coatings. Several methods have been adopted by public services confronted with winter conditions, to prevent or mitigate the local effects of corrosion. Paint systems for hot-dip galvanized steel and aluminium have been globally in use for a long time, and issues regarding the lifetime and maintenance procedures are well known. Two-component or "Duplex" systems apply water-soluble coatings immediately after hot-dip galvanizing of steel components, rather than after the galvanizing has degraded, and offer several advantages (Lugschitz et al., 2004-08).

The durability of a corrosion protection system for galvanized steel and aluminium alloys, from which many components of power network equipment are manufactured, is also influenced by the corrosivity of the concerned environment.

M. Farzaneh and W. A. Chisholm, *Techniques for Protecting Overhead Lines in Winter Conditions*, Compact Studies, https://doi.org/10.1007/978-3-030-87455-1_6

Durability is determined as the expected life of the corrosion protection system from application until the time of the first major maintenance. Based on the literature review and field experiences, there are several practical techniques for evaluating natural corrosion processes. The key point is the correct and timely diagnosis to find and use the most effective technique to protect all key equipment against corrosion. The choice of a coating and its application to a piece of equipment is highly dependent on environmental parameters and the equipment to be protected.

Maintenance intervals between coating applications are defined by the state of the structural components as well as the corrosivity of the atmospheres. Experience has shown that the best time to paint a tower is when all its profiles have a Grade 1 corrosive attack. This allows the preparation of the surface for paint application not to be very demanding, which ultimately results in fewer labour hours and lower costs.

6.1.1 Surface Preparation

Once the towers affected by corrosion are identified, the structures must be cleaned and prepared for the coating application. Depending on the level of the degradation, different methods are used, according to ASTM D6386 (ASTM SC D01.46, 2016).

The solvent cleaning is one of the applied processes. It removes soluble contaminants from the metallic structures with stiff bristle brushes, followed by scrubbing the surface with solvent.

Then, manual cleaning using steel bristle brushes eliminates superficial contaminants. Metal scrapers clear away sticky pollutants. Emery-type sandpaper is used to remove debris from hard-to-reach areas. The highly degraded parts are changed out. Some stages of the towers surface preparation are shown in Fig. 6.1.

Fig. 6.1 Tower surface preparation

The final step in surface preparation is washing the tower with an organic bonding chemical mixture diluted in water. Its proportion varies from 0.5:25 to 1:25 depending on the salt concentration which is verified by the method described in ISO (2006). If, after this process, the salt concentration is still higher than 80 mg/m^2 (SDD of 8 μg/cm^2), the tower must be washed again.

6.1.2 Anti-corrosive Coating Application on Towers

As detailed in Oberst (1998), Soto and Silva (2017-09), the selection of the most suitable anti-corrosion coating procedure involves some important steps, such as choosing the type of coating and how to apply it. In the case of a galvanized tower, the organic medium and long oil alkyd resin is the anti-corrosive material used to coat the structures. The thickness of the first undercoat was 25 μm and the second layer 305 μm, with time between applications of at least 12 h, which would enable the base to get dry.

To perform the coating application as shown in Fig. 6.2, the environmental conditions in the work area should not exceed 85% RH, wind speed lower than 35 km/h, metal surface temperature between 5 and 40 °C, and ambient temperature below 40 °C. Some anti-corrosive coatings require a certain level of absolute humidity (g/m^3) in the air to cure properly, and unfortunately the absolute humidity in winter months seldom achieves the required level. A process to apply the coatings in winter conditions should be reviewed with the coating supplier and tested before initiating a large project.

The coating is inspected two months after application. ISO 19840 (ISO, 2012) and ASTM, D3276 (2015) provide details of a suitable inspection process. If unacceptable irregularities are found, the application is repeated.

Fig. 6.2 Anti-corrosive coating application

6.1.3 Anti-corrosion Protection of Conductors

In the case of aluminium alloy conductors subjected to aggressive atmospheric conditions, corrosion occurs in the form of small pores which, over time, grow until they become visible to the naked eye. Left alone, these pores will increase in size until the strand breaks. This leads to premature failures in the operation of the conductor as its breaking strength is reduced.

All the metal elements of power lines installed in areas with aggressive atmospheres that include winter conditions will eventually be affected by corrosion. The conductors, which are mainly made of aluminium or aluminium alloy conductors, are no exception. Conductors may be fully or partially greased as an anti-corrosion measure when installed in aggressive atmospheres. Figures 6.3, 6.4, and 6.5 show corrosion on the inner and outer wires of a conductor.

These photographs show the importance of the need to protect conductors exposed to corrosive atmospheres. There are several possibilities to protect conductors from corrosion. One of the most economical and effective ways to protect and extend the useful life this is to apply anti-corrosive greases. Despite the evidence, it is common to observe today that some conductor manufacturers, some

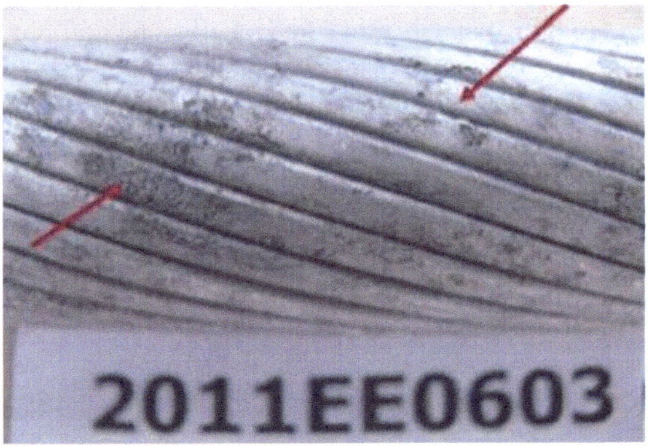

Fig. 6.3 Corrosion of the outer wires of a conductor

Fig. 6.4 Corrosion of the inner wires of a conductor

Fig. 6.5 Severe corrosion of the outer wires of a partially greased conductor

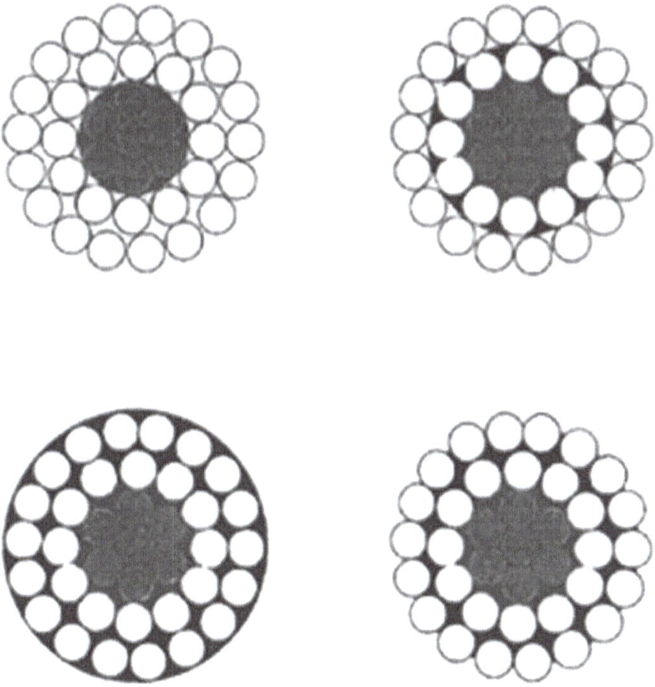

Fig. 6.6 Illustration of British standard BS EN 50182: 2001 for greasing of conductors

electric utilities, and others in the telecommunications sector still specify ungreased or partially greased to be installed in corrosive atmospheres. Figure 6.6 shows a British standard illustrating the different levels of grease application for conductors.

Fig. 6.7 Conductor with low-quality grease that migrated and accumulated at the bottom

The use of grease on the outer surface of the conductor, as shown in Fig. 6.6 at lower left, may lead to audible noise and this consideration must be balanced against the protective benefits against corrosion. High-temperature operation of greased conductors may cause the grease to soften and flow out the bottom, then form solid drops, which can also aggravate any AN problems on a line.

Some greases prove to be unsuitable for corrosion protection, as shown in Fig. 6.7. Many commercial greases protect against frictional wear, but few are also suitable for anti-corrosion applications. The desirable characteristics of an anti-corrosion grease should include low acidity and high dropping point. The low acidity ensures that the grease does not generate corrosive processes between the conductor wires and galvanized steel terminals. The dropping point of a grease is the temperature at which it passes from semi-solid to liquid state. The dropping point of anti-corrosion grease should be high enough, compared to the maximum conductor operating temperature, that the grease will not melt and flow to the lower parts of the conductor as shown.

6.1.4 Anti-corrosion Protection in Optical Ground Wires (OPGW)

Like ground wires and conductors, OPGW are attacked by corrosion depending on the corrosivity of the atmospheres in which they are installed. The following photographs show the aggressive corrosion of some greaseless wires installed in areas with high corrosivity. The classifications are presented in Fig. 6.8. Examples of OPGW degradation without and with grease in an area of high corrosivity are given in Figs. 6.9, 6.10, and 6.11.

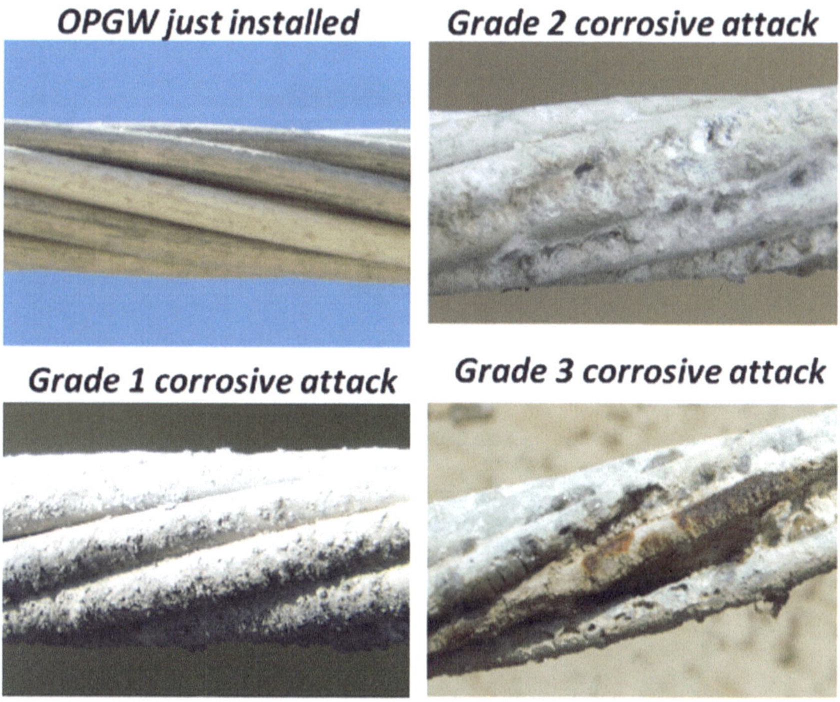

Fig. 6.8 Photographic classification guide for corrosion of the OPGW without grease in areas of high corrosivity

Fig. 6.9 OPGW without grease installed in an area of high corrosivity

Fig. 6.10 OPGW installed with anti-corrosive grease in an area of high corrosivity (left, before and right, after removing the grease—note the good condition of the wires)

Fig. 6.11 OPGW installed with anti-corrosive grease in an area of high corrosivity (upper, before and lower, after removing the grease—note the good condition of the wires)

6.1.5 Anti-corrosion Protection in Metallic Part of Insulators

Insulator manufacturers around the world are aware of the vulnerability of the metal parts of insulators to corrosion, which is why certain precautions are taken in their design, such as hot-dip galvanizing of metal parts or the installation of zinc rings as sacrificial anodes.

The reality is that these protections wear out quickly in some areas of high corrosivity. Companies may resort to painting these metal parts with little success. A paint with adequate anti-corrosion performance may pose an electrical flashover hazard across the dry arc distance of live insulators (Farzaneh & Chisholm, 2009).

In addition to the aggressive atmosphere where the tower insulation is performed, corrosion rates of metal parts of the insulators can be accelerated by induced current. DC current is a severe problem but AC currents can also increase corrosion rate. This situation forces to evaluate in detail each one of the insulators. It is very common that within the same ceramic insulator string, several levels of condition are found on caps and pins, which means that the whole chain must be changed.

An important innovation that has gained in strength to improve the performance of power lines located in areas of high corrosivity consists in the use of high-performance silicone rubber insulators and stainless steel fittings and terminals. Insulators with these characteristics offer a much longer life than conventional glass or porcelain insulators, which in some cases do not exceed 15 years due to corrosion (Asto Soto & Silva, 2019).

Figures 6.12, 6.13, 6.14, and 6.15 show examples of cap and pin corrosion on ceramic disc insulators

Fig. 6.12 Failure of paint applied to the glass insulator caps

Fig. 6.13 Porcelain insulator pin strongly affected by corrosion despite being fitted with a zinc ring as a sacrificial anode

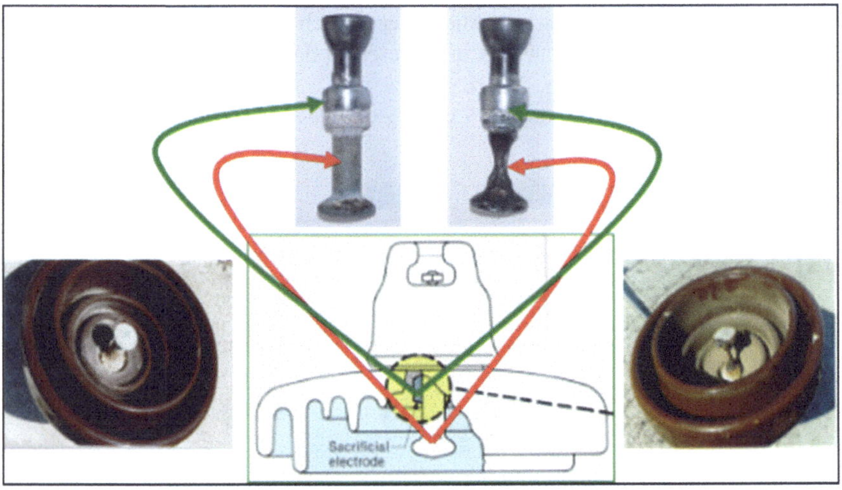

Fig. 6.14 Explanatory diagram of the possible ineffectiveness of the zinc rings installed on the pins of the insulators as sacrificial anodes

Fig. 6.15 Large amount of glass insulator strings removed from lines due to the risk of failure because of corrosion deterioration of insulator pins

6.1.6 Anti-corrosion Protection in Fittings

Fittings are critical elements in electrical substations or in a power line. Indeed, the failure of one of these elements would inevitably generate the immediate disconnection of the circuits as well as a severe safety hazard.

Unfortunately, for the corrosion protection of steel hardware elements, hot-dip galvanizing may not provide long-term protection as shown in Fig. 6.16. There are references to electrical utility companies located in tropical countries with assets very close to the sea where the fittings have a useful life of less than 15 years, which requires long outages and significant investments to replace these elements. Generally, galvanized hardware gives more than 40-year life in regions with winter conditions, although the corrosion damage is similar.

Sometimes hot-dip galvanized fittings are additionally protected with paints, seeking to improve their performance in corrosive atmospheres. However, this is an ineffective solution, it happens that the coating applied to the hardware breaks for any reason, and the point where it breaks, called a "holiday", becomes a location with accelerated corrosion. Examples of accelerated corrosion from coating failures are shown in Figs. 6.17 and 6.18.

Inherent corrosion protection can be achieved with the use of stainless steel fittings, as shown for a composite insulator corona ring in Fig. 6.19 and an end fitting in Fig. 6.20.

The extra material cost of stainless steel may have a strong business case in areas of high corrosivity because it delivers long, trouble-free service.

Fig. 6.16 Hot-dip galvanized fitting with a high degree of corrosion damage

Fig. 6.17 Hot-dip galvanized and painted suspension hardware; note the corrosive attack on the U-shackle and the onset of corrosion through the paint on the other hardware

6.1.7 Corrosion Protection of Signage

Steel is perhaps the most widely used material of information or signal plates for power transmission towers. The disadvantage of steel is that in corrosive

Fig. 6.18 Hot-dip galvanized suspension fittings with extremely severe corrosive attack

Fig. 6.19 Composite insulator with stainless steel fittings

atmospheres its useful life is very short. Therefore, some companies use fibreglass plates with high-performance adhesives, with very good results.

A corroded carbon steel plate case is shown in Fig. 6.21 and a fibreglass case in Fig. 6.22.

Fig. 6.20 Stainless steel fitting installed in an area of high corrosivity

Fig. 6.21 Information plate made of carbon steel with a severe corrosive attack

Fig. 6.22 Information plate made of made of fibreglass installed in an aggressive area

6.1.8 Anti-corrosion Protection of Stockbridge Type Aeolian Vibration Dampers

Good long-term performance of Stockbridge Aeolian vibration dampers as pictured in Fig. 6.23 is especially important in winter conditions, where low ambient temperature causes high tension and steady wind speed leads to many accumulated cycles. Perhaps due to the variety of elements that make up this type of damper, it is often the first or second component of a power transmission line to degrade when exposed to corrosive atmospheres. The other fast-wearing component seems to be the overhead ground wires exposed to frequent summer lightning flashes that lead to holidays in galvanizing or aluminium cladding. To improve the lifetime of Stockbridge dampers, some manufacturers use stainless steel both in the screws and for the messenger cable to which the weights are attached. However, no progress has yet been observed in the design of bolts and washers which also cause corrosion problems.

6.1.9 Testing of Coated Conductors

As seen in this chapter, in very harsh environments, some conductors can be protected against corrosion by specific coatings. The application of such coatings can change the electrical contact resistance between the conductors and influence the minimum or maximum slip loads that can be achieved inside a conductor fitting.

Fig. 6.23 Stockbridge type Aeolian vibration damper with severely corroded messenger wires

The verification of these aspects of the coatings is established with electrical resistance tests described later in Sect. 6.7.3

6.1.10 Concluding Remarks

Based on the literature review and field experience, there are several practical techniques for evaluating the natural corrosion process of metallic elements of power system transmission lines. The key point is the correct and timely diagnosis to find and use the most effective way to protect each part against corrosion.

The choice of an anti-corrosion coating and its application to an element of a power network equipment strongly depends on the environmental parameters and the line element to be protected. It is precisely according to these factors and the element to be protected that one decides to apply a type coating (zinc, epoxy, alkyd, metallic elements, RTV, or HTV silicone, etc.). Acceptance testing for coatings is discussed fully in Sect. 6.7.

Testing the field performance of a coating over several summer and winter periods is the best method for evaluation even though tests take several years. The adoption of new products should be sequential, some products may be useful in some countries but not in others, and factors like long exposure without rain in winter, road salt and sand, extreme cold temperature, and ice accretion make the difference.

6.2 Anti-icing Coatings

The accumulation of ice and snow on overhead power network equipment is a serious problem which may lead to major service outages. One of the most practical solutions to combat icing problems would be the use of anti-icing coatings to prevent or reduce icing loads. This chapter aims to present and discuss different types of anti-icing coatings designed to prevent or reduce the ice and snow accumulation on power network equipment.

This section considers practical and durable coatings as well as surface engineering techniques that may increase the reliability of overhead power network equipment in various aggressive environments and adverse winter weather. It also addresses certain environmental concerns caused by corona noise and partial discharge on the overhead power lines and recommends appropriate methods to evaluate the efficacity and durability of suitable coatings.

6.2.1 Classification of Coatings

Summarizing what has been presented above, from an applicative point of view, the following coating materials are used to prevent or reduce atmospheric ice (glaze ice, snow, rime, or frost) from accumulating, based on ice adhesion strength reduction or on prevention of supercool droplets freezing (Farzaneh, 2019; Farzaneh & Volat, 2008). A comprehensive review of these coatings was carried out by CIGRE WG B2.44, resulting in the publication of Technical Brochure 631 in 2015 (CIGRE WG B2.44, 2015).

Regarding the application on conductors and ground wires, to date many of the studied coatings are still at the laboratory level. Some of them have been tested at pilot scale, in large synthetic ice facilities or on test spans on field in different countries, notably in Italy, Norway, Japan, China, and Iceland. Despite extensive investigation in many countries and good progress in development of icephobic coatings, it appears that there are still limitations for applications to overhead power network conductors and ground wires. This is mainly due to the efficacy, stability, and durability of the coatings developed so far. As a result, commercial products widely applied to service conductors and ground wires are still lacking.

Conversely, coatings are already widely applied to insulators, mainly for anti-pollution purposes as described completely in CIGRE TB 837 (CIGRE WG B2.69, 2021-06). Based on the long experience on material formulation, coating application and service data, coatings for insulators are applied and tested also for anti-icing purposes or to avoid icing flashovers upon ice accretion. Therefore, more interesting application cases are reported for anti-icing insulator coatings.

6.2.2 Anti-icing Coatings for Conductors and Ground Wires

6.2.2.1 *Passive Coatings*

As previously mentioned, passive coatings are based on the exploitation of the physico-chemical properties of icing surfaces that influence the adhesion force between atmospheric ice and substrate or the nucleation of ice crystals, including surface energy and surface roughness. Passive coatings require no external energy to be effective (Farzaneh & Volat, 2008) as indicated above.

Polymeric Coatings

Teflon(R) and some silicone-based polymers or heterogeneous polymers considerably decrease ice adhesion force (CIGRE WG B2.44, 2015). Their efficacy is questionable with respect to glaze or dense ice accretions, considering that accreted ice cannot detach from their surface under its own weight or from the action of the wind (Farzaneh & Volat, 2008; Laforte & Beisswenger, 2005-06). Instead, some positive action is detectable with respect to wet snow, whose internal cohesion is lower, as reported in Balordi et al. (2019-06), Marcacci et al. (2019) after some campaigns under real snow events in Italy.

Hydrophobic and Superhydrophobic Coatings

Recent development in research concerning the surface wetting and ice adhesion and new advances in material and surface science have resulted in the development of advanced superhydrophobic coatings with self-cleaning and icephobic properties (Balordi et al., 2019; CIGRE WG B2.44, 2015; Farzaneh, 2019). These coatings with nano/microsurface roughness can prevent or reduce considerably ice adhesion strength on surfaces. Application of these coatings to conductors, ground wires and supporting structures can potentially reduce mechanical loads due to ice and snow accretions, thus increasing the reliability of power transmission in winter conditions. Experiments on in-service ground wires and test spans are being conducted in Italy (Balordi et al., 2019-06; Marcacci et al., 2019).

Viscous Coatings

Ice adhesion on an exposed surface can be reduced by applying a viscous or liquid coating such as industrial lubricants, oils, and grease, leading to ice release by gravitational, vibrations, or wind force. Laboratory investigations demonstrated that lithium grease and industrial lubricants can reduce ice adhesion strength on an aluminium surface by a factor of 63 (Farzaneh, 2019). However, this type of protection is temporary, and thus there is a need of reapplication of the coating, involving accurate field observation to predict when a severe storm hits. Finally, as these products are not biodegradable, they may constitute a threat to the environment. However, they may only be used as a last resort solution to protect strategic line sections when severe icing storms are forecasted (CEATI, 2002-04; Farzaneh & Volat, 2008). The greases are also widely applied to steel portions of conductors and ground wires for anti-corrosion purposes but seldom used on the outer aluminium wire surfaces.

Active Coatings and Devices

LC-Spiral Rod, described previously, has been implemented successfully in Japan science for about 3 decades to prevent accidents caused by the sudden fall of large chunks of snow. Due to the cost, this device is applied on short line segments, for example in correspondence of road crossing or pedestrian paths. The generated heating can be adjusted by the winding pitch using a wrapping machine. Based on some experiments in a natural Hydro-Quebec site (Leblond et al., 2009), this technology, efficient for wet snow melting, seems to be also effective under icing conditions with favourable weather and current conditions.

6.2.3 Coatings for Insulators

Insulator flashover under both pollution and/or icing has received a great deal of attention from many researchers, and many studies have been carried out in several laboratories (Farzaneh, 2014-10; Farzaneh & Chisholm, 2009, 2014).

Optimizing the maintenance of outdoor insulation is more of an art than science learned through lifelong experience. Methods like high-pressure washing, silicone dielectric greases, and RTV silicone coatings have been applied to porcelain and glass insulators. Silicone rubber housings (RTV, HTV, LSR) have proven to be more effective, offering much better long-term capacity to provide anti-pollution properties. More recently, RTV silicone rubber coatings have successfully been applied to porcelain and glass insulators to improve pollution performance. Despite all these possibilities, improving the flashover performance of insulators remains a challenge for power network operators and designers. Therefore, the search continues for new technologies that could offer technical and economic solutions. This specifically includes coatings for insulators to improve their flashover performance under pollution and ice, or a combination of these conditions. These coatings could provide self-cleaning, prevent dry band arcing, and improve the voltage gradient, or reduce ice adhesion and accretion.

A common way of improving the contamination flashover performance of insulators is to provide the insulator external surfaces with hydrophobic properties. Hydrophobic surfaces, which have a low surface energy, cause water to bead into distinct water drops that inhibit the formation of continuous conducting films. This, in turn, hinders the pollution flashover process, resulting in an improved flashover performance. A variety of hydrophobic surfaces have been fabricated, from organic polymers to inorganic materials. A brief history of these coatings is reported in (CIGRE WG B2.29, 2010). Hydrophobic greases were a traditional coating, applied and renewed by hand, for repelling water. They have been supplanted by semi-permanent coating materials such as RTV silicone that improve hydrophobicity and play an active role in encapsulating surface pollution.

6.2.3.1 *RTV Silicone Coating*

Room temperature vulcanized (RTV) silicone rubber (SiR) coatings were developed for improving contamination flashover performance of outdoor insulators. The main application of RTV coatings has thus far been applied to insulators as a measure against pollution flashovers. They are listed as maintenance or palliative measures both in CIGRE TB 361 (AC) (CIGRE WG C4.303, 2008-06) and CIGRE TB 518 (DC) (CIGRE WG C4.303, 2012-12). RTV coatings have generally a longer lifespan than that of a petroleum-based grease coating. However, an inspection programme is required to determine the condition of the coating and its need for replenishment or replacement. Such a programme normally comprises a visual inspection and a hydrophobicity measurement at an inspection interval of 5–6 years. CIGRE proposes that as a rule, coatings are used as a palliative measure and are not considered as a candidate solution for new insulation projects. However, in case of severe environmental conditions under DC, their use may be unavoidable (CIGRE WG C4.303, 2012-12).

RTV coatings have also been used as a countermeasure under icing conditions. In cold fog conditions (very light icing), an increase of about 25% in flashover strength per metre of leakage distance is gained when silicone coating is applied to the porcelain insulators (Chisholm, 2005; Farzaneh, 2008, 2019; Farzaneh & Chisholm, 2009, 2014). However, for heavy icing causing icicle bridging of insulator shed spacing, RTV is no longer effective and cannot reverse the situation. In fact, ice is retained for a longer period on RTV-coated porcelain during the dangerous melting phase (Farzaneh & Chisholm, 2009). Silicone rubber insulators show the same problem of increased ice retention (Gutman et al., 2008-08). Thus, the effectiveness of conventional RTV coating for preventing insulator icing flashovers is questionable.

An example of comparative results on uncoated and RTV-coated porcelain disc insulators is presented in Fig. 6.24. It shows the appearance of the ice accretion after 30 min and 3 h.

These pictures show that there is only a slight difference between the bare and RTV-coated porcelain cases. The ice layer on the bare insulator appears to be smoother than that on the RTV-coated insulator. The icicles themselves are more

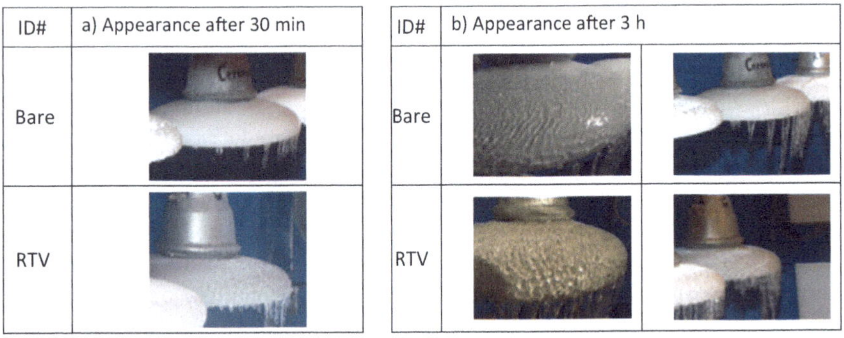

Fig. 6.24 Comparative icing appearances on bare and RTV-coated insulators **a** after 30 min, **b** after 3 h (Hayashi et al., 2013-08)

numerous, longer but thinner on the RTV-coated discs. It is believed that these effects are caused by the surface tension which is lower for RTV.

6.2.3.2 *Passive Self-Cleaning Coatings*

The low adhesion force between a liquid droplet and a surface can manifest itself in some interesting characteristics, such as self-cleaning effect. Generally, a surface is referred to as self-cleaning when it can somehow "clean" itself. On a hydrophobic surface, self-cleaning is usually done by the absorption and transportation of dirt particles in a liquid droplet moving on the surface. Therefore, a self-cleaning surface will remain contamination free in outdoor environments when it is exposed to precipitation. Clearly, self-cleaning can be of great benefit for outdoor insulators and therefore, several self-cleaning coatings have been developed with electrical insulators in mind.

One example of such coatings has been tested and applied to service insulators (Blackett-Voltshield, 2009). The paper describes the development testing of an anti-pollution treatment for glass and glazed porcelain insulators that makes the surface "non-stick" and improves the performance and durability. It was applied to capacitive voltage transformers to reduce discharges and their associated noise, because of condensation on the external insulating surfaces. Other application examples include 33 kV transformer bushings and the low-voltage DC supply of the London underground network. On transmission voltages, only a limited number of 400 kV line insulators were treated to investigate the performance of this coating in a polluted environment.

6.2.3.3 *Passive Coatings for Anti-icing*

Although there are several active anti-icing technologies that are somewhat effective, they are not applicable to the insulators due to the safety and reliability issues they may arise. Over recent years, much emphasis has been placed on the development of coatings that act as passive anti-icing surfaces.

Superhydrophobic coatings, with self-cleaning and icephobic properties are attractive to improve the flashover performance of insulators under pollution and icing conditions. A review of such coatings with potential application to outdoor insulators is reported in CIGRE TB 631 (CIGRE WG B2.44, 2015).

Surfaces with a combination of microstructure and low surface energy are known to exhibit superhydrophobic properties. Even slight amounts of water drops can entrain dirt particles adhering to the surface and clean the surface completely. To achieve effective self-cleaning on an insulator, the surface must not only be very hydrophobic but also have a certain roughness. Surface structures composed of protuberances and depressions required for self-cleaning purpose must have spacing in the range of 50 nm–200 μm and a protuberances height ranging from 50 nm to 100 μm (Baalmann et al., 2002).

An example of the effectiveness of a superhydrophobic coating (SHC) in reducing ice accretion is shown in Fig. 6.25 which shows that even after 3 h of icing exposure, no icicles were formed. This contrasts with the bare porcelain disc where icicles were present even after 30 min of exposure.

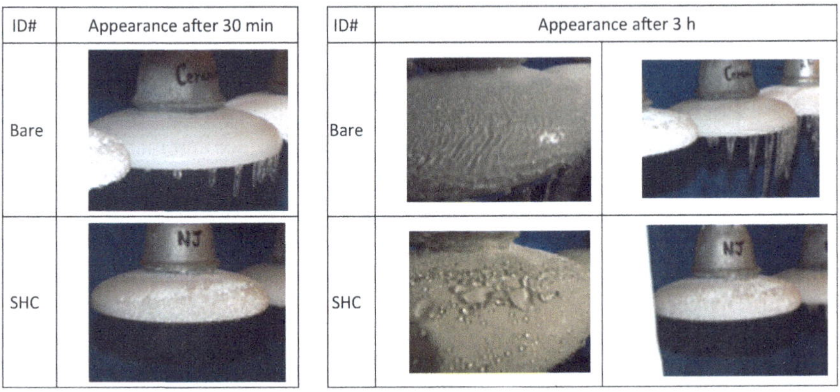

Fig. 6.25 Comparative icing appearances on bare and SHC-coated insulators **a** after 30 min, **b** after 3 h (Hayashi et al., 2013-08)

The absence of icicles on the superhydrophobic coating in Fig. 6.25 was caused by the releasing and falling of the deposited ice from the insulator surface before icicles could form.

Figure 6.26 shows an example of the effectiveness of PDMS/nanosilica in reducing ice accretion on insulators.

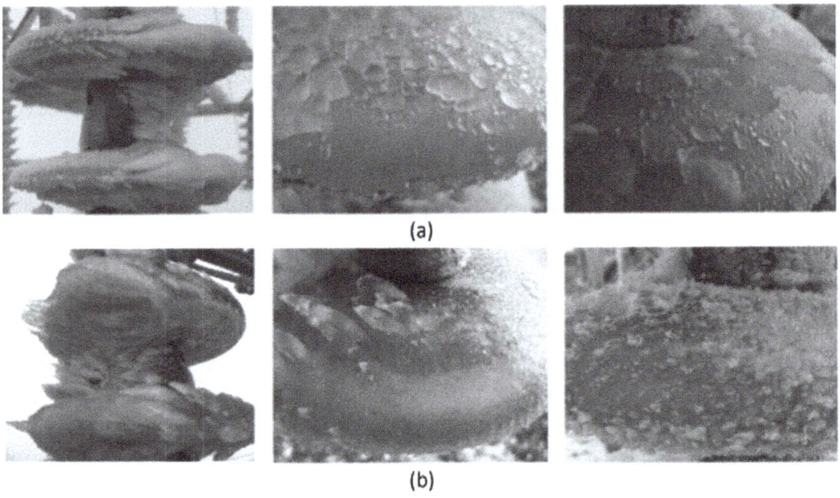

Fig. 6.26 Icing appearances on insulator surface after 18 h of ice accumulation. **a** Insulators coated with superhydrophobic coatings **b** uncoated insulators

From the photos in Fig. 6.26, compared to the uncoated insulator, much less ice was formed on the coated insulator after 18 h of icing exposure at Xuefeng Mountain Natural Icing Test Base.

6.2.3.4 Summary of Passive Coatings for Insulators

There are many academic research projects attempting to develop different types of passive coatings. A review of some of these coatings is presented in Guo et al. (2011). Table 6.1 summarizes some features of these coatings including coating deposition, as well as their chemical components and their performance under winter conditions.

6.2.4 Active Anti-icing Coatings for Insulators

6.2.4.1 *Semi-conductive Glaze*

One of the most effective treatments to improve insulator flashover performance in winter conditions is to modify the glaze of a porcelain insulator with tin, antimony, and other metals to achieve a small amount of current flow, roughly 1 mA, under normal operating voltage.

The idea of using semi-conductive coatings for insulators was proposed for the first time by Forrest in 1942 (1942), and considerable work was done on this material in the 1970s. The original concept was to increase the temperature of the insulator surface by passing a certain current through the semi-conductive insulator layer. With Joule heating, condensation of water film on the insulator surface can be avoided, and the risk of creation of conductive layer on the insulator surface can be eliminated. In cold climate regions, the Joule effect may also affect the ice formation (Forrest, 1942).

Porcelain insulators with *semi-conductive* glaze show higher flashover voltages than conventional porcelain insulators in many adverse conditions, due to drying of the surface by Joule heating as mentioned.

A more important effect may be that a suitable glaze provides an electrically conductive path around possible dry bands, giving more uniform voltage distribution along the insulators. The semi-conducting glaze delays the onset of partial discharge activity which, once initiated, proceeds quickly to flashover in materials such as ice or snow that are insulators at −2 °C, but equivalent to liquid (and metal) when their temperature reaches the melting point.

Semi-conductive glazes have been developed and applied to suspension disc insulators, to bell-shaped suspension insulators and most successfully to station post insulators. They are considered here even though the glaze is part of the original manufacturing process, and it is not common to refire high value used porcelain insulators and bushings to renew their surface glaze.

The benefits of semi-conducting glaze to improve the flashover performance of insulators under and icing environments are well established (Farzaneh & Chisholm, 2009; Radojcic et al., 2013–09). The benefits derived from the heating and grading are provided by the leakage current in the semi-conducting layer. Although

Table 6.1 Summary of passive coatings for insulators

Main chemical components	Deposition technique	Properties and applications	Durability	Comments	Ref
RTV silicone rubber	Spray coating or brushing	Hydrophobic	10–15 years	Well-established technology for polluted conditions Not efficient under icing conditions	CIGRE WG D1.14 (2011-10)
PDMS resin + silicon oil + SiO_2 nanoparticles + Pt (as catalyst)	Spin coating	Hydrophobic ARF between 30 and 40	Not tested	Ice adhesion strength was calculated from direct synthetic atmospheric icing on the insulator	Zhu et al. (2013)
Organopolysiloxane gum + silicate filler + aluminium hydroxide + organosilane or organosiloxane oligamer	Not a coating (silicon rubber composition)	Superior electrical performance	Improved stability in salt fog test 10 000 h of accelerated weathering	For detailed chemical composition refer to the patent	Cheng et al. (2009)
Various silicon-based or fluorine-based monomers such as hexamethyldisiloxane, tetraethylortho silicate, vinyltrimethylsilane, octofluorocyclobutane	Low-pressure plasma polymerization	Hydrophobic Applicable on various types of insulators	Not tested	Only suitable for small insulators due to vacuum requirement	Baalmann et al. (2002, 2003)
Organofluorine-functional silane and/or siloxane with a mineral acid	Dip coating	Applicable on ceramic, glass, or plastic Water, oil, and dirt repellent Increased flashover voltage	Excellent pH stability, heat resistance, and UV stability	At least one salt of aluminium, tin, iron, or titanium should be used in the dip solution as the catalyst	Jenkner et al. (2002)

(continued)

Table 6.1 (continued)

Main chemical components	Deposition technique	Properties and applications	Durability	Comments	Ref
Various fluorinated homopolymers and copolymers	Various deposition methods (plasma etching, spin coating, solvent casting, plasma functionalization, dip coating, spray coating or CVD)	Superhydrophobic and self-cleaning Applicable on ceramic, polymer or glass Prevents dry band arcing in contaminated environments	Excellent UV resistance	Various methods presented to improve the UV resistance of superhydrophobic coatings	Blackett-Voltshield (2009)
Silicon rubber and SiO_2/ZnO	Spray coating	Superhydrophobic Self-cleaning	Superhydrophobic properties are retained after 10 days in acidic and basic solutions Excellent thermal and UV resistance		Momen and Farzaneh (2012-05)
Silicon rubber and stearic acid	Spray coating	Superhydrophobic	Not tested		Momen and Farzaneh (2011-06)
Various organosilicons, fluorocarbons, solvents, and catalysts	Sol–gel process	Applicable to glass, polymer, ceramic, and metal	Proved to be stable in multifactor ageing test UV resistant		Xiu et al. (2009)
OH-PDMS and modified silica nanopowder	Spray coating	Superhydrophobic Significantly lower accumulated ice weight compared to RTV coating	Not tested	Icing test was performed in a climatic chamber on a treated insulator	Li et al. (2014-08)

these insulators work well under AC, their application to DC is not recommended because of corrosion issues and glaze destruction (Radojcic et al., 2013-09).

Here follows an example of using semi-conducting glaze insulators for anti-ice performance. Porcelain long-rod insulators for 33 kV lines with ordinary and semi-conducting glazes were used for the test under energized condition (20 kV_{rms} line to ground) to simulate normal operation conditions. Both specimen insulators have the same dimensions with section length of 485 mm and creepage distance of 740 mm. Figure 6.27a shows the test arrangement. Icing test was started after a 1 h energization to settle the temperature on the semi-conducting glaze surface. Temperature on surface of specimen insulators was measured by infrared thermography after 1 h energization. Figure 6.27b shows the measurement result. It can be seen that the surface temperature of semi-conducting glaze insulator was approximately 2 K higher than the ordinary one by Joule heat effect in still air.

Table 6.2 shows the icing appearances on the specimen insulators during the test. Precipitation was 1 mm/h to simulate a Canadian exposure case. Fewer icicles on semi-conducting glaze insulator were observed than the ordinary case. For

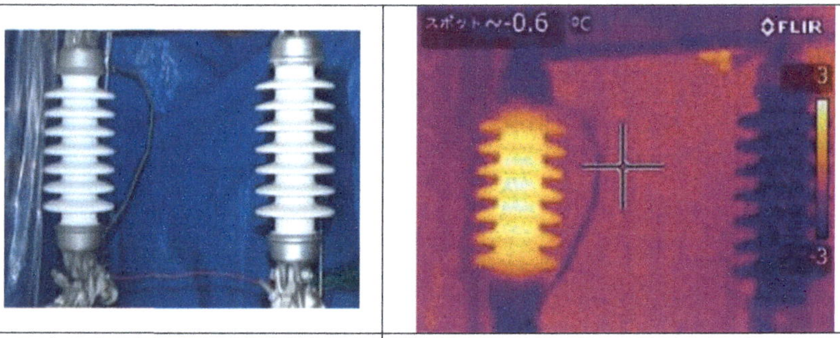

a) Right side: Ordinary glaze, Left side: Semiconducting glaze)

b) Temperature (°C) on specimen insulators (Right side: ordinary glaze, Left side: semiconducting glaze

Fig. 6.27 Specimen insulators for 33 kV line

Table 6.2 Icing appearances after 3 h

	Semi-conducting glaze	Ordinary glaze
Appearance	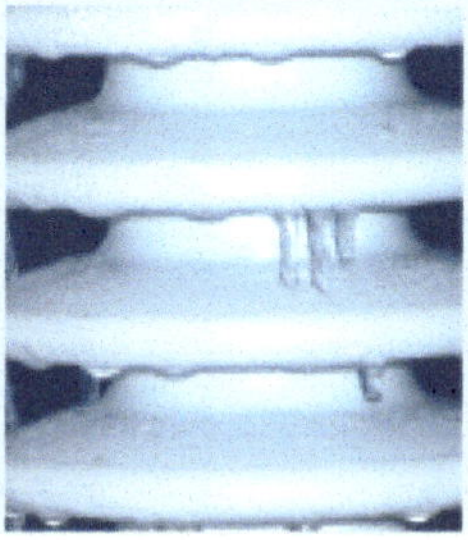	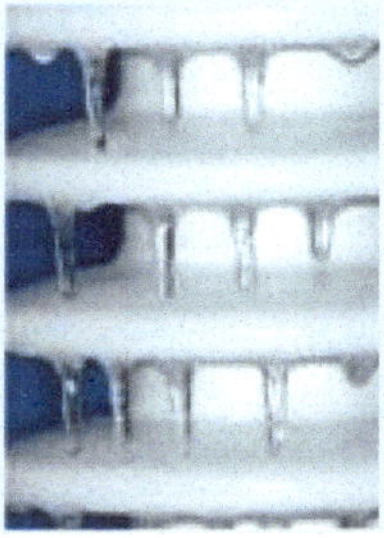

semi-conducting glaze surface, it was observed that icicles fell off easily (see Table 6.4) due to water film on the surface melted by thermal energy of 1 mA glaze current.

6.2.4.2 *Semi-conductive RTV Coatings*

In principle, semi-conductive polymeric coating can be achieved in two ways, either the coating is made from intrinsically conductive polymers such as polyaniline, polypyrrole, and polythiophene or by addition of conductive particles to an insulating polymeric matrix in an amount that allows percolation of those particles and creation of conductive paths through the coating. Coatings developed for outdoor service need proven durability against environmental conditions. So far, intrinsically conductive polymers are not suitable for electrical outdoor applications. However, insulating coatings that do meet outdoor service requirements can be modified by filling them with conductive particles.

Silicone rubber (SiR) is considered as the most promising polymeric matrix. It has been successfully applied to the outdoor insulators. Inadvertently, it was established that carbon black pollution from a manufacturing plant (Columbian Chemicals, Hamilton Ontario Canada acquired by Birla Carbon in 2011) temporarily rendered RTV silicone rubber surfaces superhydrophobic, which was desirable. The superhydrophobicity lasted for a limited period before the silicone fluid from substrate encapsulates the pollution. The surface layer did not affect the level of current in the insulator and depressed flashover voltage in laboratory cold fog and icing conditions, which was not desirable.

To deliver a desired surface resistivity like that achieved with successful semi-conductive glazes, conductive fillers such as carbon black and carbon fibres are added into the silicone rubber (SiR). The resulting materials are widely referred to, although inaccurately, as "semi-conducting" coatings as their electrical conductivity is dependent on the type and the volume concentration of the conductive particles. At low concentrations, the conductive particles are mostly separated by the insulating polymer. As the concentration of particles increases to the level of "percolation", conductive networks form within the polymer matrix. The hydrophobicity and hydrophobicity transfer of the carbon-filled SiR are not affected much by the small percentage of carbon. The added carbon is proved to be effective in decreasing the adhesion force between the ice layer and the material surface. Furthermore, Joule heating generated by the partially conducting coating can inhibit the formation of ice.

Figure 6.28 (CIGRE WG B2.44, 2015) shows an example of using a semi-conducting RTV as the anti-icing coating. Two insulator strings, each consisting of seven-unit insulators, were hung in an artificial climate chamber for the icing test. In the figure, the left insulator string is uncoated. In the right insulator string, the semi-conductive silicone rubber coating was applied only to the bottom side of the insulator to eliminate the power loss when there is no precipitation. The ambient temperature was kept at −7 °C, and insulator strings were energized at 63.5 kV. After two hours of ice accretion, the icing appearance was shown in Fig. 6.28. For the semi-conductive-coated string, there were no icicles and no ice

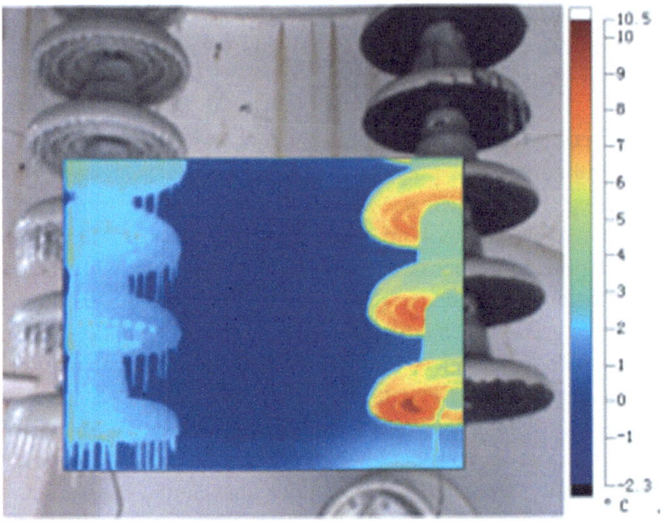

Fig. 6.28 Thermal picture of the specimens in the icing test. Left side: uncoated sample; right side: coated sample with anti-icing semi-conductive silicon rubber (Wei et al., 2012-12)

covering on the surface of the insulators. The coated string also shows an obvious surface heating effect. As shown in the thermal picture, the temperature of the uncoated insulator string surface was less than 2 °C. However, the bottom side of the coated insulators can reach as high as 10 °C.

The ice melting concept was not verified for all types of ice formation (Wei et al., 2012-12). Figure 6.29 shows how the anti-icing performance was significantly affected by the resistance of the bottom coating (Wei et al., 2014-02).

Generally, a commercial semi-conducting glaze has a coating resistance of approximately 10 MΩ to give a glaze current of 1 mA with 10 kV service voltage in a typical string of discs. The coating resistances needed for good anti-icing effect

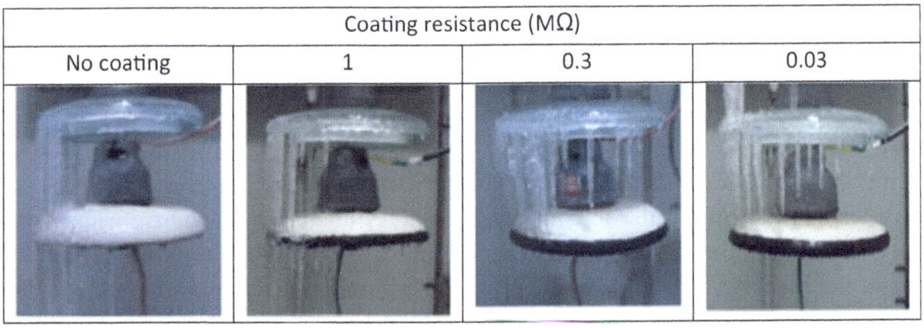

Fig. 6.29 Anti-icing performance of insulators of different resistance

are considerably lower, and it is essential to arrange that the current only flows when there is ice on the top surface of the treated disc.

Since 2009, semi-conductive rubber coating has been used in a few short sections of Guangdong and Yunnan grid transmission lines. These lines are mostly at high altitude and in areas that tend to have smog and high humidity. Two methods were adopted for spray-painting insulators in the field, either directly on the tower or at the ground, as seen in Fig. 6.30.

Comparative results of ice accretion on 35 and 110 kV insulators in service are presented in Figs. 6.31 and 6.32.

The dark insulators in Figs. 6.31 and 6.32 are covered with semi-conductive silicone rubber (SSiR) only on the bottom surface. In these natural ice comparison tests, there is clearly less ice and fewer or no icicles on the dark SSiR-coated insulators.

Fig. 6.30 Site construction work

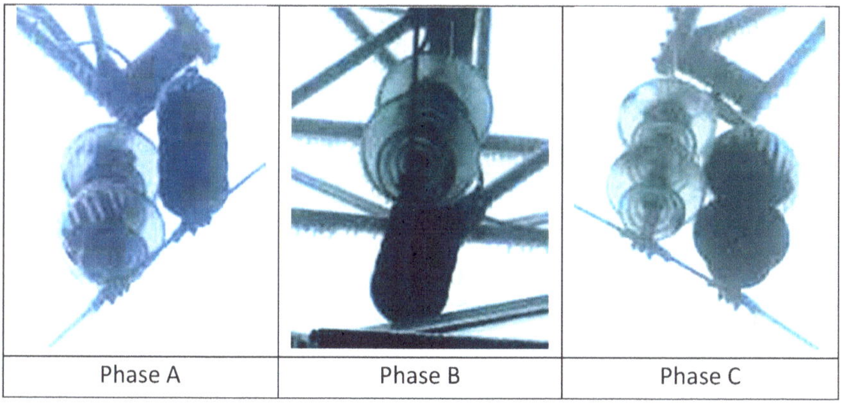

Fig. 6.31 Appearance of icing on glass and SSiR-coated glass insulator strings on 35 kV Lianhe Line in Guandong in icing conditions

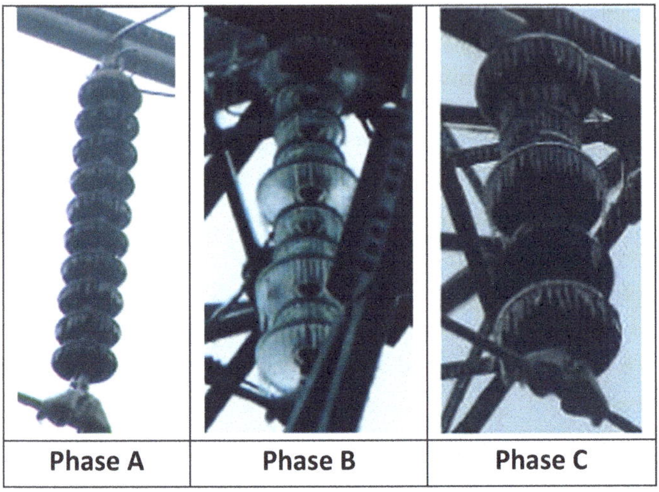

Fig. 6.32 Appearance of icing on glass and SSiR-coated glass insulator strings 110 kV Yangdian Line in Guandong

6.2.5 Coatings for Other Power Network Equipment

Ice accretion can affect normal functions of various exposed structures such as substations, transmission line antenna, and wind turbines. For wind turbines, most icing prevention methods are heating systems (Parent & Ilinca, 2011). However, anti-icing requires much more energy than de-icing because of the continuous heating required. In theory, the surface temperature of the blade must be kept above 0 °C whenever there is icing. Moreover, when ice melts on the heated elements, water can run back and freeze again in the area not covered by the heating element. To avoid this, the water must evaporate, which implies for the heated element, temperature needs to be relatively high (Parent & Ilinca, 2011). The use of a passive system such as icephobic coating alone on blades has proven to be ineffective. Laforte (2001-09) and later researchers Parent & Ilinca (2011), Adomou (2011-08) found that icephobic coatings degrade with time and lose their ability to repel ice. The combination of a coating with a heating system has proven to be the best method to prevent ice formation on wind turbine blade (Parent & Ilinca, 2011; Yamauchi et al., 2020-07). In fact, the use of coatings on aerodynamic and structural surfaces can either enhance the effectiveness of standard anti/de-icing systems, or lead to substantial reduction of the energy consumption of present systems.

No clear distinction is made in the literature regarding icephobic and superhydrophobic coatings with respect to their application to wind turbines. However, the mechanism of water/surface interaction in icing conditions is different depending on which type of coating is used (Adomou, 2011-08; Antonini et al., 2011; Kulinich et al., 2011; Yamauchi et al., 2020-07). Current research on coatings is heading towards nanocomposite type coatings. Details of those coating are often proprietary or are somewhat similar as the ones described in this chapter.

For antennas, NTT (2015) shows that the use of superhydrophobic films was efficient in reducing the adherence of snow on the antenna. This coating can be used as a stand-alone technique in wet snow condition. A hydrophobic coating was also used in Japan for antennas. Yamauchi has found that the addition of fines particles (1 μm) of polytetrafluoroethylene (PTFE) dispersed in a resin of PVDF (polyvinylidene fluoride) gives a coating that shows high water repellence on the surface of an antenna used for radio communication.

6.2.6 Concluding Remarks

A review of some of the existing coatings that can be applied to conductors, ground wires, and outdoor insulators shows that commercial products for anti-icing properties require further development. These coatings belong to the active or passive categories depending on whether they need electrical energy to be activated. Progress in the development of active coatings has stalled since 2010.

The wide acceptance of silicone rubber (SiR) for outdoor anti-pollution purposes does not extend to the domain of winter conditions. Ice adheres better to SiR than to ceramic or metal surfaces. However, SiR is also a promising polymeric matrix to support nanostructured material and conductive fillers that could deliver better, rather than similar or worse properties to ceramic and metal surfaces in winter conditions.

Most studies have dealt with superhydrophobic and icephobic coatings with application to power network equipment. Some of them can be applied to both insulators and conductors, or to other equipment such as wind turbines and transmission line antennas.

While some of superhydrophobic and icephobic coatings have a good potential for success, none of them is yet ready for application during manufacturing or to be applied to energized equipment.

6.3 New Materials and Methods for Ice Prevention

The main purpose consists of identification of new practical and durable coatings as well as surface engineering techniques to increase the reliability of overhead power network equipment in various environmental conditions. It also addresses some environmental concerns caused by corona noise and visual aspects of overhead power lines. The main power network features considered are insulators, conductors and ground wires, and transmission line structures.

Anti-icing coatings are designed to prevent or reduce the accumulation of ice and snow on power network equipment. After a reminder of certain fundamental aspects of icing, a review of developed coatings including concepts, materials typology, elaboration, and application methods is carried out, and the test methods for characterizing the performance of these coatings are described.

Regarding outdoor insulators, some coatings such as RTV silicone have already been widely used primarily to reduce pollution flashovers. However, the application of RTV coatings to outdoor insulators in icing conditions is not recommended. Several anti-icing coatings intended to prevent icing flashovers have also been developed and tested. However, these coatings are not yet ready for applications due to their unsatisfactory stability and durability. Further R&D efforts are needed to find efficient and cost-effective techniques to protect insulators under icing conditions.

Finally, to test and characterize the performance and durability of anti-icing coatings, several methods, in the absence of appropriate standards, which have been adopted in different laboratories, have been presented and discussed. Although a number of these methods, with some modifications, can be applied for testing these coatings, refinement of some of these techniques, as well as the development of new methods, is required.

In the previous section, active methods for removing ice on overhead conductors were reviewed. It is obvious that a new coating or surface treatment exhibiting perfect icephobic properties would be far more attractive than any active method especially if the overall cost of such a coating procedure is less than the cost of fabrication, maintenance, and energy needed to operate a mechanical or a thermal device. Such perfect icephobic coatings do not exist yet. Outdoor experience confirms that atmospheric ice eventually adheres to virtually all materials when icing conditions persist.

Mulherin and Haehnel (2003) summarized the main criteria to choose a suitable icephobic coating material. These are as follows:

1. Efficacity, i.e. the coating must significantly reduce the adhesion strength of ice
2. Durability and longevity
3. Acceptable life cycle cost
4. Ease of application.

While the first criterion will always be the most important one, the other three can be assigned different priorities depending on the specific application project. To illustrate the relative importance of criteria 2–4, a comparison is made in the two following examples. Among the different debris that are propelled upon NASA Space Shuttle launches, ice chunks coming off feedline brackets represent a considerable hazard (Roe, 2005). In this situation, the longevity of a proper icephobic coating would not be an issue since the latter should keep its properties only during the prelaunching and launching stages ($\sim$ few days). On the other hand, icephobic materials used to coat high-voltage power lines and/or the towers supporting them should remain active for about seventy years. This is a huge technical challenge considering that the coated material must withstand electrical, mechanical, and thermal constraints generated during the normal operating conditions or during specific events like short circuits or lightning strikes (Volat et al., 2005).

In this section, different materials used to reduce ice adhesion are reviewed. Several publications can be found on the subject, but they are either specific to one application (Boluk, 1996; Frankenstein & Tuthill, 2002; Volat et al., 2005), or too old to be considered (Sayward, 1979–04). In practically all the cited works, ice adhesion strength values were measured in direct shear mode. However, to avoid potentially misleading data comparison between different sources, no specific adhesion strength values will be given here, only qualitative information about the relative increase or decrease in adhesion strength. The rationale is that comparisons are impossible since the measurement methods were not always the same, the ice deposition process usually differed from one work to another, and coating roughness was not constant either.

Since a perfect icephobic coating does not exist, the emphasis will be put on more advanced and promising techniques developed recently by researchers to reduce ice adhesion on a given substrate. This information should help OHL engineers to explore various options and eventually adopt a technique that is optimal for the specific application of interest. Materials involving low surface energy polymers or molecules using techniques such as painting or spraying will be reviewed first, followed by a more in-depth study of recent advanced solutions to obtain highly hydrophobic or icephobic surfaces. These will include the production of heterogeneous surfaces, the use of hydrophobic self-assembled monolayers, the production of diamond-like coatings (DLC) as well as the elaboration of superhydrophobic coatings. Other miscellaneous techniques or concepts will also be reviewed. Finally, possible options and critical views on icephobic materials as well as potential future developments in this area will be presented.

6.3.1 Deposition of Hydrophobic or Icephobic Paints and Polymers

Finding a material of minimum surface energy is the main design objective for icephobic coatings. Based on this statement, polydimethylsiloxane (PDMS) or silicone, and Teflon® or polytetrafluoroethylene (PTFE), as displayed in Fig. 6.33 with parameters in Table 6.3, are the best candidates known to date as they possess very low surface energy.

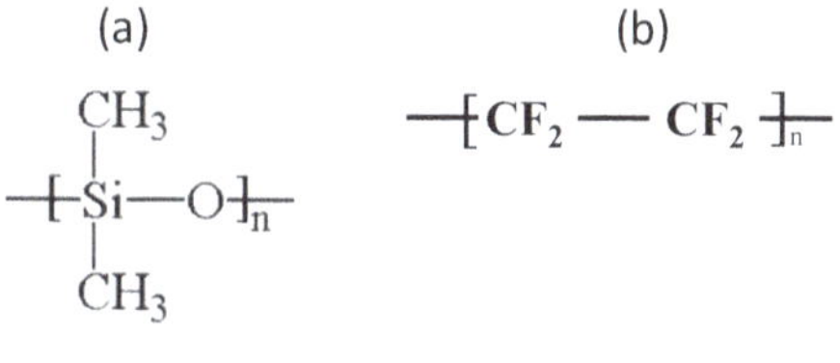

Fig. 6.33 **a** Polysiloxane and **b** polytetrafluoroethylene structures

Table 6.3 Interaction energies between hydrophobic polymer model molecules and water calculated by the SCF method (Murase et al., 1994)

Interactions forms		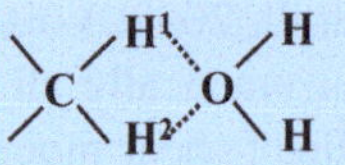	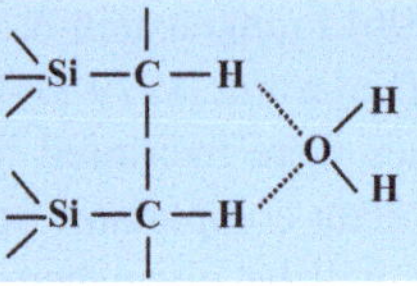	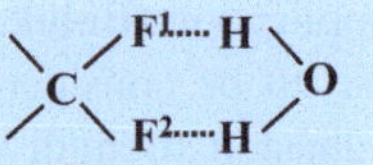
OH and FH bond length (nm)		0.252	**0.329**	**0.187**
Interaction energies (kJ mol^{-1})	E^1_1	−14.95	−15.64	**−50.89**
	E^2_1	−5.70	−12.30	**−48.51**
	E_2	−4.07	−4.64	−0.40
	E_3	−0.81	**+1.75**	**−39.89**
	E_4	−1.40	**+1.79**	**−35.81**

E_1: O_{H2O}—$H_{HC,DMS}$ F_{FHC}—H_{H2O}
E_2: O_{H2O}—$mol_{HC,\ DMS}$
E_3, E_4: H_{H2O}—$mol_{HC,\ DMS,\ FHC}$

Most research efforts have dealt with these two families of polymer materials (Anderson & Reich, 1997-01-06; Bascom et al., 1969; Crouch & Hartley, 1992; Hacker et al., 2000-10; Igoshin & Berdnikov, 1989; Jellinek et al., 1978; Kumar et al., 1999; Laakso et al., 2003-04; Lacroix & Manwell, 2000-06; Laforte et al., 2002; Maissan, 2001-03-19; Mulherin & Haehnel, 2003; Reich, 1994-01-10; Smith & Garti, 2000; USA Department of the Army, 1999). In these studies, ice adhesion was usually assessed in direct shear mode, and coating deposition techniques were usually very simple, involving paints (alkyd or acrylic) with special solvents (Crouch & Hartley, 1992) or the use of polymer resins (Hacker et al., 2000-10; Smith & Garti, 2000). Overall, silicone-based polymers performed slightly better than the PTFE-based ones.

In 1978, Jellinek et al. (1978) reported that a PDMS film had a surface tension of 21 mN m^{-1}, which is characteristic of the methyl groups lying just at the surface of the polymer film, as displayed in Fig. 6.34, with the siloxane-carbonate block polymer.

Additionally, Jellinek inferred that the dimethylsiloxane content ($-O_2Si(CH_3)_2$ groups) must lie in a certain range of weight per cent and chain length, and that the glass transition temperature T_g must remain low, T_g being a measure of segment mobility. When a polymer has a low T_g, the molecular segments can change place by actively sliding or jumping. The polymer is then said to be flexible or soft and corresponds to an amorphous state. Ice adhesion can be lowered by the addition of silicone oil which enhances the softness of the PDMS surface by acting like a lubricant and a plasticizer. Also, it was shown that low T_g polymers exhibit a low interfacial shear strength which remains constant for $T_g < -50$ °C (Anderson et al., 1994). In short, the dissimilar rheological–mechanical properties of ice- and polysiloxane-based polymers resulted in very low ice adhesion. That does not mean

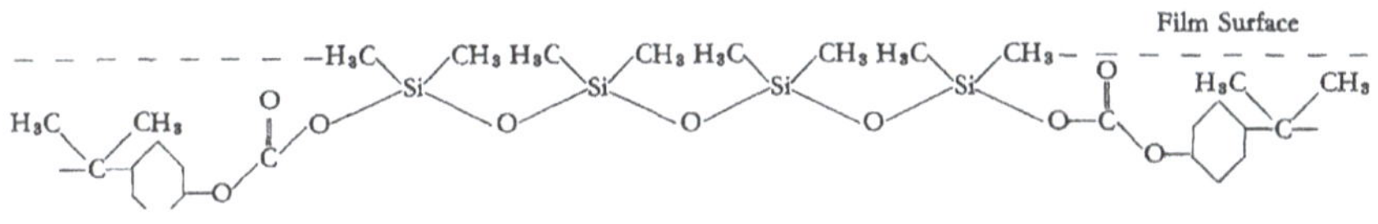

Fig. 6.34 Siloxane-carbonate block polymer (Jellinek et al., 1978)

that this dissimilarity is necessary: polyfluorocarbon (PFC)-based polymers can achieve similar performance without this contrast.

For a PFC-based polymer, the $-CF_3$ group type exhibits exceptionally low critical surface tension (6 mN m^{-1}) (Jellinek et al., 1978) and can have high T_g values (>150 °C). PFC-based materials exhibit better substrate adhesion and better mechanical properties (for wear, for instance) than their PDMS counterparts. Therefore, for temporary applications, PDMS-based coatings would be a preferable option. It is somewhat peculiar that PTFE, having strong hydrogen-type bonding with water molecules, as well as a high T_g value, is one of the most hydrophobic and icephobic substances known to date. This is explained by its low permittivity ($\varepsilon = 2.04$) which drastically reduces the electrostatic force that is the most important force involved in ice adhesion.

Mulherin and Haehnel (2003) tested 16 commercial materials labelled as "icephobic" and concluded that they indeed reduced the amount of energy needed to remove ice from the surface of these materials, but no material prevented ice build-up. In fact, the tested materials offered little benefit over Teflon and polyethylene (UHMV) coatings and claddings. Thus, to improve icephobicity, researchers needed to formulate more advanced materials based on the physics of ice-solid substrate interactions. The rest of this section will be dedicated to these more advanced and recent techniques as well as to some potential candidates to fight ice adhesion.

6.3.2 Heterogeneous and Composite Coatings

Several researchers (Crouch & Hartley, 1992; Hacker et al., 2000-10; Mulherin & Haehnel, 2003) have found that better reduction in ice adhesion strength could be obtained by mixing polysiloxane and fluorocarbon materials than by using homogeneous coatings with either a PDMS or a PFC structure. However, this finding has not yet been explained.

In 1994, Murase et al. (1994) published a very important paper on heterogeneous polymer coatings aimed at decreasing ice adhesion. Three different types of heterogeneous polymers were studied: one of them, the organopolysiloxane grafted fluoropolymer (FX), is displayed in Fig. 6.35. The other two are polyperfluoroalkyl (meth)acrylate combined with hydrophobic silicon dioxide (NX) and an organopolysiloxane modified with a lithium compound (SIII).

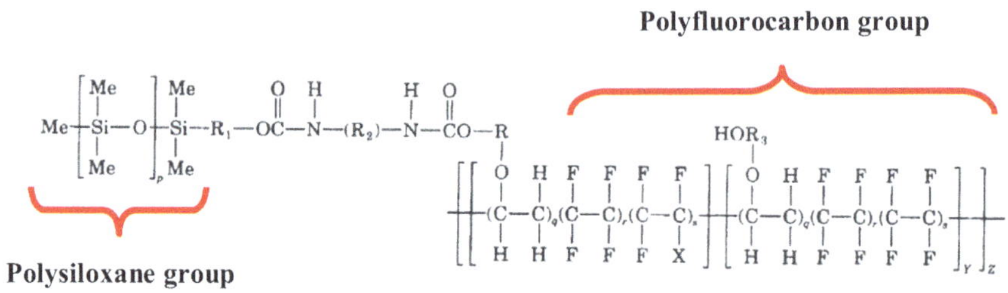

Fig. 6.35 Organopolysiloxane grafted fluoropolymer (FX) (Murase et al., 1994)

Snow accretion did not occur on the NX coating, and ice adhesion was reduced two times compared to PTFE. On the other hand, ice adhesion was 25 times lower for the SIII compound compared to PTFE while snow accretion was only divided by two.

To explain such a decrease in ice or snow adhesion in the presence of a heterogeneous polymer material, Murase et al. (1994) performed molecular orbital energy calculations using the SCF method with the MOPAC programme. Three molecules were analysed: ethane (LHC or C_2H_6), dimethylesiloxane (DMS), and hexafluoroethane (FHC or C_2F_6). The hydrogen bond lengths (O–H and F–H) as well as the different interaction energies are displayed in Table 6.3.

These values indicate that hydrogen bond energies and lengths vary considerably depending on the kind of molecular group involved. There is a slight repulsion between a water molecule and the siloxane group, for example $E_4 = 1.79$ kJ mol^{-1}, while a strong attraction was calculated for the fluorocarbon group and a H_2O molecule, comparing $E_4 = -35.81$ kJ mol^{-1}. Also, the water molecule orientations at the surface of the fluorocarbon group and at the polysiloxane group are reversed. Therefore, by creating various disparities at the molecular level in terms of energy bonding and water molecule orientations, the ice-material interface is weakened, with the probable creation of a wide range of dislocations and slips in the structure of the quasi-liquid layer.

A US patent (Byrd, 2004) from the Boeing company describes the use of a polysiloxane (amide-ureide) coating for icephobicity applications. The concept is the same as the one described by the Murase group: the heterogeneity of this polymer creates a synergistic effect which leads to low ice adhesion strength. Another US patent (Mizuno et al., 2003) describes a heterogenous coating using a tetrafluoroethylene and a silicone resin with an appropriate organic solvent to form a water-repellent coating.

6.3.3 Self-Assembled Monolayers (SAMs)

It was emphasized in Crouch and Hartley (1992) that $-CH_3$ and $-CF_3$ groups must be as close as possible to the coating surface to obtain maximum hydrophobicity.

An attractive technique to obtain such oriented layers is to employ self-assembled monolayers (SAMs). SAMs are molecular assemblies that are formed spontaneously by the immersion of an appropriate substrate into a solution of active surfactant in an organic solvent, as shown in Fig. 6.36.

In Fig. 6.36, an organosilicon type of anchoring group was chosen to illustrate this deposition technique since this group covalently binds to several important metal or non-metal oxides such as Al_2O_3, TiO_2, SnO_2, SiO_2, and glass. The spacer group consists of an alkyl chain while the surface group can be chosen following the desired surface property. In the case of icephobicity, CF_3 or CH_3 is the obvious choices for R.

The use of SAMs to obtain icephobic surfaces is very attractive. Highly functionalized groups are in direct contact with water molecules of the ice surface. Strong covalent bonds anchor the alkyl chain. The overall deposition technique is easy to implement and inexpensive. At least, three studies have been reported in this area.

Somlo and Gupta (2001) have shown that ice adhesion decreases when a SAM of dimethyl-n-octadecilchlorosilane is formed on an aluminium alloy surface. More recently, Petrenko and Peng (2003) has applied mixtures of self-assembling 1-dodecanethiol and 11-hydroxylundecane-1-thiol with various degrees of hydrophobicity/hydrophilicity on Au surfaces and has shown a good correlation between the contact angle of water and the ice adhesion strength. Nanostructured superhydrophobic surfaces have also been prepared by self-assembly of hydrophobic n-octadecyltrimethoxysilane ($H_3C(CH_2)_{17}Si(OCH_3)_3$) and (3,3,3-trifluoropropyl) trimethoxysilane ($F_3C(CH_2)_2$ $Si(OCH_3)_3$) monolayers from the gas phase on porous alumina, ZnO nanowire and GLAD (glancing angle deposition) surfaces (Kemell et al., 2005).

Despite this progress, systematic knowledge is still lacking to explain how the microstructure and surface chemistry of SAMs influence their hydrophobic and icephobic properties (Kulinich & Farzaneh, 2004). The case depicted in Fig. 6.36 is

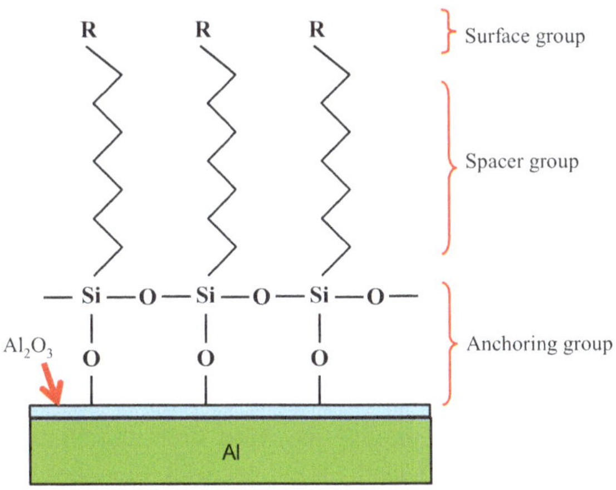

Fig. 6.36 Schematic representation of SAMs grafted onto an aluminium substrate

an ideal one: the surface is perfectly flat, and all the surface-active groups are ideally positioned. In the presence of a porous or rough surface, chains of molecules become bent or collapse, resulting in a decrease in hydrophobicity due to surface exposition of alkyl groups such as $-CF_2$ or $-CH_2$ (Kulinich & Farzaneh, 2004). Another aspect that requires more study is the fact that the efficacity of a coating appears to be related only to one or two layers of molecular chains. In other words, if some patches of the substrate to be protected are damaged, the ice adhesion will likely increase. However, this technique is very promising to fight ice adhesion. The challenge is to develop robust SAMs that will remain active on complex surface topographies and will exhibit good ageing properties in outdoor exposure.

6.3.4 Diamond-Like Coatings (DLC)

Plasma-enhanced chemical vapour deposition (PECVD) can produce diamond-like coatings (DLC) through the ionization and radicalization of a feeding gas like C_xF_y (Ji et al., 2002). Highly reactive radical species such as $CF_2^{\cdot}$ or $CF_3^{\cdot}$ are produced during the PECVD process and are subsequently deposited on the desired surface forming a dense, hard, and highly adhesive coating, with a long-life expectancy.

Recent studies have demonstrated the potential of ultra-thin films based on DLC associated with fluorocarbon gas (Cicala et al., 2003-10; Ji et al., 2002-02; Kiuru, 2004-08-20; Koshel et al., 2003; Woodward et al., 2003). Obtained by a common PECVD technology, these DLC coatings exhibit high hydrophobicity with strong adhesion to aluminium and porcelain, as well as good mechanical properties. This technique may also be used to coat microtextured surfaces to make them super-hydrophobic. However, no ice adhesion tests have been done so far with these films. Even though coatings with very good mechanical properties can be fabricated with the PECVD technique, several issues or limitations must be considered. These include high cost, the small film thickness (around 1 μm), and difficulty to plate large and complex-shaped substrates.

6.3.5 Superhydrophobic Coatings

Small-scale rugosity can have a spectacular influence on hydrophobic properties for a given surface. Numerous examples can be found in nature where the surface of some plants (Barthlott & Neinhuis, 1997-04; Neinhuis & Barthlott, 1997-06; Wagner et al., 2003-04) or animals (Cong et al., 2004; Mock et al., 2005-02; Parker & Lawrence, 2001-11) exhibits the so-called superhydrophobic behaviour with an example in Fig. 6.37.

On superhydrophobic surfaces, water drops tend to behave like pearls and can roll off the surface if it is slightly tilted or naturally curved. This phenomenon allows some plants or animals to remove dirt from their surface by collecting surface deposits through the water drop motion and release. Beetles in hot desert regions use this superhydrophobic trick to gather life-saving water drops from the

Fig. 6.37 Water drops sitting on a lotus plant leaf (Wagner et al., 2003–04)

morning dew (Parker & Lawrence, 2001-11). Some of the superhydrophobic coatings made by scientists in trying to mimic nature will be reviewed. Here, we define a superhydrophobic surface as one that achieves a contact angle $\Theta > 150°$. This is the level where repelling effects on water droplets are observed.

The influence of surface roughness on ice adhesion was discussed, using Wenzel (1936) and Cassie and Baxter 1944) models, in Sect. 4.4.4.1.1. These models were shown to be relevant to superhydrophobicity properties by He et al. (2003-05). Using photolithographic techniques, these researchers created micropatterned surfaces to validate the models using a simple geometric layout, as shown in Fig. 6.38.

From Eq. 6.1, a geometrical parameter A is defined, and new contact angles are proposed in Eqs. 6.2 and 6.3.

$$A = \frac{1}{((b/a) + 1)^2} \tag{6.1}$$

$$\cos \theta_{\mathrm{r}}^{\mathrm{cas}} = \mathrm{A}(1 + \cos \theta_{\mathrm{flat}}) - 1 \tag{6.2}$$

$$\cos \theta_{\mathrm{r}}^{\mathrm{wen}} = \left[1 + \frac{4\mathrm{A}}{(\mathrm{a/H})}\right] \cos \theta_{\mathrm{flat}} \tag{6.3}$$

Then, using different values of b and a, they plotted the theoretical contact angle obtained for the two models versus the b/a ratio in Fig. 6.39. The two models are completely different, and the plots intersect at only one point, at $b/a \approx 0.5$.

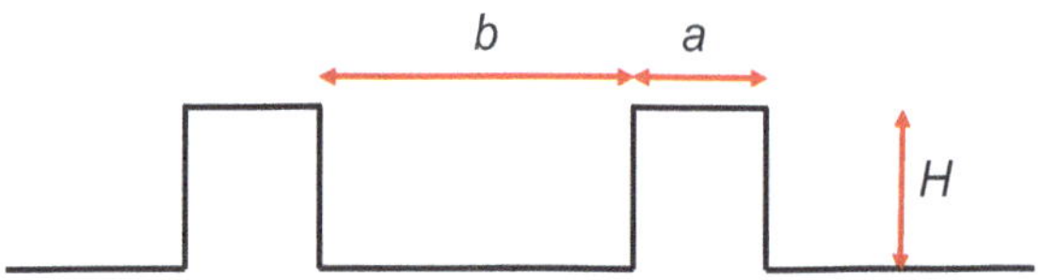

Fig. 6.38 Two-dimensional geometrical representation of the patterned surface (He et al., 2003–05)

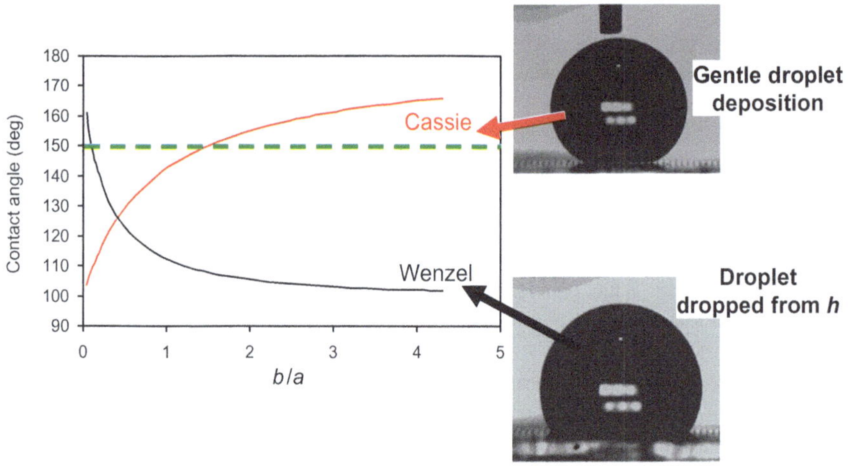

Fig. 6.39 Theoretical contact angles versus geometric factor *b/a* for the Wenzel and Cassie regimes (He et al., 2003–05) $a = 25\ \mu m$, $\Theta_{flat} = 100°$ and $a/H = 0.83$

The most important insight of this work is its experimental verification that the Wenzel and Cassie regimes coexist. Gentle water drop deposition gives Cassie regime, as seen in Fig. 6.39. A droplet impact from a certain height resulted in a Wenzel type regime, visualized as filling the individual areas *bH in* Fig. 6.38 with liquid. These results are crucial for the consideration of microtextured materials for icephobic applications. During an ice storm, impinging supercooled water droplets on rough surface may result in liquid inclusion into surface asperities and its subsequent and rapid freezing inside the three-dimensional structure. If ice growth occurs, this could be problematic since mechanical interlocking could take place, giving better ice adhesion than on an untreated surface.

Numerous studies can be found in the literature on the making of superhydrophobic micro and nanotextured surfaces. However, few considered icephobic applications. An early review on the subject can be found in Nakajima et al. (2001-01). Researchers have created a primary porous structure either by etching a given substrate (Wu & Shi, 2005-08), depositing nanoparticles of oxides (Soeno et al., 2004-10), using nanolithography (Shiu et al., 2004-01), or by electroplating polymers (Yan et al., 2005-05; Zhang et al., 2003-09). To enhance the hydrophobicity, subsequent coating with a low surface energy compound has been realized using various techniques such as PECVD, deposition of self-assembled monolayers (SAMs) (Kemell et al., 2005; Kulinich & Farzaneh, 2005-08), and passivation with stearic acid (Saleema et al., 2006-05-07).

Some work describes methods to reproduce the orientation and size of natural micro or nanofibres. A superhydrophobic poly(vinyl alcohol) forest-like nanofibre structure was produced by extruding the polymer through holes with diameters comprised between 20 and 500 nm in an alumina template (Feng et al., 2003-02). Lau et al. (2003-10) produced a carbon nanotube forest using the PECVD technique which was subsequently coated with PTFE to obtain superhydrophobicity. A PTFE superhydrophobic surface was made by extension of the fibres comprising the base material (Zhang et al., 2004-06).

Production of nanofibre networks using the electrospinning process has recently drawn some attention. Block copolymer poly(styrene-b-dimethylsiloxane) fibres with diameters in the range 150–400 nm were electrospun to produce superhydrophobic materials (Ma et al., 2005-04). In follow-up work, poly(caprolactone) was electrospun and then coated with a thin layer of hydrophobic polymerized perfluoroalkyl ethyl methacrylate (PPFEMA) by iCVD (Ma et al., 2005-10). The electrospinning of poly(AN-co-TMI) with a perfluorinated linear diol (fluorolink-D) and tin(II) ethyl hexanoate in DMF (Acatay et al., 2004-09) can also be cited. In each of these three cases, superhydrophobicity was reached when the microstructure of the materials was composed of both fibres and polymer beads.

Finally, the work of the Saito group in Japan was aimed at producing highly hydrophobic microtextured coatings made of PTFE particles blended in a PVDF binder applied by a spray drying technique (Saito et al., 1997-04, 1997; Takai & Yamauchi, 1998). Snow accretion on antennae was dramatically reduced by such a coating. The work of Saito is the only one to date linking ice or snow adhesion to superhydrophobic coatings, which means that such coatings must be studied in icing conditions. Their fragile structure, their adhesion to a given substrate as well as their behaviour after several ice removals are major hurdles that scientists will have to face in future developments.

6.3.6 Joule-Effect Methods with Semi-conductive Coatings

The role of carbon black concentration in SiR insulator coatings has been described above.

Hydrophobic composites for coating metal-PTFE can be produced by electrodeposition or electroless plating. PTFE micro or nanoparticles are dispersed in each electrolyte and are entrapped during the reducing process of the metallic ions. Using the electroless technique, Ni-P-PTFE coatings exhibiting a water contact angle of 110° were produced, as shown in Fig. 6.40 (Ming der Ger et al., 2002-01).

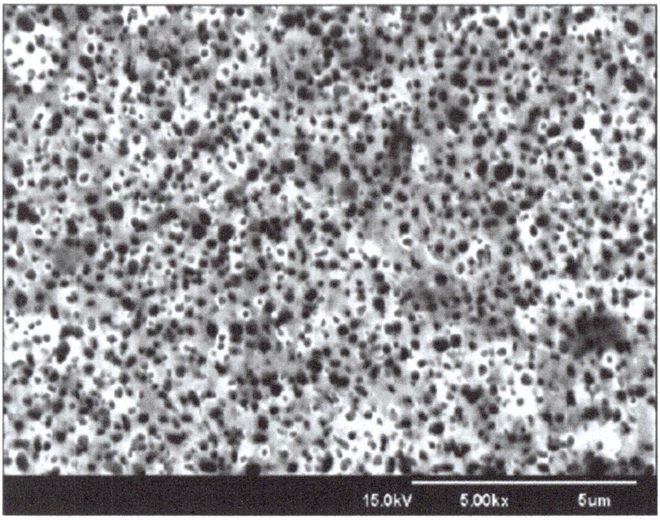

Fig. 6.40 SEM photomicrograph of a composite Ni–P-PTFE coating (Ming der Ger et al., 2002-01). The black spots correspond to the PTFE particles

Similar results were also obtained by others with the same technique as reported in (Zhao et al., 2005-02). Ultra-dispersed 300 nm PTFE particles within a nickel matrix were produced using electrodeposition (Wang et al., 2004-02). In this study, water contact angles ranging from 123° (28% vol. inclusion of particles) to 155° (47% of particles) were observed.

No studies on ice adhesion on electro-composite coatings can be found in the literature. Even though the presence of high surface energy area (metal) may be problematic, the composite nature of such surfaces may lead to interesting new properties. Electrochemical deposition techniques are inexpensive and can produce thick adhesive films (between 1 and 100 μm). A further advantage is that large and complex parts can be plated with these techniques.

6.3.7 Ice De-bonding Induced by Pressure Build-Up Within Rough Surfaces

Various aspects of icing were studied on porous pavements (Penn & Meyerson, 1992). One novel aspect of this study was the pressure increase measured during freezing inside an enclosed space. Because of the unique expansion of ice at freezing point, ice formed in blind pores pushes tight against the walls, creating strong mechanical interlocking. However, if air is entrapped within the pores, the pressure build-up can be significant and lead to crack initiation and propagation as well as ice de-bonding. Pressure increase was also measured during freezing as a function of the initial volume of water trapped in the enclosed space. It was found that when 2% of the water was frozen the pressure was 36 atm, and that it reached 420 atm with 20% of the ice formed. The pressure generated within the specimen is manifested as internal or residual stress. To use the pressure, or residual stress, to their advantage for ice de-bonding, the researchers drilled holes into a pavement portion, which was then covered with ice. After top-down freezing, the ice caps popped up as shown in Fig. 6.41 after a certain time depending on the size of the holes.

In another work at a different scale, Bascom et al. (1969) found that replicas of ice sheared from four different PDMS coatings showed bubble-like features as shown in Fig. 6.42.

The replicas in Fig. 6.42 strongly suggested that tiny air bubbles had been trapped between the ice and the coating, resulting in lower ice adhesion properties.

Although surface roughness can lead to potential strong mechanical interlocking and therefore to high adhesion strength, the two examples just described can be the basis of new coating materials having an optimum roughness to create enough internal stress to crack and possibly de-bond ice.

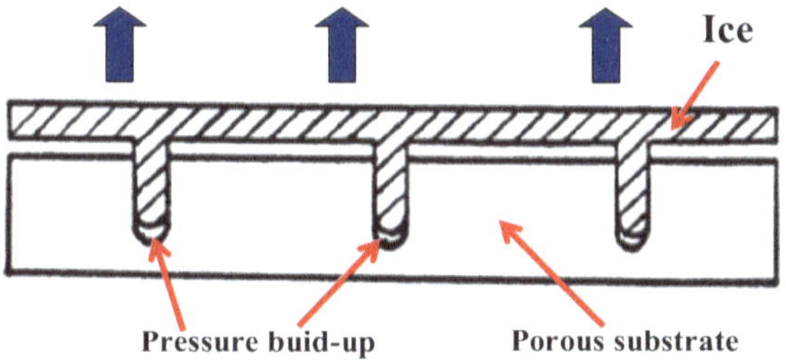

Fig. 6.41 Ice de-bonding on a porous pavement section (Penn & Meyerson, 1992)

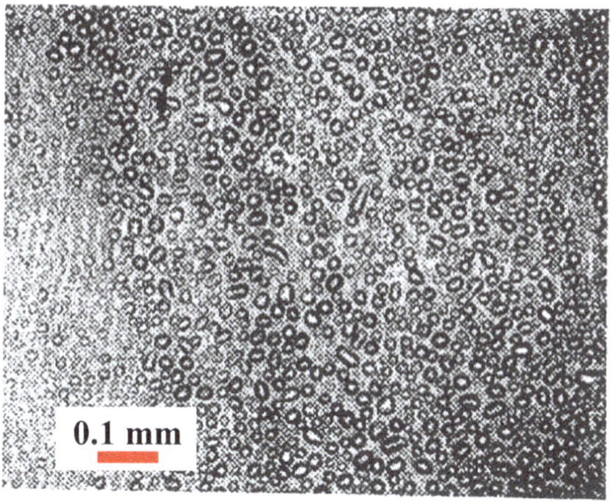

Fig. 6.42 Replicas of ice sheared from polysiloxane polymer coatings (Bascom et al., 1969)

6.3.8 Concluding Remarks

Many scientific publications can be found on materials with high hydrophobic and superhydrophobic properties. In contrast, the materials science literature aimed at developing icephobic coating have few contributions. The reasons for this are the lack of understanding of the physics behind ice-solid interactions, the availability of alternative methods such as active methods and hybrid methods (active and passive), or simply because such new materials are difficult to fabricate.

Recent advances in ice adhesion physics, materials science (functionalized nanomaterials, advanced deposition techniques, new polymers, etc.), and in analytical tools (STM/AFM) have spurred new interest in developing materials having enhanced specific properties. Icephobic materials are no strangers to such developments. It would be extremely difficult, if ever possible, to formulate a truly icephobic material (with zero ice adhesion strength). However, by drastically reducing its adhesion strength, ice may be removed using a minimum amount of work or heat or can even detach itself under its own weight.

To decrease ice adhesion on a given solid, its surface must be modified or coated with a material being able, at the molecular or crystal level, to disrupt the structure of the ice immediately adjacent to the solid. Several strategies may be exploited to reach such a goal:

1. Coat the substrate with a low surface energy material
2. Coat the substrate coating with a low dielectric material
3. Create, at the molecular or nanoscale level, a heterogonous surface promoting a non-uniform stress distribution in the ice adjacent to the solid
4. Promote the presence of tiny air pockets at the interface between the ice and the coating to disrupt bonding by creating stress concentrations
5. Achieve an optimum degree of roughness at several different scales to promote development and propagation of cracks in the ice.

Surface roughness may have a detrimental effect, through mechanical interlocking, on the decrease of ice adhesion with a coating. Therefore, it will be important for the engineer to test the new coatings in simulated or real weather conditions and on real surfaces with a variety of conditions.

The time for merely testing commercial hydrophobic coatings to assess their icephobic properties has come to an end. New pressures call for the development of innovative advanced coatings to drastically reduce ice adhesion. Environmental issues such as the hazard presented by de-icing fluids in the aeronautic sector, energy consuming mechanical, and thermal de-icing methods in the case of wind turbines for example, potentially huge safety and economical costs generated by ice or snowstorms as well as all the cost to operate active devices, are the main drivers to explore novel and specific icephobic coatings to protect overhead lines in winter conditions.

From a practical point of view, most important factors in a successful new icephobic treatment are as follows:

- The durability or longevity of the coating
- The cost and ease of application
- Adhesion of the coating itself to the substrate metal conductor
- Ageing due to corrosion, which may be normal or accelerated by acid rain, airborne contaminants, or the nearby presence of seawater
- UV exposure, both from sunlight and from local partial discharge activity
- Thermal exposure
- Resistance to abrasion wear during installation for instance
- Resistance to erosion/ablation due to the combined effects of rain and wind

6.4 Methods for Coating Preparation and Application

Coating applications will differ significantly depending on the material of the tower components, surface condition and previous coatings applied. Furthermore, applications will vary significantly depending on where the work is performed. Coatings can be applied in factories and on site.

When it comes to location of the coating application, tower members can be coated in the factory and then assembled on site. This is the preferred option when quality of application, durability of coating and environmental aspects are all considered.

When it comes to field application, pollution of environment is usually covered by national regulations. In absence of such regulations, special care should be paid to industrial waste, dust noise, odours, organic solvents, etc. All waste should be collected and treated in accordance with applicable regulations.

Table 6.4 Advantages and disadvantages of shop application (ISO, 2019)

Advantages	Disadvantages
Better control of application	Possible limitation of the size of the tower components
Controlled "ambient conditions for coating" would include temperature and humidity	Possibility of damage due to handling, transport, and erection
Easier to repair damage	Possible contamination of the last coat
Greater output	Possible risk of paint stream electrical flashover
Better waste and pollution control	

To ensure maximum performance of the coating system, most coats or, if possible, all the coats should be applied in shop. The advantages and disadvantages of shop application are shown in Table 6.4 (ISO, 2019).

The on-site coating application is also strongly influenced by the daily weather conditions, and this has an influence on the expected lifetime.

For field application, weather conditions during surface preparation have a significant effect on the quality and durability of the coating. In general, the quality of cleaning and preparation on site is not as good as in the specialized workshop.

The total cost for a field application is about 50–60% more than a workshop application and applied coating has a shorter lifetime. In case of inclement weather conditions, schedules may need to be readjusted. In addition, field application requires the additional effort and cost of site preparation including protecting the area against contamination, avoiding paint spillage/overspray, and waste disposal considerations.

6.4.1 Pretreatment

Surface pretreatment is defined as one or a series of operations including cleaning, removal of loose material and physical and/or chemical modification of the surface onto which coatings are applied. The main reasons for undertaking surface pretreatment prior to applying a coating are as follows:

- To remove or prevent the later formation of weak layers on the surface of the substrate
- To maximize the degree of molecular interaction between coating and substrate surface
- To create specific surface microstructure on the substrate.

The basic surface pretreatment methods are as follows:

- Degreasing
- Abrasion
- Chemical treatment.

The application of water-soluble DUPLEX coatings on freshly galvanized steel components (Lugschitz et al., 2004-08) is considered to be an example of chemical pretreatment.

6.4.1.1 Surface PreparationGeneral

Surface preparation is essential to ensure proper adhesion and durability of the coating. Other surfaces of major interest are hot-dip galvanized surfaces that were previously painted.

A brief overview of surface preparation methods according to ISO 12944-4 (ISO, 1999) is given in Table 6.5. This overview does not include all possible methods and instead has a focus on those that have some field experience.

6.4.1.2 *Preparation of Hot-Dip Galvanized Surfaces*

Hot-dip galvanized surfaces are steel surfaces coated with zinc or zinc alloy by means of immersion in a molten bath. One commonly used standard for hot-dip galvanized coatings of steel tower components is ISO 1461 (ISO, 2009).

Surface preparation for the purpose of anti-icing is the same as surface preparation for the purpose of anti-corrosion protection. Various standards are available for advice on hot-dip galvanized surface preparation; one example is ISO 12944-4 (ISO, 1999).

It is recommended that tower manufacturers develop and follow surface preparation and application procedures in cooperation with their coating suppliers/service providers. Coating a hot-dip galvanized surface should be done within a defined time frame after surface preparation to ensure absence of zinc corrosion products.

Damages to the zinc coating shall be repaired according to relevant working procedures, for example, by zinc spray or by a suitable zinc-rich paint. Any other contaminates like grease, oil, residual flux or similar must be removed by suitable methods, and guidance is given in Table 6.5.

Zinc coating can be treated by sweep blast cleaning using non-metallic abrasives. After the cleaning process, the zinc coating surface must be free of damage and irregularities that would impair the adhesion of subsequent coatings.

Examples of irregularities that can influence the subsequent coating adherence (ISO, 1999) are as follows:

- Runs or areas with excessive thickness
- Pinholes
- Lack of adhesion between zinc and the steel substrate
- Zinc dips
- Zinc ash.

After sweep blast cleaning, the surface should have a uniform and dull appearance. The minimum thickness of the retained zinc coating must meet specifications; the simplest requirement would be that the minimum thickness of the retained zinc coating meets the general requirements for hot-dip galvanized tower components.

The roughness of the surface must be agreed upon, and it is recommended that it is according to the specifications given by the coating manufacturer. Surface roughness is often expressed as the arithmetical mean roughness (Ra) in micrometres. It is common that the coating (paint) manufacturer specifies the surface roughness required to achieve satisfactory adhesion of the coating.

In the case of weathered hot-dip galvanized surfaces, zinc corrosion products form on the surface. Such a surface should be prepared by choosing a suitable

Table 6.5 Procedures for removal of extraneous layers and foreign matter (ISO, 1999)

Matter to be removed	Procedure	Remarks
Grease and oil	Water cleaning	Freshwater with added detergents. Pressure (<70 MPa) may be used. Rinse with freshwater
	Steam cleaning	Freshwater. If detergents are added, rinse with freshwater
	Emulsion cleaning	Rinse with freshwater
	Alkaline cleaning	Aluminium, zinc, and certain other types of metal coatings may be susceptible to corrosion if strong alkaline solutions are used. Rinse with freshwater
	Organic-solvent cleaning	Many organic solvents are hazardous to health. If cloth/rags are used to clean, replace at frequent intervals else oily and greasy contaminants may not be removed. Additionally, a smeared film may appear after the solvent evaporation
Water-soluble contaminants, e.g. salt	Water cleaning	Freshwater. Pressure (<70 MPa) may be used
	Steam cleaning	Rinse with freshwater
	Alkaline cleaning	Aluminium, zinc, and certain other types of metal coating may be susceptible to corrosion if strongly alkaline solutions are used. Rinse with freshwater
Mill scale	Acid pickling	The process is normally not performed on site. Rinse with freshwater
	Dry abrasive blast cleaning	Shot or grit abrasives. Residuals of dust and loose deposits will have to be removed by blowing off with dry oil-free compressed air or by vacuum cleaning
	Wet abrasive blast cleaning	Rinse with freshwater
	Flame cleaning	Mechanical cleaning will be required to remove residues from the combustion process, followed by removal of dust and loose deposits
Rust	Same procedures as for mill scale, plus:	
	Power-tool cleaning	Mechanical brushing may be used in areas with loose rust. Grinding may be used for firmly adhering rust. Residuals of dust and loose deposits will have to be removed
	Water blast cleaning	For removal of loose rust. The surface profile of the steel is not affected
	Spot blast cleaning	For localized removal of rust
Paint coatings	Stripping	Solvent-borne pastes for coatings sensitive to organic solvents. Residues to be removed by rinsing with solvents. Alkaline pastes for saponifiable coatings. Rinse thoroughly with freshwater. Stripping is restricted to small areas

(continued)

Table 6.5 (continued)

Matter to be removed	Procedure	Remarks
	Dry abrasive blast cleaning	Shot or grit abrasives. Residuals of dust and loose deposits will have to be removed by blowing off with dry oil-free compressed air or by vacuum cleaning
	Wet abrasive blast cleaning	Rinse with freshwater
	Water blast cleaning	For removal of poorly adhering paint coatings. Ultra-high-pressure (>170 MPa) cleaning may be used for firmly adhering coatings
	Sweep blast cleaning	For roughening coatings or removal of the outermost coating layer
	Spot blast cleaning	For localized removal of coatings
Zinc corrosion products	Sweep blast cleaning	Sweep blast cleaning on zinc may be performed with aluminium oxide (corundum), silicates, or olivine sand
	Alkaline cleaning	5% (m/m or mass percent) ammonia solution in combination with a synthetic-fabric pad with embedded abrasives may be used for localized spots of zinc corrosion. Alkaline cleaners may be used for larger surfaces. At high pH, zinc is susceptible to corrosion

method. Oxidation products, salts, and other contaminants should be removed. After removing the contaminants, a proper surface roughness should be achieved to ensure good adhesion of subsequent coatings.

6.4.1.3 *Preparation of Previously Painted Surfaces*

It is highly recommended that a work procedure be established before the start of the work and that the personnel in charge of the work have adequate relevant experience. Furthermore, it is recommended that the work procedure be written in accordance with the recommendations of the coating manufacturer or in cooperation with the coating manufacturer.

Prior to any coating an evaluation of the existing paint system must be conducted. There are several standards covering the topic, one possible choice is ISO 4628 which consist of the following parts, with a common title *Paints and varnishes—Evaluation of degradation of coatings—Designation of quantity and size of defects, and of intensity of uniform changes in appearance*:

- Part 1: General introduction and designation system (ISO, 2016)
- Part 2: Assessment of degree of blistering (ISO, 2016)
- Part 3: Assessment of degree of rusting (ISO, 2016)
- Part 4: Assessment of degree of cracking (ISO, 2016)

- Part 5: Assessment of degree of flaking (ISO, 2016)
- Part 6: Assessment of degree of chalking by tape method (ISO, 2011)
- Part 7: Assessment of degree of chalking by velvet method (ISO, 2016)
- Part 8: Assessment of degree of delamination and corrosion around a scribe or other artificial defect (ISO, 2012)
- Part 10: Assessment of degree of filiform corrosion (ISO, 2016).

After evaluation of the condition of the existing coating an appropriate surface cleaning method from Table 6.5 can be selected.

The towers already painted are in almost all cases coated in the field. When performing surface preparation and applying the new coating, environmental considerations and local regulations should be considered.

6.4.2 Deposition Methods

6.4.2.1 Laboratory Methods

Spin coating has been used for several decades for the application of thin films. A typical process involves depositing a small puddle of a fluid resin onto the centre of a substrate and then spinning the substrate at high speed. This is a useful laboratory technique but not relevant to protecting overhead lines in winter conditions.

Plasma treatment plays some important roles in the development of icephobic materials. Plasma etching is mainly used to generate roughness on the surface. However, it may change the surface chemistry to some extent. On the other hand, plasma polymerization and plasma sputtering are generally used to develop thin films from low surface energy materials, but they may also increase surface roughness.

In the plasma sputtering process, the target coating material and substrate are placed as cathode and anode in a plasma reactor, respectively. An inert feed gas such as argon is injected between the electrodes, and a voltage in the range of 100 V to several kV is applied. Positively charged ions from plasma strike negatively charged particles with high energy, thus disintegrating atoms or molecules from cathode surface. Ejected atoms then settle on the surface of the substrate, forming a thin film from the target material. Compared to conventional surface treatment methods, sputtering is a simple, environmentally friendly, and time-saving process which results in more adhesive films. It is also not limited by the melting point of the coating materials and can form multicomponent thin films (e.g. alloys) on various substrates. Many studies have been done on the development of one-step methods for fabrication of superhydrophobic surfaces. Some studies have investigated the effect of preroughening of samples in plasma sputtering process. For example, "bohemitage" or anodization of aluminium samples results in micro–nanoroughness on the surface. Just like polymerization, preroughening of the samples has proven to be an efficient approach in the development of superhydrophobic surfaces.

- Reactive sputter deposition
- Cathodic arc deposition
- Chemical vapour
- Thermal chemical vapour deposition
- Plasma-enhanced chemical vapour depositions (PECVD).

Plasma polymerization deposits polymer films through dissociation and excitation of a monomer gas in plasma and subsequent deposition and polymerization of the excited species on the surface of a substrate. In comparison with conventional surface treatment methods, plasma polymerization can be a very advantageous approach. It can form ultra-thin films with suitable mechanical properties, require lower temperatures, can be used for monomers that do not polymerize under normal conditions, is environmentally friendly, and is significantly faster. Plasma polymerization processes can be divided into two general categories: low pressure (pressures less than atmospheric pressure) and atmospheric pressure. Low–pressure plasma polymerization tends to generate more homogeneous films with fewer impurities. It is also considered to be a green process due to significantly less amount of process gases involved. On the other hand, atmospheric pressure plasma polymerization can be advantageous due to its lower energy consumption, shorter processing times, lack of need for vacuum equipment and considerably higher growth rates. Pulsed plasmas or modulated glow discharges are used frequently in plasma polymerization applications. In a modulated glow discharge, plasma is turned on with a low duty cycle. In the off period, enough time will be given for the polymerization process. Pulsed plasma can lead to unique morphologies or enhance superhydrophobic behaviour. For example, in plasma polymerization of 1H, 1H, 2H, and 2H peruorooctyl acrylate (PFAC), deposited layers in pulsed and continuous deposition were shown to be very similar in chemical composition. But morphology is significantly different, leading to extremely high CA (contact angle).

Several materials can be used in plasma polymerization, such as oxides, nitrides, and oxynitrides of silicon, crystalline materials such as polycrystalline silicon, epitaxial silicon, and refractory metals and their silicides. These materials play a crucial role in development of microelectronics. However, in superhydrophobic applications, it is essential to consider materials with low surface energy. For improvement of hydrophobic behaviour, some specific materials are suitable due to their low surface energy, such as fluorocarbons and organosilicons. Some studies even report and increase in roughness after plasma polymerization. Some of the categories cannot be utilized in plasma polymerization. Several studies have been done on the development of superhydrophobic fluorocarbonic surfaces. Furthermore, a few studies have investigated the development of plasma polymerized superhydrophobic surfaces by other precursors, such as TiCl4, CH4, or acetylene.

6.4.2.2 *Dip Coating*

In a dip coating process, a substrate is dipped into a liquid coating solution and then is withdrawn from the solution at a controlled speed. Coating thickness generally increases with faster withdrawal speed. The coating thickness is determined by the balance of forces at the stagnation point on the liquid surface. A faster withdrawal speed pulls more fluid up onto the surface of the substrate before it has time to flow back down into the solution. The thickness is primarily affected by fluid viscosity, fluid density, and surface tension. Fluid viscosity in turn is strongly affected by fluid temperature.

6.4.2.3 *Brush or Roller Coating*

Paint brush and roller coating should always be undertaken using an appropriately sized, good quality synthetic, or natural fibre brush compatible with the product. This application technique is relatively slow and is generally used for coating small areas with decorative paints and for surface tolerant primers, where good penetration of rusty steel substrates is required. It is particularly suitable for the application of stripe coats and for coating complex areas where the use of spray methods would lead to considerable losses due to overspray and associated dry spray problems. Brush application requires considerable care when applying non-convertible coatings over one another, such as chlorinated rubber over chlorinated rubber, or vinyl on top of vinyl. In these cases, the solvents in the wet coat readily redissolve the previously dry bottom coat. For tower painting, specialized wool mittens allow the worker to immerse the entire hand into a large (>20 l) container of paint. The worker must take some care not to allow any continuous stream of paint to drip from these gloves near energized conductors. Paint streams have flashed over at voltage gradient of 50 kV/m (Farzaneh & Chisholm, 2009), where the typical 230 kV line voltage stress is 70 kV/m.

6.4.2.4 *Spray Coating*

The spray coating technique is well established for industrial coatings. This high-throughput large area deposition technique ensures ideal coatings on a variety of surfaces with different morphologies and is often used for in-line production. Moreover, the fluid waste is reduced to minimal quantities, and the deposition can be easily patterned by simple shadow masking. Also, the spray coating technique can access a broad spectrum of fluids with different rheologies, offering the opportunity to tune the system to deposit virtually any kind of solution and obtain the desired film properties.

The steps to optimize a spray process are detailed and shown in Fig. 6.43.

Good success has been achieved, with excellent surface finish and minimum overspray, with the use of high-volume, low-pressure (HVLP) spray systems to deposit SiR coatings. The HVLP equipment may run at pressure as low as 0.5 bar (7 psi) where conventional paint spraying systems require five times this pressure. A multistage turbine compressor provides the high airflow. In an HVLP system, the supply of paint is also pressurized, rather than feeding from syphon or gravity cup as illustrated in Fig. 6.43.

Coating systems for icephobic properties are typically designed as for a low or very low pollution class that would be pollution class C1 or C2 as defined in ISO 12944-2 (ISO, 2017). Examples are given below for coating systems that could serve just a single purpose. If the coating system is to have a dual purpose, giving

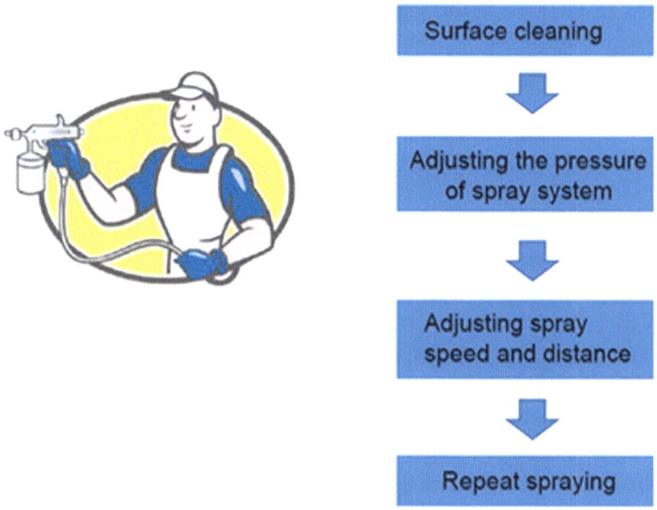

Fig. 6.43 Exploration of spray coating process

both corrosion protection and icephobicity, the coating system should be designed accordingly considering the service conditions. Further information can be found in various standards, including ISO 12944-5 (ISO, 2019).

Some coating systems are listed in Tables 6.6, 6.7, and 6.8 as examples. Surface preparation is listed as a part of the coating system as it is a key factor in the performance and life of the coating. The term "anchor profile" in these tables refers to the distance from the highest peak to the lowest valley on the surface and may be

Table 6.6 Example of a coating system for galvanized steel with alkaline washing (Statnett, 2018)

Application	Surface preparation	Coat	NDFT
Galvanized steel	Cleaning with alkaline detergent followed by hosing with freshwater	1 coat epoxy primer 1 polyurethane topcoat	50 μm 70 μm Total: 120 μm

Table 6.7 Example of a coating system for galvanized steel with sweep blasting surface preparation (Statnett, 2018)

Application	Surface preparation	Coat	NDFT
Galvanized steel	Sweep blasting with non-metallic and chloride-free grit to obtain an anchor profile of 25–45 μm	1 coat epoxy primer 1 polyurethane topcoat	50 μm 70 μm Total: 120 μm

Table 6.8 Example of a coating system for galvanized steel with sweep blasting surface preparation and one coat painting (Statnett, 2018)

Application	Surface preparation	Coat	NDFT
Galvanized steel	Sweep blasting with non-metallic and chloride-free grit to obtain an anchor profile of 25–45 μm	1 coat of painting suitable for application directly on blast cleaned hot-dip galvanizing	80 μm

measured with a depth gauge, composite plastic tape obtaining a reverse image or surface profile comparator.

6.4.2.5 Electrostatic Sprayers and Electrodeposition

An exploration of spray coating process variables for application of both anti-corrosion and anti-icing coatings should include electrostatic spray gun technologies. The head of the sprayer is held at negative potential (up to 100 kV and 100 μA) relative to the grounded component. The negatively charged atomized paint is drawn to and even wraps around the back side of the grounded surface. Demonstrations show that the back side of a grounded 25 mm cylinder is also painted while the spray pattern coats the front side with similar thickness and surface gloss. Electrostatic spray gun systems require some electrical safety awareness, grounding and bonding training that is already a common part of modern utility overhead line safety management processes.

Powder coating is another popular spray coating system option for larger components that can be baked to the necessary curing temperature. The process is specialized, and the powder is usually charged electrically before application.

Electrodeposition is a fascinating phenomenon by which it is possible to provide a shiny coating on one metal surface from another metal simply by donating electrons to ions in a solution. Direct current is made to flow between two electrodes immersed in a conductive, aqueous solution. Electrodeposition is exceptionally versatile, and valuable applications keep being developed. This approach can be most suitable with practical applications, such as coating efficiency, film thickness uniformity on irregular complex shapes, cost reduction, and mass-productivity in mind. This process is useful for applying materials onto any electrically conductive surface. The materials which are being deposited are the major determining factors in the actual processing conditions and equipment which may be used.

Metal matrix composites with PTFE particles as the reinforcing phase can be produced by electrodeposition. In this approach, PTFE micro or nanoparticles are dispersed in an electrolyte and are entrapped during the reducing process of metallic ions. The hydrophobic behaviour of such surface has been investigated in a few studies (ASTM SC G01.05, 2019), and superhydrophobic characteristics were observed in some cases. Although no ice adhesion tests have been done on such surfaces, the composite nature of the surface, the low cost of electrodeposition process, and the possibility to treat large and complex parts can prove to be beneficial in icephobic applications.

6.4.3 Concluding Remarks

Many utilities have considerable experience with application of anti-corrosion coatings on overhead line components. Most advise colleagues in regions where winter corrosion effects are strong to monitor the loss of galvanizing and to initiate recoating when corrosion is at an early stage. Proper surface preparation is needed to obtain a long service life.

The use of anti-icing coatings is an interesting and attractive way for preventing or reducing icing loads and the associated phenomena and damage on overhead electrical network equipment such as conductors, ground wires and insulators. In the laboratory, a variety of techniques have been developed for applying coatings onto surfaces to protect them against icing. Coating application is preceded by a surface pretreatment including cleaning, removal of loose material and then physico-chemical modifications. Some of these techniques, such as spray coating are well established for industrial applications while others, like plasma sputtering, are still at early stages of development.

Despite research and good progress in the development of passive anti-icing coatings, there are still limits for the application of icephobic coatings and treatments to towers, conductors, and ground wires. Progress in the past decade has been slow, especially for processes to apply active coatings, leaving these approaches as laboratory concepts. Finally, icephobic coatings that continue to function in areas of high corrosivity, and at the same time provide anti-corrosion characteristics, have not yet been identified.

6.5 Methods for Handling Coated Power Line Components

6.5.1 Handling and Installation of Coated Insulators

Utilities generally obtain experience with handling insulators with anti-pollution coatings first. Successful experiences can guide the successful transport and installation of insulators with anti-icing coatings that may be more fragile.

Silicone coatings for anti-pollution or ice repelling duty are thin layers that may be damaged during shipping or handling. Therefore, precautions are necessary from the factory packing stage to the storage and handling end.

Standard wooden crates lined with soft cushion material (soft plastic, polyamide, or polyester sheets) need to be added as shown in Fig. 6.44. Other packaging materials including individual cardboard boxes (Fig. 6.44) are sometimes used. In this case, the number of empty boxes is often inconsistent with field scrap disposal guidelines.

Every utility has its own handling and installation guidelines, but special instructions are provided by the manufacturers to avoid damage of the coating

Fig. 6.44 Typical coated insulator packaging methods

during installation (Fig. 6.45). Small cuts and tears in the coating do not present much risk of reduced service life in many applications.

Figure 6.46 gives examples of users handling methods. In the example shown at the right of this figure, we can see that the crate, while protecting most of the string,

Fig. 6.45 Example of coated insulator damage

Fig. 6.46 Handling examples and protections

Fig. 6.47 Damaged units while they were still wet during handling. Crates were stored in water

can still damage the coating of some insulators. In this case, foam or soft protective material may be used to limit the risk of damage.

Silicone by nature will allow water to permeate through the coating and in stagnant water this can lead to a weakness of the coating if handled wet. Whenever possible storage in dry places is recommended. Packaging that holds water, or pallets that sit in puddles of water, should be avoided. Figure 6.47 shows the risk to damage the coating if it is handled wet.

6.5.2 Handling and Installation of Coated Conductors

There is little difference between stringing operations of uncoated and coated conductors. Not all coatings have the same surface adhesion. Depending on the coating, it might be necessary to handle the conductor more carefully and use plastic lined stringing equipment to limit coating damage. Figure 6.48 shows a coated conductor on a drum. The vertical flange is covered with a polymer protector on the inside, to avoid scratches on the conductor surface.

In this case, the conductor has been coloured to reduce visibility, but the same considerations would apply to a conductor that has been treated with an icephobic or anti-corrosion coating at the factory.

Manufacturers should issue clear handling instructions, and it is recommended that the contractor is made familiar with them in an early phase of a project.

Surfaces of the conductor that need to be electrically connected with fittings, like suspension clamps, earthing clamps, parallel groove clamps, etc., require to be cleaned and the coating removed.

Whether or not the coating needs to be removed in places where the conductor is in contact with other fittings, like for example wedge tension clamps, needs to be determined by tests.

The conductor coating can be removed mechanically and/or with solvents. If brushes or other mechanical means are used, it is a good practice to conduct a trial

Fig. 6.48 Coloured conductor on a drum with a flense covered with a polymer protector (Nexans, 2016)

coating removal on the ground, determine the tools and time needed, and plan for the additional time needed for installation. In cases where solvents are used, local environmental regulations must be followed and an assessment or test made to determine how a solvent influences the frictional characteristics of the surface.

6.5.3 Handling and Installation of Coated Fittings

Fittings must have all the surfaces that meet the conductor uncoated, so that the coating does not change the surface friction properties. In addition, all electrical contact surfaces should be uncoated if the coating is non-conductive. Masking may be needed to ensure that all surfaces in contact with the conductor should remain uncoated; see Fig. 6.49. Nuts and bolts are often left completely uncoated.

In the case of non-conductive coatings, the contact surfaces must remain uncoated and must be masked prior to coating. See Fig. 6.50.

Most fittings have markings that include the name of the manufacturer, product codes, model numbers, batch information, and manufacturing dates. It is important that markings remain legible even after the coating application.

Like for towers and tower members, if the coating of the line is planned before installation, the coating application is much easier if it is done by the fitting manufacturer or any other shop prior to installation.

Handling and installation of fittings treated for corrosion or icephobocity are the same as for regularly coated fittings apart from packing procedures. Fittings are usually packed in crates, and there can be even several types of fittings in the same crate.

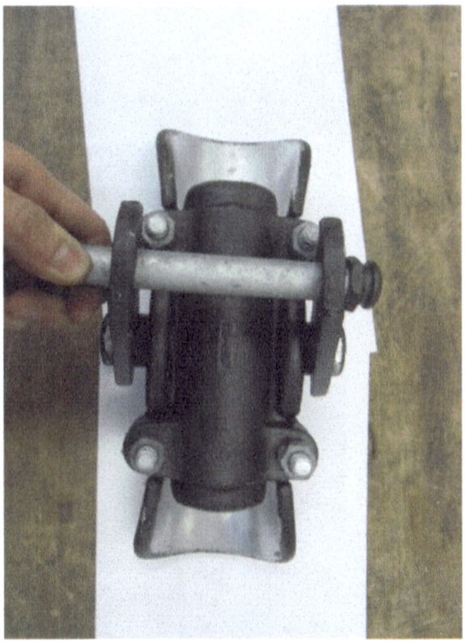

Fig. 6.49 Coated suspension clamp

Fig. 6.50 Fittings with uncoated contact surfaces

Fittings packing should be such that the coating is not damaged during transport. Care should be taken to use solutions with smallest possible environmental impact and that the packing material is disposed of in a responsible manner. Example of packing for coated fitting is shown in Fig. 6.51.

Fig. 6.51 Example of possible fittings packing

6.5.4 Handling and Installation of Coated Tower Members

There are no special considerations for handling and installation of painted steel tower members. However, as the coated surface is more prone to damage than hot-dip galvanized members, certain practice rules are advisable including:

- Coating should cure before transportation and handling
- Packing materials, procedures, and containers should be optimized to mitigate the risk of damage during transport.

An on-site repair/touch-up procedure should be specified and agreed upon.

6.5.5 Concluding Remarks

Utility experiences to manage damage to coated insulators have a learning curve, just like the experiences with brittle fracture of long polymer insulator FRP cores when workers treated them like vaulting poles. Coatings for protecting power network equipment in winter conditions will benefit from handling guidelines in place for successfully installing insulators with anti-pollution coatings.

6.6 Characterization of the Coating Functional Properties

The aim of surface treatment or applying a coating to power network equipment is to provide its surface with specific functional properties. The most important functional properties in winter conditions are as follows:

- Hydrophobicity and superhydrophobicity
- Icephobicity
- Self-cleaning
- Anti-corrosion
- Visual impact.

Once the component is coated, the properties of the coating should agree with the results reported in Table 6.9.

Test procedures ensure reproducibility of sample preparation, testing conditions and the measurement procedure. Conductor samples tested are mainly obtained from two sources: power utilities and directly from the manufacturers. For transport, conductor surfaces are covered to prevent scratches, and mechanical fittings are used to retain the compact shape of the conductor. Conductors recovered from service are generally coated with aged greases. Coating samples from towers can be obtained in situ at tower base, and insulators removed from service can be used to assess coating integrity more conveniently than coatings on energized, in-service conductors.

Table 6.9 Physical and chemical properties of the coating

Properties	Recommended property after coating
Hydrophobicity	The detection method is implemented in accordance with DL/T864-2004, giving the desired water contact angle
Adhesion	The detection method is implemented in accordance with GB/T 9286–1998, and the result is $\leq$ 1 level
Hardness	The detection method is implemented in accordance with, and the result is $\geq$7H
Chemical resistance	Soaking in 3% HCl, 3% NaOH, and 10% NaCl for 60 d, 30 d, and 300 d, respectively, and the coating has no peeling, blistering, or discoloration Alternately boiling in 0.1% NaCl (conductivity 2 mS/cm) for 100 h has traditionally identified good surface preparation and coating adhesion using the same criteria
Salt fog resistance	After 1080 h salt spray test, the gloss retention rate of the coating is 98%

6.6.1 Characterization of Hydrophobicity

The theoretical background to wetting phenomena was described in Chap. 4. In this section, several methods generally used to characterize the wetting behaviour of a surface are discussed.

The wetting behaviour of a surface is commonly quantified by the concept of contact angle. Therefore, contact angle measurement is the most common characterization process used in the literature. Other methods are also available, which will be discussed briefly.

It should also be noted that the hydrophobic properties of a surface may change over time as the material ages or responds to the stresses placed on it. Some materials permanently lose their initial hydrophobicity whereas other materials may recover their hydrophobicity after stress events. It is therefore also important to determine the degree by which a coating can recover its hydrophobicity.

6.6.1.1 *Contact Angle*

The contact angle is the result of interface tensions between a liquid and a solid surface surrounded by a vapour. It is defined by the angle between the solid/liquid interface and liquid/vapour interface.

Contact angle can be measured by means of two routes. These two routes are explained briefly in the next part of this section.

Static Contact angle Measurement (CA)

An indication of the surface wetting properties of a given material may be obtained by placing a water droplet on a flat section of the material and measuring the static contact angle between the water droplet and the surface as shown in Fig. 6.52.

Generally, to measure the contact angle, the shape of the droplet is somehow estimated using a mathematical equation. Some methods, such as Young–Laplace estimation, try to assess the shape of a complete droplet considering all the factors contributing to the physical equilibrium of a resting droplet, such as droplet weight or air pressure. Some other methods, such as tangent method, only fit a specific part of the droplet (e.g. the vicinity of the three-point interface), and produces a faster and easier, but less accurate calculation. Generally, a small droplet (less than 10 μL)

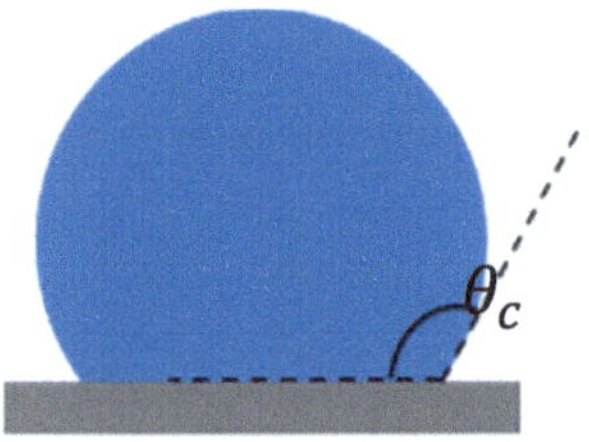

Fig. 6.52 Definition of the static contact angle (θ_C)

is deposited on the surface, and the shape of the droplet is estimated using one of the methods available.

Traditionally, a superhydrophobic surface is defined as a surface on which the water contact angle is higher than 150°. However, some recent studies (Di Mundo & Palumbo, 2011; Müller & Oehr, 2011; Strobel & Lyons, 2011) have suggested that the concept of static contact angle should be completely revisited. They argue that the *static contact angle* is a random number between the advancing and receding angles and is not an accurate indication of wetting characteristics. To truly characterize the wetting behaviour of a surface, the dynamic wetting behaviour should be studied.

6.6.1.2 *Dynamic Contact Angle Measurement*

As mentioned in Chap. 4, the dynamic (advancing θ_a and receding θ_r) contact angles are defined as the contact angles on moving liquid fronts (Radojcic et al., 2013-09). Subsequently, contact angle hysteresis (CAH) is defined as the difference between the dynamic contact angles. To measure θ_a and θ_r on a surface, two methods, shown in Fig. 6.53, are commonly used:

(a) The advancing and receding contact angles can be measured by moving the droplet relative to the substrate and thus directly produce the advancing and receding contact angles on two sides of the drop.
(b) Another method, which is widely accepted in the literature, is to deposit a small drop on a surface and gradually increase its volume. The advancing contact angle can be acquired by measuring the contact angle right before the droplet baseline expands. Similarly, the receding contact angle can be determined by reducing the drop volume and measuring the contact angle right before the droplet baseline shrinks in Fig. 6.53b.

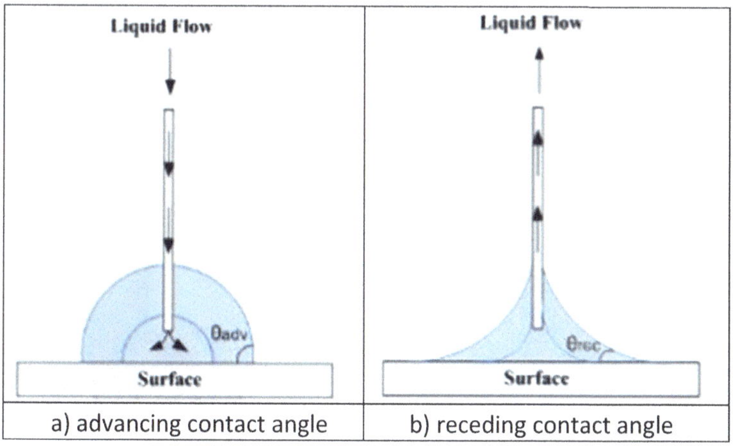

Fig. 6.53 Schematic diagram showing the dynamic contact angle measurement

6.6.1.3 *Other Measurements of Hydrophobicity*

According to IEC Technical Specification 62073 (IEC, 2003), "Guidance on the measurement of wettability of insulator surfaces", there are two other methods, besides contact angle measurement, to estimate the wettability of an insulator surface:

1. The surface tension of an insulator may be measured by spraying the surface with a range of organic liquid mixtures with known surface tensions. The time needed for the sprayed on liquid to break into distinct droplets can indicate the surface tension.
2. Another method is to compare the visual appearance of a wetted section of an unknown surface with a series of standard photographs. This is a simple and practical approach where a common spray bottle is used to spray the area of interest with a fine mist of uncontaminated water (Fig. 6.54). The wetted surface is then inspected and categorized according to standardized photographs and descriptions.

6.6.1.4 *Hydrophobicity Recovery*

Hydrophobic recovery is a phenomenon that is exhibited in many of the treatment methods, whereby the water contact angle increases towards its original value after a period, with total recovery to the pretreated state which sometimes happens within a few hours. This effect has an impact on the applicability of these techniques by determining the "shelf life" of the treatment. X-ray photoelectron spectroscopy (XPS) and contact angle analysis can be used to characterize hydrophobicity recovery. Some tests can verify and compare the recovery of hydrophobicity of different materials. Examples for two of them, recovery after immersion and recovery after contamination (hydrophobicity transfer), are presented below.

Fig. 6.54 Spray bottle method for determining surface wettability

6.6.1.5 *Water Immersion Test*

In one specific study by Gutman et al. (2012), a superhydrophobic surface along with a high-temperature vulcanized (HTV) rubber sample was prepared and separately immersed in tap water for 10 days. Water conductivity was measured to be 240 μS/cm, and its temperature was kept constant at 25 °C. After the 10-day period, the samples were removed from the water and left to dry naturally. Once completely dried, they were sprayed with tap water, and their wettability class was determined. The determined wetting class over a period of 11 days is shown in Fig. 6.55.

Figure 6.56 shows photographs of the samples after 11 days. The superhydrophobic nanocomposite surface maintained its higher wettability class during the immersion test better than standard rubber.

6.6.1.6 *Hydrophobicity Transfer*

Several studies have used different characteristics of the surface to estimate the hydrophobic recovery of a coating. One example presented in Gutman et al. (2012) uses the same principle utilized for similar applications in (Gutman et al., 2011-08). Two series of samples were prepared: ten samples of a superhydrophobic nanocomposite (70 × 40 × 15 mm) and ten samples of HTV silicon rubber as the control group (120 × 50 × 10 mm). The HTV samples originated from Gutman et al., (2012) where they are described in detail. The method consisted in measuring resistance of the polluted and wetted test samples over time to estimate the rate of

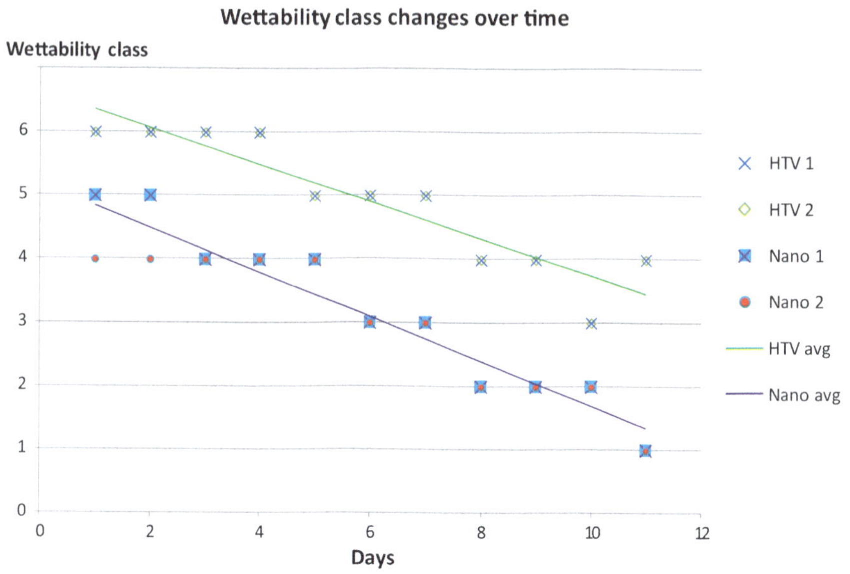

Fig. 6.55 Wettability class measurements over time

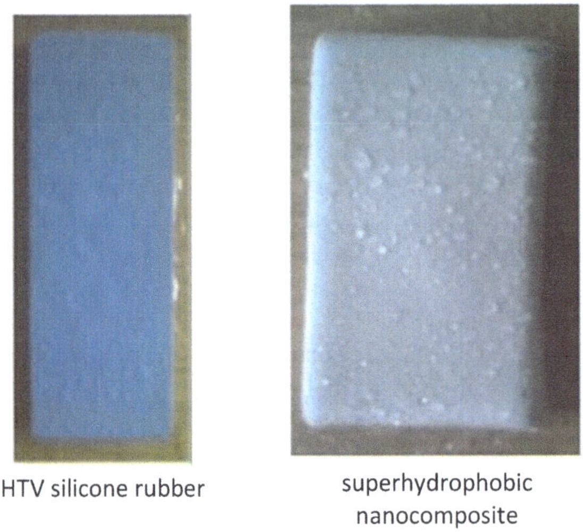

Fig. 6.56 Hydrophobicity after 11 days of water immersion

hydrophobic recovery. Two different pollution levels were chosen. The simple device used for the measurements is shown in Fig. 6.57.

The results of the resistance test are presented in Fig. 6.58. This test method suggests that hydrophobic recovery of the superhydrophobic nanocomposite coated samples was much faster than that of the HTV samples.

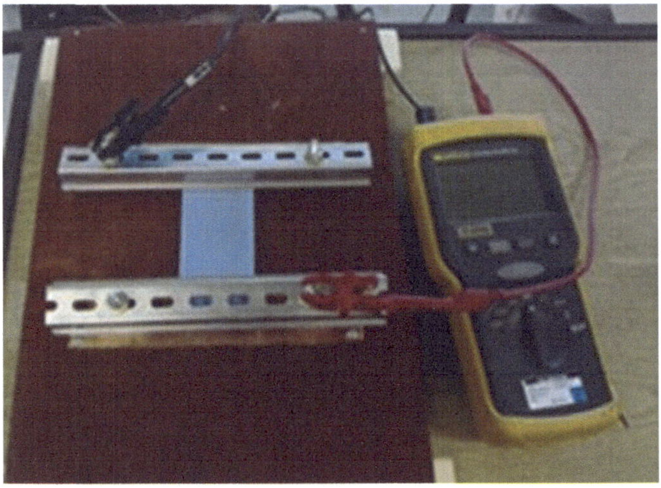

Fig. 6.57 Test set-up for resistance measurement

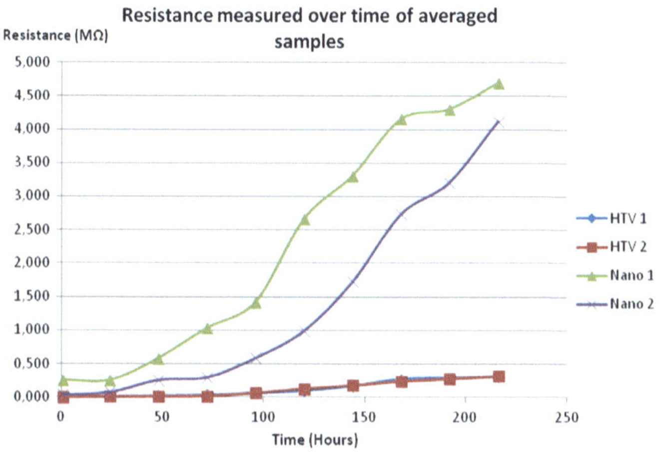

Fig. 6.58 Resistance measurement over time (averaged values)

6.6.2 Characterization of Icephobicity

The icephobic properties of a coating are measured in terms of how effective it is to prevent ice accretion on the coated surface and how strongly ice adheres to the surface once formed. These characteristics are evaluated with ice accretion and ice adhesion tests.

6.6.2.1 *Ice Accretion Test*

As ice adhesion is related to the process and type of ice accretion, it becomes important to consider a representative ice type for the evaluation of icephobicity for coated surfaces. Different types of ice include glaze ice, dry and wet snow, rime ice (soft and humid), and hoar frost. The process of accretion and characteristics of these ice types are described in CIGRE WG B2.29 (2010).

Artificial Ice Accretion Test

Several ice accretion tests methods have been proposed for testing the ice flashover performance of insulators, which are described in IEEE publications and standards (Farzaneh et al., 2003; IEEE DEIS, 2009-10) as well as in CIGRE publications (CIGRÉ TF 33.04.09, 1999, 2000).

Most of these tests have a few common features that include

1. **Preparation**: The insulators are prepared and conditioned before the start of the ice accretion phase to obtain the correct surface conditions representative of the service conditions and to ensure a repeatable ice accretion process. In this respect, the insulator temperature, surface contamination, installation configuration, and surrounding e-field are important aspects that require consideration.

2. **Exposure**: After the preparation phase, the insulators are ready for **exposure** to the conditions resulting in ice accretion. During this time, the test subject is kept under specific environmental conditions and exposed to a water spray or a flow of chilled water.
3. **Quantification**: The severity of the ice accretion is **quantified** when the exposure phase is concluded. This typically consists of measuring the ice thickness on a round metal rod or tube.
4. **Evaluation**: The final stage of the test is the **evaluation** of the insulator flashover performance. The ice-covered insulator is subjected to melting conditions while energized to determine the flashover voltage.

Practically, the same test procedure can be used to characterize the ice accretion properties of treated insulators and conductors (Li, J. et al., 2014-08). In this case, the test is terminated after step 3 to compare the amount of ice accreted by direct weight measurement. For insulators it is also possible to use the test in its entirety to determine the effect of differences in ice accretion on the flashover voltage.

The artificial ice can be accumulated on the samples in indoor tests, cold rooms or icing wind tunnels, but it is also possible to do this outdoors when temperature and wind conditions are appropriate (CIGRÉ TF 33.04.09, 1999, 2000). In this section, two recent examples of conductor ice accretion tests are also presented.

In the first example, short conductor samples, as shown in Fig. 6.59, are used. The middle section of the conductor is coated and then subjected to the artificial outdoor ice accretion test (Radojcic et al., 2013-09).

For the exposure test, the conductor samples are installed on a rotating drum forming a squirrel cage, as illustrated in Fig. 6.60. The drum with conductor samples should be rotating during the exposure test to ensure that all samples are

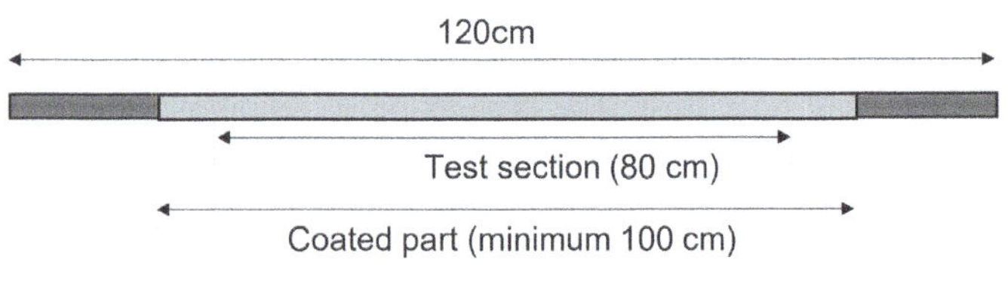

Fig. 6.59 Test conductor sample

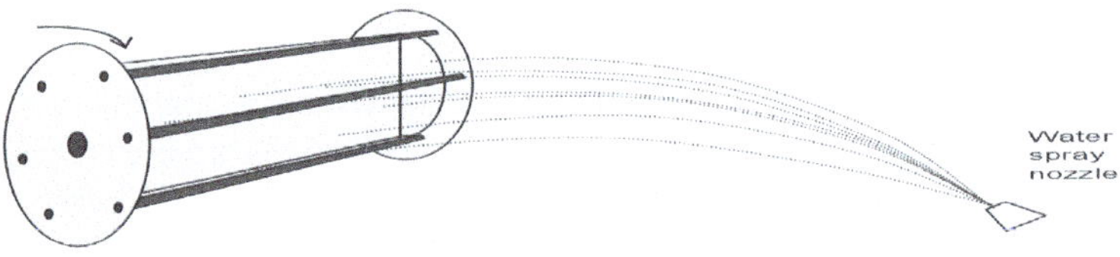

Fig. 6.60 Illustration of the rotating drum with conductor samples set-up for icing test

equally exposed to the water spray. This rotation, at about 6 rpm, also facilitates the accretion process and prevents the formation of long icicles. Removable covers are applied to the conductor ends so that only the test section of the conductor is exposed to the freezing water spray. It was possible to remove the conductor samples from the test rig for weighing without disturbing the ice accretion. An uncoated test conductor was included on the drum to serve as the reference sample.

Before the accretion phase, both the conductor and test area were prepared to create the right conditions for an effective and repeatable ice accretion. For example, for tests in the climate chamber (45 m^3), the ambient temperature in the hall was kept between −7 °C and −4 °C, and the conductor samples had been precooled to −7 °C for a period of 24 h. This test may also be performed outdoors if the ambient temperature is below −10 °C, but then the test locations should be selected so that the test samples are not exposed to direct sunshine. Examples of the indoor and outdoor testing set-ups are given in Fig. 6.61.

For this test two icing types, namely glaze and rime ice were deposited, respectively, by spraying the conductors with a high-pressure washer and a spray gun which sprays a mixture of air and water. The amount of ice accumulated on test samples and the reference conductor is determined by weighting them after ice accretion. Examples of results achieved are presented in Figs. 6.62 and 6.63 (Safaee, 2008).

In the case of the glaze ice test, it was found that the weight of ice accreted was about the same on the RTV and superhydrophobic-coated conductors. The adhesion strength of the ice to the superhydrophobic-coated conductor was however significantly lower than that of the RTV-coated conductor. Figure 6.62 also shows that ice on the superhydrophobic coating had a bumpier appearance compared to that of the RTV-coated conductor.

In the case of rime ice test, it was found that a thinner layer of ice was accreted on the superhydrophobic coating than on the other conductors tested. There were however no obvious differences in the appearance of the ice layers between the superhydrophobic and RTV-coated conductors in Fig. 6.63.

Fig. 6.61 Artificial ice accretion tests performed in laboratory (left) and outdoors (right)

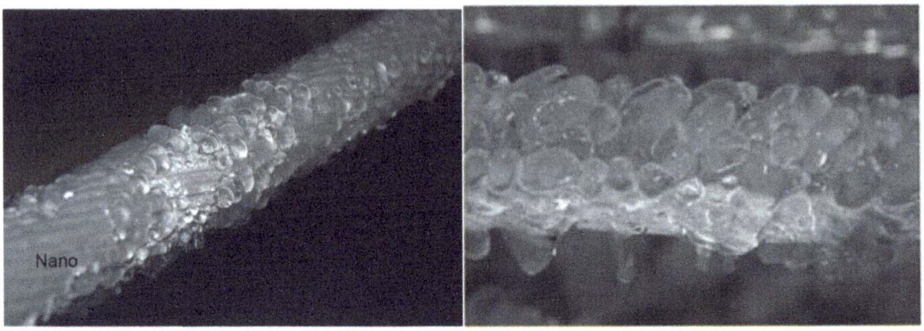

Fig. 6.62 Illustration of bumpy appearance of ice accretion on superhydrophobic conductors

Fig. 6.63 Ice accretion on RTV-coated and superhydrophobic-coated conductors

For the second example of ice accretion test, a small environmental chamber is used to test the ice accretion properties of small conductor samples. For this test special conductor samples are used. They are prepared, before applying the coating, by replacing the inner strands of the conductor by a copper tube. For the test the conductor sample is placed in a double-sided environmental chamber as shown in Fig. 6.64.

The temperature inside the chamber can be controlled by circulating cool air between the outer and inner walls while the temperature of the conductor is controlled by circulating a coolant through the tube inside the conductor. A set of nozzles in the roof of the chamber is used to spray the conductor with cold water, or an ultrasonic humidifier can be used to generate a cold fog. The conductor is weighed before and after the test to determine the amount of ice accreted. An example of ice accretion achieved on uncoated and coated conductor samples as shown in Fig. 6.65.

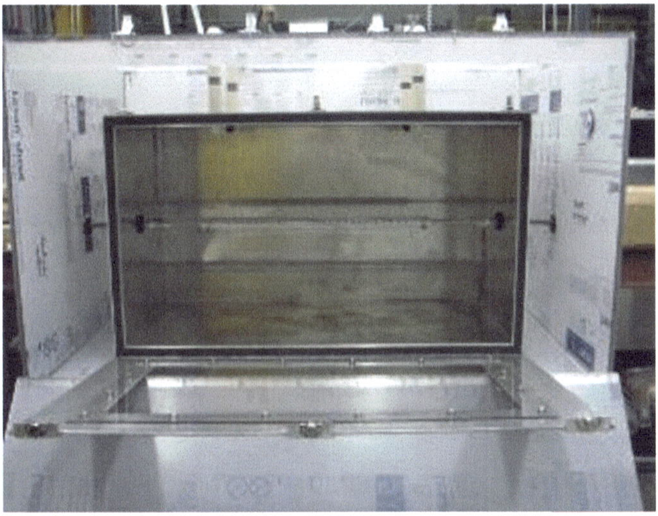

Fig. 6.64 Double-sided environmental chamber used for the ice accretion test

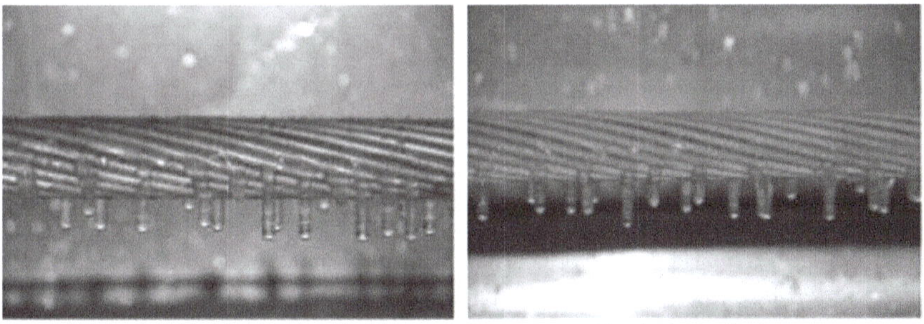

Fig. 6.65 Example of ice accretion achieved in a small chamber test

Natural Ice Accretion Test

A simple standard exposure frame can be used to expose samples to the environment during natural ice storms to determine the icephobic properties of the coatings. An example of such a test frame is shown in Fig. 6.66. Regular visual inspections were done during periods of freezing temperature. Usually, these inspections were carried out early in the morning, before the samples were exposed to direct sunlight to determine the degree of ice or frost accretion on the sample surface. Inspections were also performed after periods of snowfall. After the visual inspections, the samples were also sprayed with fine water mist to determine if the water droplets would freeze onto the coating surface. Pictures were taken as visual record of the extent of ice accretion.

Fig. 6.66 Example of a natural ice accretion test frame

a) The Deadwater Fell test site viewed from the North end of the 190 m test span

b) The tested Cu conductor fitted with vibration monitors and load cells

Fig. 6.67 Example of natural exposure test site

A similar exposure rig can be used to evaluate ice accretion on conductors in natural conditions. An example of a full-scale conductor natural exposure test site is given in Fig. 6.67a. The tests were carried with uncoated ACSR and coated copper conductors (Fig. 6.67b), with representative findings given above in Chap. 3.

6.6.2.2 *Ice Adhesion Tests*

Ice and wet snow adhesion to outdoor surfaces is known to cause serious problems to power transmission lines, aircrafts, boats, etc. Ice adheres strongly to almost all surfaces, and yet, test methods to evaluate the strength of ice adhesion to surfaces are scarce. A few test methods do exist, but they are complicated and time consuming. Moreover, they are quite expensive. An overview of the methods that are commonly used is presented in this section.

Ice Shear Stress Test on Flat Samples

The following method is used in China for the measurement of the ice adhesion stress on insulating materials (China Electricity Council, 2019).

Ten identical samples are coated on the surface of a base pad which has an area of 300 × 300 mm and a thickness of 10 mm. The coated surfaces need to be flat without bending, skewing, and any other deformation.

Five stainless steel ice boxes with area of 100 × 100 mm as shown in Fig. 6.68 are used for ice shear stress test.

The base pad with coating sample is fixed on a steel platform which also has a thickness of 10 mm. The ice box is placed at the centre of the coated base pad as illustrated in Fig. 6.69. It should be noted that the base pad, the pull rod of the base platform, and the pull rod of the ice box are kept parallel.

The test device as displayed in Fig. 6.69 is placed horizontally in the climate chamber, and the ambient temperature of the climate chamber is adjusted to −7 to −5 °C. After a thermal equilibrium is obtained between the device and the ambient

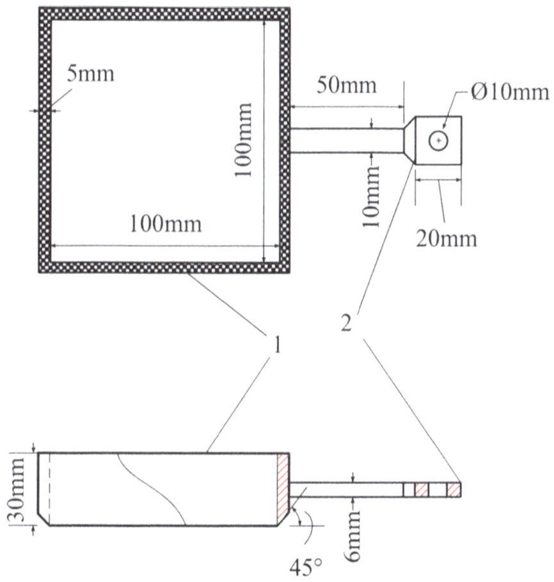

Fig. 6.68 Schematic diagram of ice box for ice shear stress test. (1) ice box; (2) pull rod of ice box

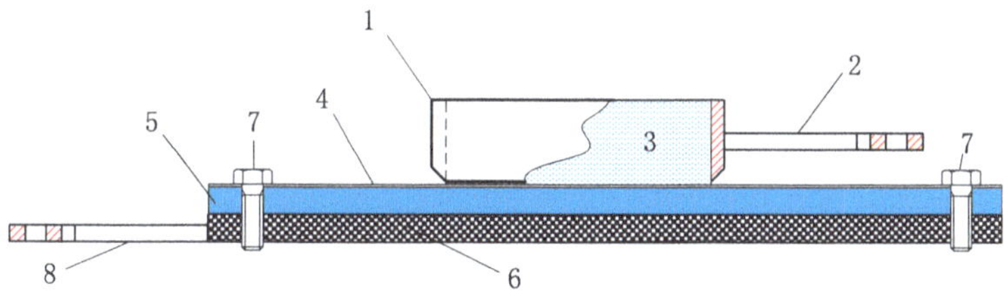

Fig. 6.69 Layout of the ice shear stress test. 1—ice box; 2—pull rod of the ice box; 3—accreted ice layer in the ice box; 4—coated anti-icing coating; 5—base pad; 6—steel platform; 7—fixing bolt; 8—pull rod of the steel platform

environment, the ice box is sprayed with de-ionized water. The precipitation rate of the spraying should be controlled to make the ice grow gradually. The spraying process is stopped when the glaze ice in the ice box reaches a thickness of 25 mm. After that, the ice accretion on the surface of the coating sample outside the ice box is removed. To harden the accreted ice in the ice box, the ambient temperature of the climate chamber is kept constant at least for another 15 min (China Electricity Council, 2019).

To ensure that the axis of the upper and lower retainers coincides with the axis of the pull rods of the ice box and the steel platform, respectively, and the pull rods of the steel platform and the ice box in Fig. 6.69 are fixed to the upper and lower retainers of the tensile testing machine, respectively. The tensile testing machine is turned on to load the sample with a speed of (50 ± 5) mm/min until the ice in the ice box is completely separated from the coating surface. The maximum load F_1 of the tensile testing machine is recorded. During the experiment, the temperature is kept at –7 ~ −5 °C (China Electricity Council, 2019).

The ice shear stress of coating surface is calculated as follows:

$$p_l = \frac{F_1}{S} \tag{6.4}$$

where

- p_1—ice shear stress (Pa)
- F_1—maximum load of sample surface under ice shear stress (N)
- S—ice area in the ice box (m^2).

The average value of five samples is measured as the ice shear stress of the coating surface.

The reduction rate of ice shear stress AC_1 is calculated according to the following formula (ISO, 2017):

$$AC_1 = \left(1 - \frac{p_1}{p_{10}}\right) \times 100\% \tag{6.5}$$

where

- p_1—ice shear stress of coating surface (Pa)
- p_{10}—ice shear stress of base pad without coating (Pa).

The reduction rate of ice shear strength AC_1 should exceed 95% when p_{10} is obtained on glass or ceramic surface.

Ice Shear Stress on Cylindrical Samples

Shear stress analysis on cylindrical samples is performed in Italy (Balordi et al., 2019) with a home-made apparatus, equipped with an electromechanical testing system INSTRON 4507, after Susoff et al. (2013-10). Figure 6.70 shows the testing layout.

Cylindrical samples, 12 mm diameter and 200 mm length, are prepared with the appropriate coating. The samples are then immersed in 40 ml of de-ionized water in a proper aluminium alloy mould and frozen at −19 °C for at least 8 h. The de-ionized water should be boiled to remove air prior to pouring into the mould. After this period, the mould is rapidly fixed into the machine, and the test is started after one minute. Test samples are pulled from the ice with a speed of 4 mm/min, and the force F(N) needed to pull the sample off the mould is recorded. The ice adhesion strength (σ) in shear (Pa) can be calculated by:

$$\sigma = F/A \tag{6.6}$$

where A is the surface area of the bar in contact with the ice (m^2).

The shear stress is averaged over five different specimens for each coating treatment.

To assess the durability of the coating, the test needs to be repeated several times on the same bars.

In this method, the adhesion strength of ice is evaluated by measuring the force required to detach the ice grown on a sample surface by mechanically pulling the ice from the surface (Fig. 6.71). A charcoal block with a hole drilled through its centre with the sample affixed at one end is used as the mould to grow the ice. The

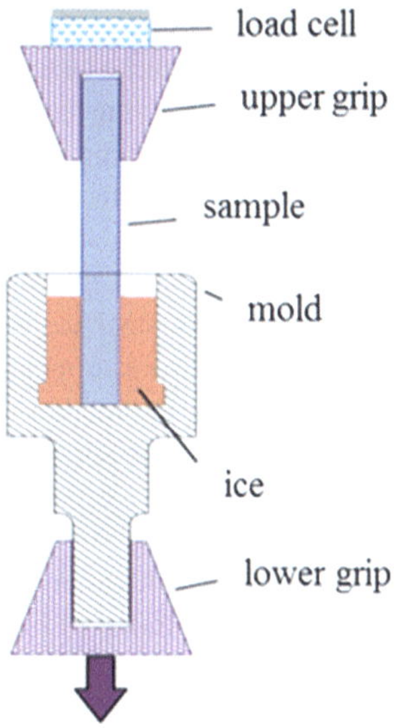

Fig. 6.70 Sketch of shear stress measurement apparatus

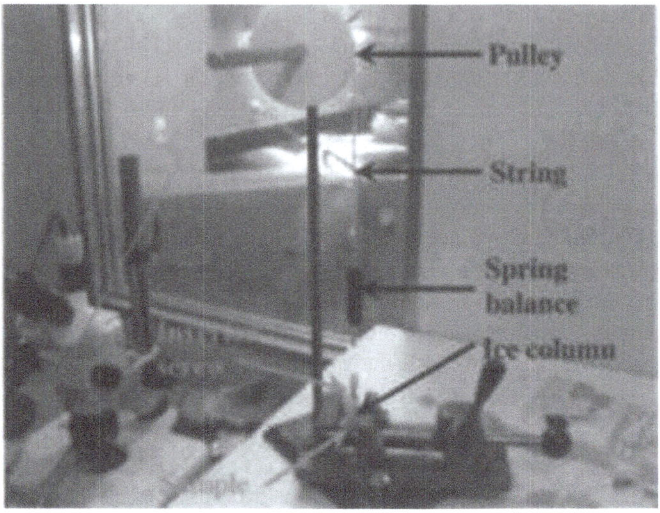

Fig. 6.71 Photograph of the apparatus used to measure the ice adhesion strength

ice is grown by freezing preboiled de-ionized water at −20 °C. The clear, bubble-free ice grown on the sample is then separated from the mould by cleaving off the charcoal surrounding the cylindrical column of ice. The apparatus used to test the adhesion strength of ice mainly consists of a stand holding a pulley and a strong base to hold the sample with the intact ice. A cord goes over the pulley to a spring balance attached to the ice on the sample. With the sample fixed to a base, the ice is slowly pulled vertically upwards by applying a force to the free end of the cord. The force needed to detach the ice from the surface is read from the spring balance providing the force of adhesion and hence the adhesion strength from Eq. 6.6. Force can be determined from $F = mg$, where m is the mass required to detach the ice from the sample surface and g is the gravitational force. This method is useful in determining the adhesion strength of ice on hydrophilic samples as ice adheres strongly on those surfaces with lower water contact angles.

A similar set-up is adopted in (China Electricity Council, 2019), and the coating samples are prepared the same way as indicated.

Five stainless steel ice boxes as shown in Fig. 6.72 are used for ice vertical adhesion stress measurement. The bottom end of the pull rod, which has a T shape, is placed under the metal cover of the ice box for the following tensile adhesion stress test. To let the spraying water get inside the ice box, there are several circular holes made on the ice box cover.

Like the ice shear stress test, the base pad coated with coating sample is fixed on a steel platform which has a thickness of 10 mm, and the ice box is centred on the base pad as illustrated in Fig. 6.73. The pull rods of the cooler and steel platform, components 2 and 8 in the diagram, must be in a straight line, and both are perpendicular to the coated base pad.

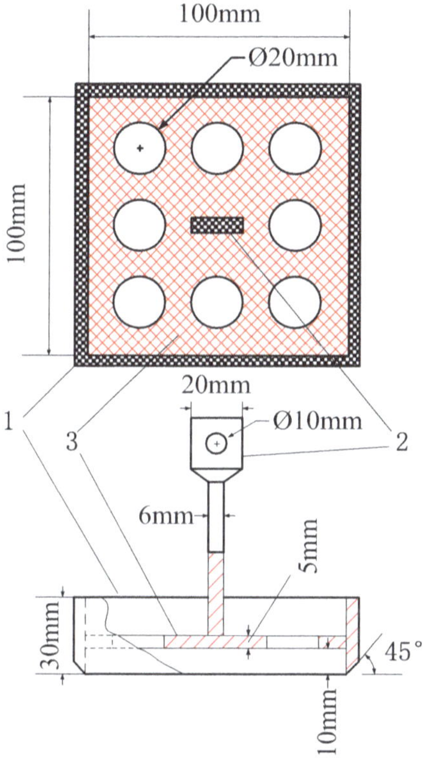

Fig. 6.72 Ice box for ice tensile adhesion stress measurement

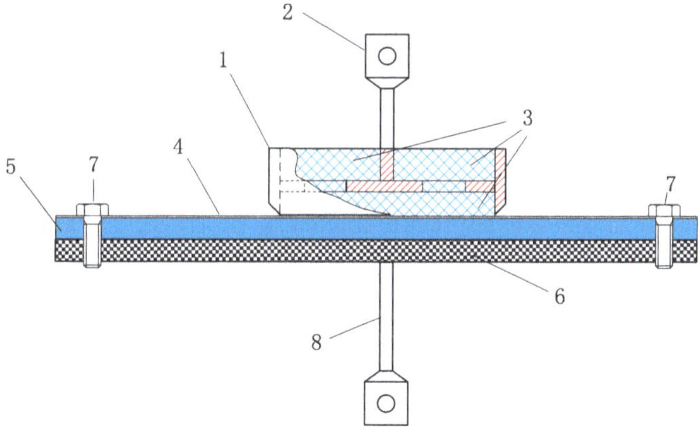

Fig. 6.73 Layout of tensile adhesion stress measurement. 1—ice box; 2—pull rod of the ice box; 3—accreted ice in the ice box; 4—coated anti-icing coating; 5—base pad; 6—steel platform; 7—fixing bolt; 8—pull rod of the steel platform

The artificial icing process is the same as that of the ice shear stress test.

The pull rods of the steel platform and the ice box are fixed to the upper and lower retainers of the tensile testing machine, respectively. The tensile testing machine is turned on to load the sample with a speed of (50 ± 5) mm/min until the ice in the ice box is completely separated from the coating surface. The maximum load F_2 of the tensile testing machine is recorded. During the experiment, the indoor temperature is kept at −7 °C to −5 °C.

The ice vertical adhesion stress of coating surface is calculated as follows:

$$p_2 = F_2/S \tag{6.7}$$

where

- p_2 is ice vertical adhesion stress (Pa).
- F_2 is maximum load of sample surface under ice shear stress (N).
- S is ice area in the ice box (m^2).

The average values of five samples are measured as the ice vertical adhesion stress of the coating surface.

The reduction rate of ice vertical adhesion stress AC_2 is calculated according to formula (Newell et al., 1968-04).

$$AC_2 = (1 - \frac{p_2}{p_{20}}) \times 100\% \tag{6.8}$$

where

- p_2 is ice vertical adhesion stress of coating surface (Pa).
- p_{20} is ice vertical adhesion stress of base pad without coating (Pa).

The reduction rate of ice vertical adhesion stress AC_2 should exceed 95% when p_{20} is obtained on a flat steel surface.

Centrifugal Chamber Methods

The ice centrifuge adhesion test (CAT) (Laforte & Beisswenger, 2005-06) consists of a two-step procedure. Test blades with one extremity either bare or coated are iced on a stand in a cold room as displayed in Fig. 6.74. The blades are then rotated in a centrifuge until one end sheds its ice deposit. An accelerometer device or perhaps a high-quality microphone is set on the centrifuge vat to detect the instant when the ice detaches and trigger a recording of the rotation speed achieved. The rotation speed of the centrifuge can be monitored with good precision, allowing a close estimate of the force on the ice coupon especially if it is small compared to arm length r.

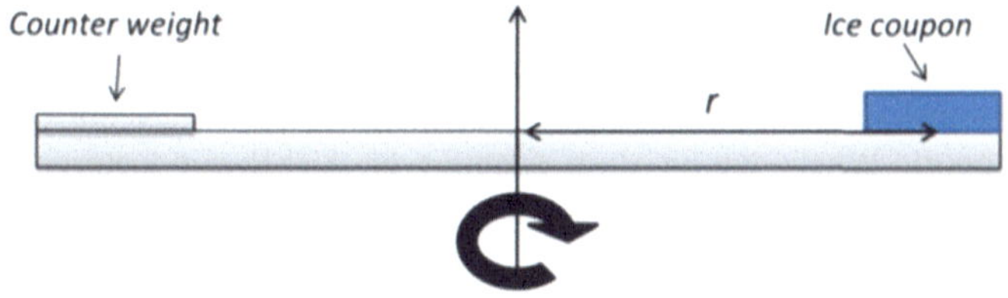

Fig. 6.74 Scheme of the centrifuge adhesion Test (Brassard et al., 2018)

A similar centrifugal apparatus was designed and developed in-house at Université du Québec à Chicoutimi (UQAC) for measuring the ice adhesion strength on various surfaces (Farhadi et al., 2011; Safaee, 2008). In this method, test samples (32 × 50 mm) with accumulated ice are mounted at one end of an aluminium beam (32 mm wide and 30 cm long). A counterweight is attached to the other end to balance the beam. The beam is then fixed in the home-built centrifuge test chamber (Fig. 6.75) maintained at −10 °C. By rotating the beam at increasing speed, a controlled ramp of the centrifugal force can be achieved. When this force reaches the adhesion force of the ice, the ice detaches from the sample surface. The exact rotation speed at the time of ice detachment is determined using the controlling software, which was also developed in-house. The adhesion force $F = mr\omega^2$ is determined after the ice detaches, where m is the mass of ice, r is the radius of the beam, and ω is the rotation speed. The shear adhesion strength ($\sigma = F/A$) of ice is then determined from the apparent area A of the sample surface which was in contact with the ice.

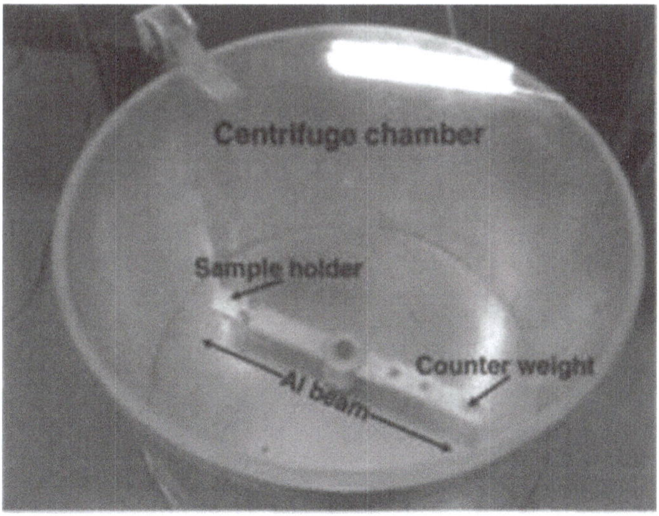

Fig. 6.75 Centrifugal chamber where the ice-covered samples are fixed on the sample holder and the Al beam rotated under computer control

The centrifuge test is easily adapted to test the adhesion of ice coupons on short lengths of ACSR conductor with various strand diameters, outer diameters, and values of rugosity. If necessary, long conductor samples can be stiffened with a parallel support. When cutting short samples of ACSR conductor, usually the strands are bound on each side of the hacksaw cut in gear clamps to maintain the normal strand orientation and spacing.

Sliding Weight Method

For testing the ice adhesion in field conditions, a sliding weight method was proposed by Scandinavian power companies. Figure 6.76 shows a tool used for comparing the adhesion force between two different surfaces (Radojcic et al., 2013-09; Wang et al., 2010). It is important to standardize the impact energy on the ice coverage and to measure it in field outdoor conditions. Thus, a weight of 1.42 kg which can slide on a rod is used. The rod is placed, resting by its own weight, on the conductor. The weight is let down on the conductor, first from half the length of the rod, 15 cm, then from the full length, 30 cm. These two actions are referred to as "half punch" and "full punch", respectively. Such a simple device made it possible to quantify the difference in adhesion strength of ice for surfaces with different coatings applied. For example, the adhesion strength of rime ice on RTV-coated conductors was found to be less than that for rime ice on bare metal conductors (Radojcic et al., 2013-09).

Ice Push Off Test

At the Electric Power Research Institute (EPRI) in the USA, an ice push off test is used to assess the icephobicity of coatings. The test set-up is shown in Fig. 6.77. For this test, coated flat samples are placed on a cold plate cooler that is in line with a linear motion table that has a force gauge secured to it. Three ice forms are placed,

Fig. 6.76 Demonstration of the sliding weight method

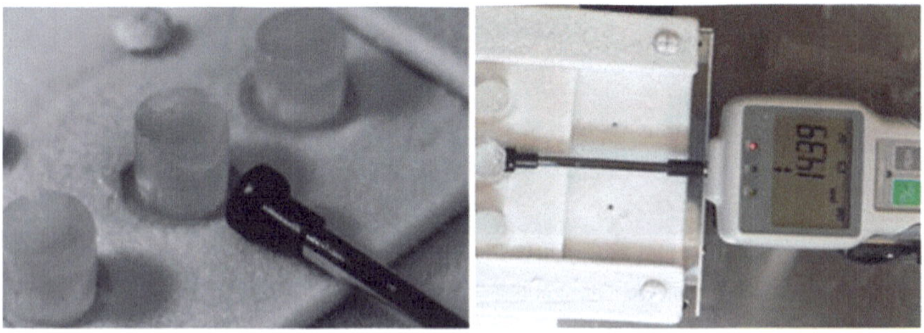

Fig. 6.77 Ice push off test (courtesy of EPRI)

equally spaced directly in line with the force gauge's probe, and then filled with distilled water. The cold plate is then powered on and set to a target temperature of −10 °C.

Once the ice samples have frozen, the ice forms are carefully removed, exposing an ice column. Then, the motion table, controlled by a servo motor and drive, moves the force gauge secured to it at a velocity of 1 mm/s. The force gauge's probe pushes against the ice column, and the peak force is measured when the ice column detaches from the coated surface. The test is then repeated with the two remaining ice columns.

The peak force (F) is then divided by the cross-sectional area of the ice column (A) resulting in the shear strength σ using Eq. 6.6.

Conductor Ice Pull Off Test

Using a rod, wire, and manufactured conductor to which the coating has been applied, a simple method has been developed to measure the ice adhesion to wires. For this test a wire is immersed into a container with water, which is then frozen. Figure 6.78 shows the container with the conductor frozen into the ice.

Fig. 6.78 Image of wire immersed within an ice block (courtesy of LaFarga)

Fig. 6.79 Side view of the ice adhesion test

The wire is positioned at the centre of the container and kept in place with a rubber lid. The wire passes through a hole in the centre of the lid, as shown in Fig. 6.79. It is important to measure accurately the contact surface between the wire and the ice prior to closing the lid.

Ice adhesion strength on the conductor is then measured by pulling the wire from the ice, as shown Fig. 6.79. The pull-out force is measured with a standard dynamometer. Figure 6.79 also provides details that the container is clamped into a bench vice, and the wire is attached to the dynamometer through a hook.

6.6.3 Self-Cleaning Properties

The self-cleaning ability of coatings is determined by artificially applying a contamination layer onto the coating and then by measuring the washing effect by spraying the contaminated sample with a spray bottle. This method is still in development, and a repeatable method needs to be found to apply the contamination layer. It was found that the artificial contamination method proposed for hydrophobic insulators in TB 555 (CIGRE WG C4.303, 2013) was unsuitable for superhydrophobic coatings. With this method, the hydrophobic properties of silicone rubber are masked by the application of dry kaolin or kieselguhr powder. After hydrophobicity is suppressed, it becomes possible to use a flow or dipping method to apply the contamination slurry to the insulator. On superhydrophobic coatings, it was found that the dry powder did not have the intended masking effect, which made it impossible to wet the insulator with the contamination slurry.

Currently, an entirely new method to evaluate the self-cleaning properties of a coating is in development. In this method, a layer of condensed water is deposited on the coating surface to capture and bind a dry contamination mix applied by airflow. The initial results with this method look promising as it was possible to generate a contamination deposit on the samples. Some examples are shown in Fig. 6.80. After the deposition and drying of the applied contamination, the hydrophobicity spray bottle test is used on tilted samples to test the self-cleaning ability of the coating.

For testing under more realistic conditions, a natural exposure rig, as shown in Fig. 6.81, may be set up at a site in a contaminated area. Standard equivalent salt

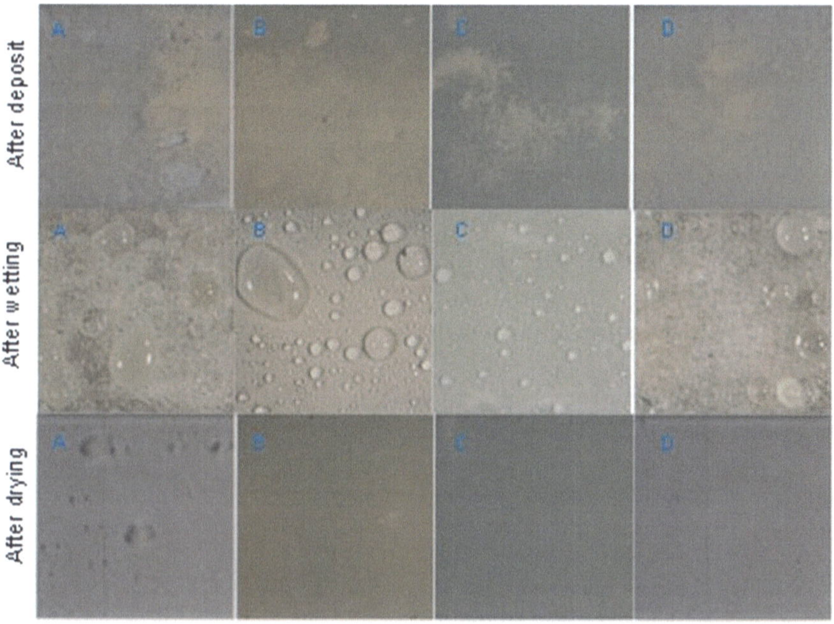

Fig. 6.80 Results of the self-cleaning test for superhydrophobic coatings

Fig. 6.81 Tier 2 natural contamination test

deposit density (ESDD(IEC, 2008)) measurements are taken at regular intervals to track the build-up of contamination on the coated insulator samples. Some modification to the standardized ESDD method may, however, be necessary for textured surfaces as the cotton strands from the pad used for swabbing may become ensnared in the surface structure. In such cases, an alternative solution could be to dunk the insulator into a container with the distilled wash water.

Neither of these contamination accumulation tests are standardized, and this is a proposed line of additional investigation.

6.6.4 Anti-Corrosion Characteristics

Coatings play an important role in corrosion control, used either alone or as part of an overall strategy of corrosion control. Generally, the anti-corrosion properties of the coatings originate from the pigments used in coating formulation. Anti-corrosive coatings can be categorized into three types according to their pigments:

- Inhibitive coatings (passive): These pigments are slightly soluble in water and react to form a passive film on the substrate, which reduces the corrosion of the latter. They are in the form of metal oxides such as molybdates, phosphates, or chromates (zinc or lead). In this category, we also find barium metaborate, silicates, and some ion exchangers based on silica and calcium salts.
- Barrier coatings: These pigments prevent or slow the moisture absorption on the substrate. The micaceous iron oxide, mica, aluminium, zinc, stainless steel, and nickel flake are part of this category.
- Zinc-rich coatings (sacrificial pigments): These are used for the cathodic protection of steel. They are in the form of zinc dust, oxide, or zinc carbonate. Because zinc will oxidize preferentially to steel, the steel will be protected.

Corrosion is usually driven by some electrochemical non-homogeneity in the metal or its environment. In this process, different areas of the metal, having different levels of free energy and therefore different corrosion potentials, become the electrodes of an electrochemical cell in contact with a common electrolyte. A corrosion test is a type of materials test in which the goal is to determine a material's sensitivity to chemical reactions that might cause damage. Typically, all conventional corrosion testing methods are expensive and time consuming. There are three major types of corrosion tests: laboratory, field, and service testing.

In the first two tests, the four major factors in accelerated weathering are moisture, oxygen, sunlight, and heat. Standard testing procedures have been set up for various laboratory tests and are given in Table 6.10. Pollution can also be considered as a fifth factor but since the type and level of pollution can vary dramatically, even on a small distance, there is no universal test. Such accelerated testing for assessment of corrosion performance should only be used for

Table 6.10 Summary of specific laboratory corrosion (accelerated weathering) tests and their corresponding standards

Organization	Salt fog	Cyclic salt fog/ultraviolet
ASTM	**B117**, "Standard Practice for Operating Salt Spray (Fog) Apparatus", 2011 **G85**, "Standard Practice for Modified Salt Spray (Fog) Testing", 2001	**D5894**, "Standard Practice for Cyclic Salt Fog/UV Exposure of Painted Metal, (Alternating Exposures in a Fog/Dry Cabinet and a UV/Condensation Cabinet)", 2013
ISO	**15110**, "Paints and varnishes—Artificial weathering including acidic deposition", 2013	**16151**, "Corrosion of metals and alloys-Accelerated cyclic tests with exposure to acidified salt spray, "dry" and "wet" conditions", 2005 **11997-1**,"Paints and varnishes—Determination of resistance to cyclic corrosion conditions—Part 1: Wet (salt fog)/dry/humidity", 2005

comparative purposes. No direct correlation with natural weathering should be made because there is no test that is proven to give results like natural weathering.

6.6.5 Visual Characteristics

Overhead lines are technical objects in the landscape and are often under discussion. Many methods and approaches have been developed worldwide to get greater acceptance from the public. One of these methods is to camouflage lines so that they blend into the environment. The visibility of overhead lines and the impression on the public can be reduced if they are coated in an appropriate colour. In some regions, one of the preconditions to get permission for the building of an overhead line is that the towers are to be coated to give a particular desired colour. Often sunlight reflection effect from newly erected galvanized towers and of conductors is a source of annoyance for some landowners. Coating of conductors for reducing ice accretion may reduce the visibility. On the other hand, if camouflage of conductors is needed, the coating may be selected with properties to reduce ice accretion.

The visual impact of conductors and insulators (and other components of the OHL) varies depending on their appearance (i.e. colour, degree to which its surface reflects sunlight and size), the background (e.g. sky, land, and forest), illumination effects caused due to service operation (e.g. corona), and visibility. Due to the large number of parameters, it was decided first to compare visibility through a simple direct comparative test where samples of coated conductors were mounted in an outdoor rig as shown in Fig. 6.82 (Gutman et al., 2013). Differences in visibility of the conductor types were documented and compared through the photographs taken under different weather conditions and averaged ranking from eight inspectors.

Fig. 6.82 Example of picture for ranking of visual impact of different conductors. A-painted; B-acid; C-blasted; D-RTV; E-new; F-superhydrophobic (nano)

The untreated, RTV-coated and blasted conductors were found to be the least visible. The dark conductors (painted or acid treated) were generally found to be the most visible. The visibility of the superhydrophobic conductor in Fig. 6.82 generally fell between these two extreme cases, except in bright sunlight when it was found to be the most visible.

The visual impact of coated conductors can also be assessed on actual transmission lines. By careful selection of locations for coated conductor sections and positioning web cameras, the visual impact of the conductor can be assessed against different backgrounds such as forest or sky under different weather conditions and times of day (Fig. 6.83). The results of this investigation showed that it was possible to quantify the visibility of insulators by an observer in an objective way (Gu et al., 2013-06).

The effectiveness of coating conductors and structures was demonstrated when a 380 kV double circuit "camouflage line" with dark green-coated towers, conductors and fittings was erected in the Austrian Alps in the 1990s. It is of course difficult to show a nearly invisible OHL with a picture. Figure 6.84 shows two 380 kV towers, one close to the other: on the left a well visible galvanized tower (not coated line) and on the right a nearly invisible camouflage line (the chimneys from the foundation can be seen and indicate the location of the tower). As explained above, the camouflage effect strongly depends on the colour of the background and the diversity of the landscape.

The introduction of coloured coated overhead line conductors (green or blue) has a major environmental advantage in terms of visual impact. However, it should be kept in mind that an "invisible" conductor could present obstacles to birds, paragliders, helicopters, aeroplane, etc. Hence, power lines located in areas with high air traffic should be marked to avoid incidents.

Fig. 6.83 Typical image of one tower with three different insulators (coatings) and three different backgrounds captured from the test site

Fig. 6.84 Non-camouflage and camouflage line in the Austrian Alps

6.6.6 Concluding Remarks

Coatings are applied to power network equipment to provide their surfaces with specific functional properties. At the present time, there are only a few standardized tests to characterize these properties. Various institutions have, however, elaborated various test methods for this purpose. In this chapter, brief descriptions for methods to measure or quantify the hydrophobicity and icephobicity of coatings are provided.

Icephobicity is characterized by measuring the ice adhesion strength on a protected surface covered with ice. This is generally done by measuring the force necessary to break the bond between the ice and the substrate. At present, there are no standard tests available; therefore, various organizations have developed different procedures to measure icephobicity.

Approaches for assessing the self-cleaning and anti-corrosion properties of coatings as well as their visual impact were discussed. Further investigations are needed to establish standard methods to characterize the functional properties of coatings.

6.7 Test Methods for Characterizing the Coating

6.7.1 Intrinsic Material Properties

In the absence of a standard for the performance evaluation of icephobic coatings, several different methods, related to the specific application, are adopted in different laboratories. Regarding overhead power lines components, an in-depth examination of these methods was carried out by CIGRE WG B2.44 and reported in TB 631 (CIGRE WG B2.44, 2015). More generally, testing methods of icephobic properties are also discussed in Brassard et al. (2018).

6.7.1.1 *Coating Thickness*

Thickness of the coating is a sum of all layer thicknesses. Thickness of a non-homogeneous layer shall be estimated as the maximal geometric size of the contained particles. Thickness of the coating shall be measured according to the ISO 2808 (ISO, 2007) or (ASTM, D1005, 2013; ASTM, D4414, 2013; ASTM, D5796, 2010; ASTM, B504, 2011) standards.

6.7.1.2 *Density*

The density of a coating is its weight per unit volume of the coating, surface, or length of the substrate. Density of the coating shall be tested according to the ISO 2811 (ISO, 2011) or (ASTM, D1475, 2013) standards.

6.7.1.3 Hardness

Hardness is a measure of how resistant a coating is to permanent deformation when subject to a compressive force. Hardness depends on various material properties including ductility, elastic stiffness, plasticity, strain, strength, toughness, viscoelasticity, and viscosity. The measuring standard techniques can be found in ASTM, D3363 (2011) or ISO 15184 (ISO, 2012).

6.7.2 Coating/Material Interface Characteristics

Ice adhesion forces can be measured in the two components, normal to the surface (tensile adhesion strength) or parallel to the surface (shear adhesion strength). The two components are generally not equal. Different geometrical configurations are therefore set up on different sample shape (e.g. planar or cylindrical samples), while ice blocks may have variable dimensions, from a few millimetres (a single drop solidifying) up to tenths of centimetres.

Other variables include a procedure to generate ice or a snow block and make it adhere to the sample. This part of the testing procedure can result in data that cannot easily be compared among different laboratories. In facts, ice adhesion stress values, as determined in all methods, vary considerably, ranging from 0 to 2 MPa (Raraty & Tabor, 1958). The latter value corresponds to the cohesive shear fracture energy of ice (Fortin et al., 2010-08). These large variations are caused by many differences, such as the way ice is formed, atmospheric or refrigerated in a cold box, as well as the substrate characteristics (Brassard et al., 2018). Anyhow, some tests can produce highly variable results, with up to 300% variation, even in the same testing laboratory.

The adhesion reduction factor (ARF), defined as the ratio between shear stress of the bare reference sample and the coated sample, was introduced in 2000 by researchers at the University of Quebec in Chicoutimi (UQAC) (China Electricity Council, 2019), to normalize ice adhesion reduction values between the different existing methods. High values of ARF can be measured on very effective icephobic coatings while lower values indicate poorly performing materials.

The capacity of the coating to delay the freezing process is normally tested by dropping some water on the cold sample and observing the solidification. In other cases, the water is dosed on the sample at room temperature, and the whole system is cooled down until the complete solidification is observed. The temperature setting, the temperature ramp, and in general the procedure details strongly influence the observed results, so that only a comparison internal to the laboratory is possible.

Electrochemical methods look at the reaction mechanism at the interface of a system such as a substrate and a coating. An electrochemical method, such as electrochemical impedance spectroscopy (EIS), gives information about the reaction mechanism of an electrochemical process: different reaction steps will dominate at certain frequencies. Information obtained includes detection of coating holidays (voids), time to corrosion failure due to water penetration and absorption, measurement of resistance and capacitance changes in coatings, corrosion rate measurements, undercoat corrosion, measurement of activity in holidays or pores through coatings, and corrosion resistance due to a protective coating on metal in various electrolytes (Lasia, 1999).

6.7.3 Electrical Characteristics

The application of a coating on an overhead conductor can change the contact electrical resistance between the conductor and its fittings and thus influence the performance of the overhead line. This may also be the case for example when applying temporary grounding connections for work on the line. In addition, the conductor surface properties influence the minimum or maximum slip loads that can be achieved inside a conductor fitting. The verification of these aspects calls for testing and methods that are common for all coatings. It is important to consider this possibility in the planning stage of a project if conductors with changed surface properties are used.

6.7.3.1 *Dielectric Tests*

Electrical characteristics of coatings applied to transmission towers, nominally for corrosion protection, can also provide a degree of electrical puncture strength that can be helpful in managing touch potentials. Generally, coatings need to have a minimum thickness to achieve adequate puncture strength, especially from ground level to the maximum reach, about 2 m off the ground. The high-visibility guards placed around guy wires to protect right-of-way users, for example snowmobile operators, should provide similar electrical strength if the guy wires are not insulated from the structure with guy strain insulators. Dielectric strength of a coating intended to provide touch potential mitigation is established using IEC 60243–1: 3.0 or the equivalent ASTM D149 test method.

The dielectric test is performed on conductor coatings which are applied to flat aluminium samples. A commercial dielectric tester is connected to the sample with the ground lead connected to the metal substrate and the high-voltage probe (via a pointed probe) to an electrode that rests on the surface of the coating. The test set-up is shown in Fig. 6.85. The HV electrode used is as specified in IEC TS 62217 (IEC, 2012) for the high-voltage test on the core material.

The dielectric tester has an automated test sequence which starts at the minimum output voltage and slowly increases the voltage until the material breaks down. The test is repeated at various locations on the coating surface to obtain statistical information on the uniformity of the results.

6.7.3.2 *Fitting Resistance Tests*

Fittings that should be tested together with a coated conductor are as follows:

- Compression dead-end clamps, mid-span joints, and jumper terminals
- Parallel groove clamps
- Earthing clamps

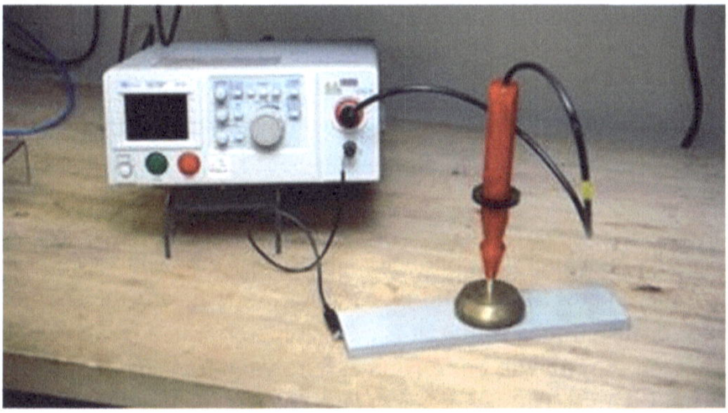

Fig. 6.85 Dielectric test for conductor coatings

- All types of repair fittings, such as splices, repair sleeves, and others
- All other electrical connection fittings

Figure 6.86 shows a typical resistance test to determine the conductor fitting electrical performance.

A typical test protocol using the four-terminal resistance metre with 0.1 μΩ resolution in Fig. 6.86 would be

1. The resistance measurement contact points are placed on the conductor each side of the fitting 25 mm from the fitting edge.
2. The current connections are at a distance not less than 50 times the conductor diameter.

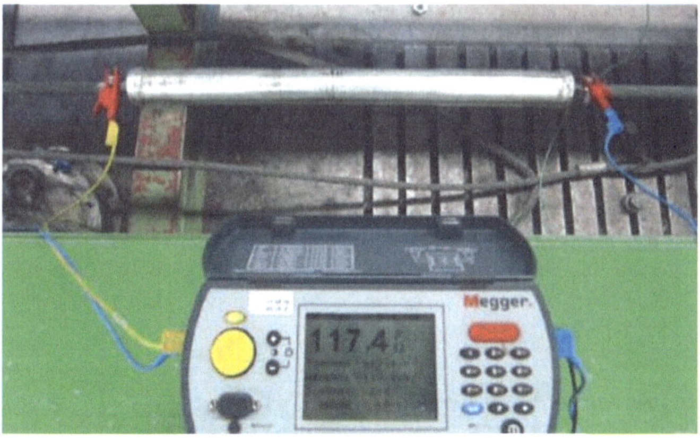

Fig. 6.86 Resistance measurement across a mid-span joint installed on a coated conductor

3. The measurement is done with DC current.
4. The measurement is repeated with the polarity reversed, and the average of the two measured values is taken as the measured value.
5. An acceptance criterion is reviewed, for example that that the resistance across the fitting is less than 75% of the resistance of the equivalent length of the conductor.

The effect of precoating on conductor installation procedures can be different depending on the type of fittings and the coating properties. If the coating needs to be removed prior installing a fitting, it is advisable to test in advance what might be a practical method of removal. For example, removal of the coating may require the use of hand-held wire brushes, fine sandpaper, or battery-powered tools that leave residue or particles that need to be removed for successful installation of fittings.

6.7.3.3 *EIS Tests*

Electrochemical impedance spectroscopy (EIS) and electrical impedance have common aspects. The reactions happening at the interface of a system can be modelled by an equivalent electric circuit. Each component of the electric circuit represents a physical phenomenon. In the EIS technique, a small amplitude signal, usually a voltage between 5 and 50 mV, is applied to a specimen over a wide range of frequencies of 0.001–100,000 Hz. The real (resistance) and imaginary (capacitance) components of the impedance response of the system are measured. Often, data obtained by EIS are expressed graphically in a Bode plot or a Nyquist plot (Jorcin, 2007-03). A suitable RLC electric circuit model is then fitted the shape of the EIS spectrum. By fitting the EIS data, it is possible to obtain a set of parameters which can be correlated with the coating condition and the corrosion of the steel substrate.

Figure 6.87 shows various equivalent electric circuits used. The best-known model used for paint is from Beaunier et al. (1976) as Fig. 6.87e. A first group is for film characteristics such as pore resistance and film capacity. The second group characterizes the mechanisms happening at the interface of paint and steel: $R_{tc:}$ charge transfer, C_{dc}: double layer capacity, and R_e in all circuit models for the electrolyte resistance.

Other electrochemical methods can also be used such as voltammetry, chronoamperometry, or chronopotentiometry. Those methods are stationary measurement which means that either the current or the potential remains constant through the test. These techniques give information more about the thermodynamics of a system. The ASTM standards for the electrochemical methods are ASTM, G59 (2014), ASTM, G106 (2010), and ISO,16773-1, 16773-2, 16773-3,16773-4 (2007).

6.7.3.4 *Insulator Surface Conductance Tests*

The surface conductance of an insulator, which is the ratio of the power frequency current flowing over a wetted insulator surface to voltage apply to the insulator (*I*/*V*), can represent the level of contamination on the insulator surface. The applied

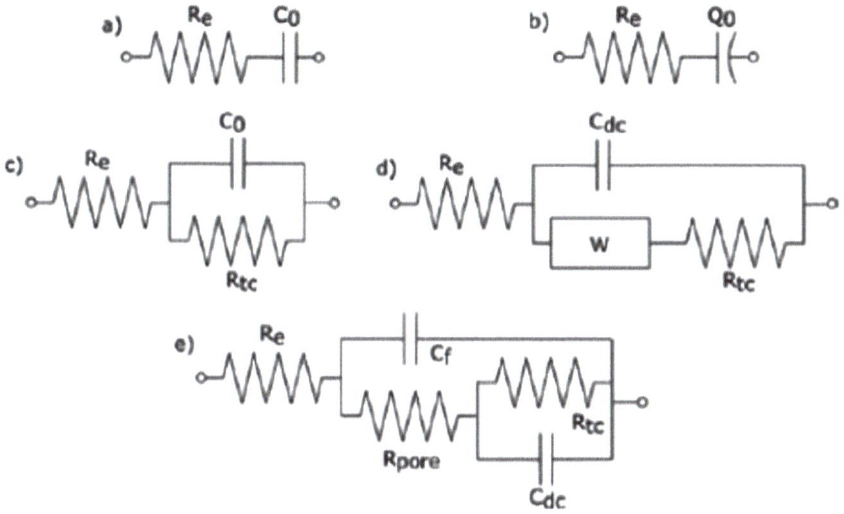

Fig. 6.87 Equivalent electric circuits for fitting results of electrochemical impedance spectroscopy (EIS)

voltage should be high enough to obtain sufficient current. On the other hand, the formation of dry bands and discharge activity should be avoided by energizing the insulator only for a short time duration. In standardized tests steam fog is used to achieve condensation on the insulator in a repeatable way such as IEC 60507 (IEC, 1991).

The use of cold fog (accretion of frost followed by melting) (IEEE DEIS, 2009-10) or simply frosting the insulator with de-ionized water spray both offer suitable technical alternatives for measuring surface conductance in winter conditions. A layer of frost tends to stabilize any pre-existing pollution, but as the melting process is not reversible, the current must be monitored for a maximum value during a melting phase.

Insulator surface conductance is not directly usable in modelling, since it is dependent on the insulator dimensions (i.e. leakage path length, diameter, etc.). It is therefore more useful to express the severity of the contamination in terms of the surface conductivity (IEC, 2008). There are also devices available to measure the surface conductivity directly, as described in IEC 60507 (IEC, 1991).

6.7.4 Thermal Characteristics

Thermal characteristics are obtained by thermal analysis which encompasses a variety of techniques used to measure change in material properties with changes in temperature. The information obtained from these techniques can be divided into three categories (Koleske, 1995) as follows:

- Intrinsic properties, such as glass tension temperature, T_g
- Processing properties
- Product properties.

Some techniques relevant to coatings for winter conditions are described in this section.

6.7.4.1 Differential Scanning Calorimetry (DSC): Heat Difference

DSC measures the difference in heat flow between a sample and a reference under controlled thermal conditions providing quantitative and qualitative data on endothermic and exothermic processes. DSC can be used to measure

- Glass transition temperature T_g
- Melting temperatures
- Heat of fusion
- Cure temperature
- Degree of cure
- Crystalline phase change.

6.7.4.2 Differential Thermal Analysis (DTA): Temperature Difference

DTA is like DSC. In this technique it is the heat flow to the sample and reference that remains the same rather than the temperature. Changes in the sample, either exothermic or endothermic, can be detected relatively to the inert reference. The data obtained from DTA is like that obtained from DSC. DTA can be used to measure:

- Glass transition temperature T_g
- Crystallization
- Melting
- Sublimation

6.7.4.3 Thermogravimetric Analysis (TGA): Mass

TGA measures change in sample mass as a function of time or temperature. It can be interfaced with a mass spectrometer RGA to identify and measure the vapours generated, though there is greater sensitivity in two separate measurements. TG can be used to measure

- Thermal stability
- Accelerated ageing
- Decomposition kinetics
- Oxidation stability.

The relevant ASTM standards for TGA techniques are as follows:

- Standard Test Method for Ignition Loss of Cured Reinforced Resins (ASTM, D2584, 2011)
- Standard Test Method for Compositional Analysis by thermogravimetry (ASTM, E1131, 2014).

6.7.4.4 Thermomechanical Analysis (TMA) or Dilatometry Thermal Analysis (DTA): Dimension

TMA measures changes related to sample dimension as a function of time or temperature. TMA can be used to measure

- Coefficient of linear thermal expansion (α)
- Transition temperature
- Determine degree of crystallinity.

The relevant ASTM standards for TMA techniques are as follows:

- Standard Test Method for Linear Thermal Expansion of Solid Materials by TMA, (ASTM, E831, 2014)
- Standard Test Method for Assignment of the Glass Transition Temperature by TMA (ASTM, E1545, 2011).

6.7.4.5 Dynamic Mechanical Analysis (DMA): Mechanical Stiffness and Damping

DMA, also known as dynamic mechanical spectroscopy, is a technique used to study the viscoelastic behaviour of polymers. In this method, a sinusoidal stress is applied, and the strain in the material is measured, allowing it to determine the complex modulus of the material. The temperature of the sample or the frequency of the stress is often varied, leading to variations in the complex modulus. DMA can be used to measure

- Glass transition temperature
- Transition corresponding to other molecular motions.

6.7.4.6 *Dielectric Thermal Analysis (DEA): Dielectric Permittivity and Loss Factor*

DEA is like DMA. An oscillating electrical field is used instead of a mechanical force. The great advantage of DEA is that it can be employed not only on a laboratory scale, but also during a process.

6.7.4.7 *Freeze–Thaw Test*

A freeze–thaw cycle procedure based on the IEEE 1783 (IEEE DEIS, 2009-10) for electrical strength of iced insulators can be developed and adapted for the deposition of ice on coated insulators. Such a method, as described below, has been developed in China (China Electricity Council, 2019):

(A) Three base pad samples, each with an area of 100 × 100 mm, are cleaned according to the selected process, and the coating material is applied with a thickness of 5 mm.
(B) The coated base pads are placed horizontally in the artificial climate chamber with a temperature of −7 to −5 °C.
(C) De-ionized water is sprayed on the samples for artificial glaze icing as described below (T/CEC 184-2018, 2018-11-06):

- Preparation of spray water: The conductivity of spray water can be measured by the actual collected water on site. In the absence of field data, the conductivity of spray water can be taken as (100 ±15) $\mu S/cm$(when converted to 20 °C), or the conductivity of other spray water can be used according to the test requirements. The conductivity of spray water can be controlled and changed by adding a certain amount of analytical pure NaCl into de-ionized water.
- Pretreatment of spray water: The spray water should be precooled whose temperature should be reduced to below 4 °C before icing.
- Control of rain process:
  - Rainfall: According to the test conditions, continuous spray or intermittent controllable spray can be used, but the continuous rainfall of ice-covered rain should be the same as that of coating artificial rain test. The spray rate can also be controlled according to the actual situation of icing.
  - Wind: Wind can be introduced in the test as a parameter, and the deviation of wind speed should be less than 10%.
  - Direction: When the raindrop (or foggy water drop) formed in the process of raining reaches the position of the test coating, the movement direction should be oblique downward, and the angle with the vertical direction should be 45° ± 10°. The vertical component and the horizontal component of the average rainfall rate measured at the coating location should be similar, that is, they are in the range of 1.0–2.0 mm/min.
- Leakage current control: During the whole icing process, attention should be paid to the detection of leakage current. When the leakage current is greater than 20 mA, the spray time should be reduced, and the freezing time should be increased. If the leakage current is greater than 50 mA, the overall icing pattern could be damaged which should be avoided.

(D) After 6 h of icing, keep freezing for no less than 15 min and then increase the ambient temperature in the artificial climate at the rate of 2–3 K/h to make the ice layer on the sample surface melt naturally.

(E) Repeat the icing and melting cycles 5 times for each sample with an interval of 2 h between each cycle. After the completion of the cycle test, the contact angle of the coated sample surface should be measured in the laboratory environment after natural drying for 24 h, which should not be less than 95% of the initial contact angle after the cycle test.

6.7.5 Mechanical Characteristics

Mechanical characterization of coatings for winter conditions shall be tested according to the following standards:

- Adhesion: (ASTM, D4541, 2002) or ISO 4624 (ISO, 2016) and (ASTM, D3359, 2008) or ISO 2409 (ISO, 2013)
- Tensile strength: ISO 527-3 (ISO, 1995)
- Scratch resistance: ISO 1518 (ISO, 2011)
- Tear resistance: ISO 1520 (ISO, 2006)
- Impact resistance: ISO 6272 (ISO, 2011), ASTM D2794 (ASTM, D2794, 2010)
- Bending elasticity: ISO 1519 (ISO, 2011).

As examples of mechanical characterization according to the above standards, the two following methods are presented for adhesion measurement of a coating on a substrate.

6.7.5.1 *Measuring Adhesion by Tensile Strength*

The adhesion of the coating to the substrate is measured according to the test method of ASTM D4541 (2002). For this test a loading feature, called a dolly, is fixed to the coating with a special adhesive. The adhesion strength of the coating is then determined with a pull-off test using a portable tester, as shown in Fig. 6.88.

The pressure required to separate the dolly from the coating is noted as well as the failure interface. The location of the failure is described by a code based on the following abbreviations:

- A: Substrate
- B: First coat
- C: Second coat
- Y: Adhesive
- Z: Dolly.

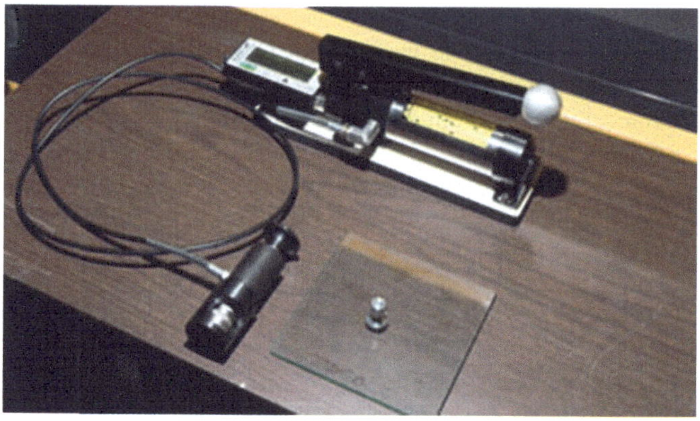

Fig. 6.88 Portable pull-off adhesion tester

In addition, the failure interface is inspected to determine the relative proportions of the different failure interfaces, expressed as % adhesion failure, % cohesion failure, and % glue failure.

This test can only be performed on visible coatings and not on transparent coatings or surface treatments.

6.7.5.2 *Measuring Adhesion by Tape Test*

The adhesion resistance of a visible coating to the substrate is measured according to the test methods of ASTM D3359 (ASTM, D3359, 2008). This test cannot be performed on transparent coatings or surface treatments.

Test method B (i.e. cross-hatch test) is used for assessing thin superhydrophobic and self-cleaning coatings (<125 μm coating thickness). Briefly the method is as follows:

- A right-angled lattice pattern is cut into the coating with a special cross-hatch cutter.
- Adhesive tape is applied over the lattice cuts and then removed.
- The amount of coating removed from the hatch pattern is then evaluated against a standard classification guide.

This appearance of the cut pattern is classified as follows:

- ASTM 5B: The edges of the cuts are completely smooth. None of the squares of the lattice is cut.
- ASTM 4B: Small flakes of the coating are detached at intersections. Less than 5% of the area is affected.
- ASTM 3B: Small flakes of the coating are detached along the edges and at intersections of cuts. The area affected is 5–15% of the lattice.

- ASTM 2B: The coating has flaked along the edges and on parts of the squares. The area affected is 15–35% of the lattice.
- ASTM 1B: The coating has flaked along the edges of cuts in large ribbons, and whole squares have detached. The area affected is 35–65% of the lattice.
- ASTM 0B: Flaking and detachment worse than grade 1B.

For the thicker RTV silicone rubber coatings (>125 μm coating thickness), test method A is used for the assessment. Briefly, the method is as follows:

- Two cuts at an angle of approximately 45° are made into the coating with a special cutter.
- Adhesive tape is applied over the cut intersection and then removed.
- The amount of coating removed from the cut intersection is then evaluated against a standard classification guide.

The visual appearance of the cut pattern is classified on a scale of 0A to 5A as for Method B. Classification 5A is the best and 0A the worst. Typical equipment used for a scratch test on insulator discs coated with SiR, and results of good and bad adherence are found in Fig. 6.89.

(CIGRE WG B2.69, 2021–06) TB 837 provides additional details about the scratch test, including a discussion about whether a single-blade or multiblade tool is more appropriate.

While performing the adhesion test, which may also follow ISO EN 2409 (ISO, 2013), observations should also be made on field- or factory-coated applications about:

- Good adherence, qualified through a scratch test (see Fig. 6.89)
- Smooth aspect without exposed areas and without silicone drops (see Fig. 6.90)
- Good control of thickness control to avoid an excessive amount of coating. Excessive thicknesses should be avoided to ensure that a glass shell would still leave a stub for visual inspection if a unit would shatter. Minimum coating thickness of 250 μm and maximum thicknesses of 400 μm can be considered as good guidelines but manufacturer specifications should always be respected.

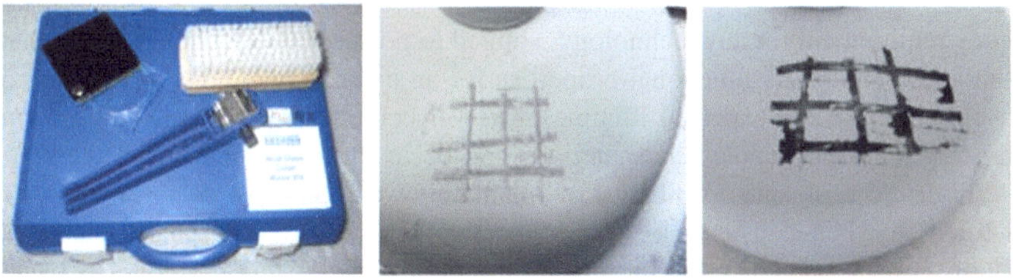

Fig. 6.89 Adherence test (left: test equipment, centre: good adherence, right: bad adherence)

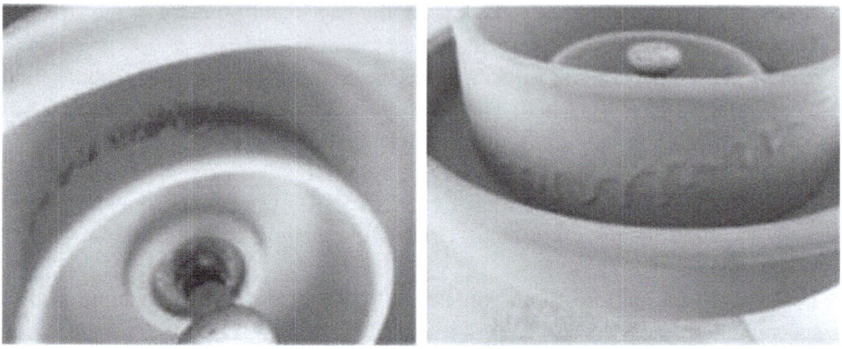

Fig. 6.90 Examples of visual defects (uncoated area, runs) in insulator coating

6.7.6 Concluding Remarks

Extensive testing is required to ensure that coatings have the correct properties to fulfil their intended function. There are many standards already available for such testing at normal temperatures. To test and characterize the performance and durability of anti-icing coatings, several methods have been adopted in different laboratories in the absence of appropriate standards. Although a number of these methods—with some modifications—could be applied to these coatings, further development and are is required.

This section has presented a summary of pertinent characteristics that should be tested. The list includes determination of the intrinsic material properties of the coatings, as well as their electrical and mechanical characteristics. However, there is still a need for further refinement of existing tests and development of new tests, as the strengths and weaknesses of these new coatings are discovered. The durability of the coatings is another important performance aspect which will be discussed next.

6.8 Test Methods for Characterizing the Coating Durability

A key requirement for any technology, applied to power network equipment, is that it should maintain its functional properties for an extended period, certainly when considering that the expected lifetime of power equipment is typically more than 50 years, and the maintenance intervals can be in the order of several years. In addition, the cost and practicality of maintaining or replacing coatings must be factored in when considering their application. The cost of the coating itself is usually low compared to the total cost of applying the coating. Besides, in many cases, it may also be difficult to obtain access to the line or substation for maintenance due to restrictions on planned power outages. Service experience shows that high-quality room temperature vulcanized (RTV) silicone rubber coatings can have a useful life expectancy of 10 years or more. Since RTV coatings are widely used as palliative, their performance can be used as benchmark for new insulator or conductor coatings.

Considering the factors above, there is a clear motivation to subject new coating technologies to extensive durability testing. Environmental, electrical, mechanical wear, and tear stresses all need to be considered.

6.8.1 Environmental Tests

6.8.1.1 *Ultraviolet Tests*

Sunlight is an important cause of damage to coatings. Short-wavelength ultraviolet (UV) light has long been recognized as being responsible for most of this damage. Accelerated weathering testers use a wide variety of light sources to simulate sunlight and the damage that it causes. UV weathering is evaluated by exposing a coated sample to accelerated weathering (UV, humidity, and temperature) for 2 000 h.

Evaluation is based on colour change and gloss retention (ASTM, D4587, 2011).

Xenon Arc UV Testing

The UV resistance of the coatings is tested in the weather-o-metre simulated environmental test according to (ASTM, G155, 2013). UV weathering is evaluated by exposing a coated sample to an accelerated weathering cycle (UV, humidity, and temperature) for 2 000 h. Each cycle consists of 4 h of dry UVA exposure at 60 °C, followed by 4 h of condensation exposure at 40 °C without UV radiation. The light spectrum of this lamp matches closely to that of the sun. The test chamber in operation is shown in Fig. 6.91a.

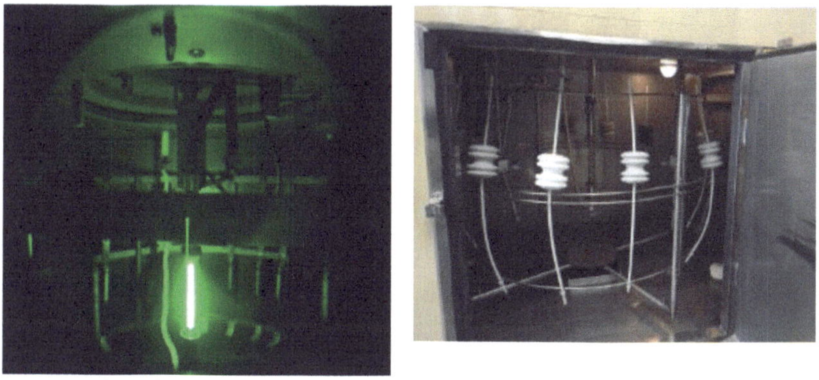

a) UV chamber in operation b) Insulator coating test on small spool insulator

Fig. 6.91 Xenon arc UV test chamber in operation (courtesy of EPRI)

The coating is visually inspected after the test according to the criteria listed in ASTM, G151 (2010) after the test for signs of deterioration such as chalking, cracking, or blistering and changes in hydrophobic properties.

Although this xenon arc test is usually performed on flat, or cylindrical samples, the chambers have sufficient space to test small, coated conductors or small spool insulator—see an example of such a test in Fig. 6.91b.

Fluorescent UV testing

Another UV resistance test for the coatings is the UV-Con test according to ASTM, G154 (2012). The test objects are installed inside a cabinet and are exposed to light from fluorescent lamps and to high humidity for 2 000 h. A photo of the UV-Con test cabinet is shown in Fig. 6.92.

After testing in the cabinet in Fig. 6.92, the coating is visually inspected according to the criteria listed in ASTM, G151 (2010) after the test for signs of deterioration such as chalking, cracking, blistering, chalking, and changes in hydrophobic properties.

6.8.1.2 *Humidity Tests*

The response of the coating to extended periods of high humidity is determined in the humidity chamber test. The coated samples are placed on a flat surface with the coating facing upward in an enclosed chamber. Cold humid air supplied by an ultrasonic humidifier is injected into the chamber to keep the relative humidity at 100% for 100 h. A photo of the chamber and test object placement is shown in Fig. 6.93.

The test sample is visually inspected for deterioration such as debonding, and the surface hydrophobicity is measured after the test.

Fig. 6.92 UV-Con test cabinet

The inside of the humidity chamber

Test object placement in the humidity chamber

Fig. 6.93 Humidity chamber

6.8.1.3 *Temperature Cycling Tests*

The test samples are subjected to temperature cycles to verify the coating's ability to handle temperature changes with a temperature cycle based on (ANSI, C29.2, 1992) with modified temperature maxima. Over a 24 h period, the temperature is increased over 4 h to a temperature of 60 °C after which it is soaked for 8 h, and the temperature is then lowered to −20 °C over 4 h and again soaked for 8 h. The cycle is repeated four times for a total test duration of 96 h.

The test sample is visually inspected, and the surface hydrophobicity is measured after the test.

6.8.1.4 *Boiling Tests*

In many international standards for polymer insulators, surge arresters, and SiR coatings, a boiling test, of 44 h or 100 h duration, is specified (IEEE DEIS, 2018-10). Many surface preparation, application and coating material systems have some difficulty in passing this test, which calls for no blisters, debonding, or discoloration after the test. These observations are intended to be made directly after removing the coated samples from the boiling tank, not several hours or days later.

The coatings are not exposed to temperature of 100 °C during their service life, and indeed other test standards allow longer exposure to water of reduced temperature such as 80 °C.

While the boiling test for adhesion of RTV coating has been criticized for this technical issue (CIGRE WG B2.69, 2021-06) and other weaknesses, it must be noted that the boiling test, with no follow-up scratch test, identified the following application process defects:

- Wiping the insulator surface with the RTV coating native solvent and paper towels (leaving fibres on the surface that were the origin of blisters), as recommended by a coating supplier

- Spraying the insulator surfaces using conventional paint guns
- Reuse of paint guns after they had been contaminated with the silicone product.

Systems that did pass a 100 h boiling test included

- Cleaning insulator surfaces with silicone-free auto body shop products
- Rinsing either with de-ionized water or a specialized insulator cleaning solution and air drying
- Spraying the insulator surfaces using a high-volume, low-pressure (HVLP) system that pressurized the coating material, as recommended by a coating supplier.

Substrates that passed the 100 h boiling test for SiR adhesion include new and aged porcelain, aged EPDM, and an insulating polymer for 600 V DC traction systems with more than 60% ATH filler.

The "systems"—cleaning, application, and material—that passed the 100 h boiling test in 1994 produced SiR coatings, in situ at substations, that are still in good condition after 27 years of service that included high levels of winter pollution and road salt. If correctly installed coatings from the 1990s were able to achieve this long service life in adverse winter conditions, there is a reasonable expectation that today's improved application processes and materials will give their desired anti-corrosion and anti-icing properties over a similar service life, especially if they also pass a boiling test.

6.8.1.5 *Heat Stability Tests*

Heat stability resistance is assessed by evaluating the behaviour of coatings on flat or bent samples, subjected to accelerated ageing by temperature. The samples are maintained continuously in a dry atmosphere at a defined temperature.

Evaluation is based on colour change and/or physical–chemical properties of coating (ASTM, D5324, 2010).

6.8.2 Other Environmental Tests

A resistance of the coating shall be verified by the sequence of related tests:

- Resistance to SO_2 shall be tested according to the ISO 3231 (ISO, 1993).
- Resistance to salt fog shall be tested according to the ISO 9227 (ISO, 2012).
- Resistance to cyclic corrosion shall be tested according to the ISO 11997 (ISO, 2005).
- Resistance to abrasive wear shall be tested according to the ISO 7784 (ISO, 2016).
- Resistance to ozone shall be tested according to the ISO 1431-1 (ISO, 2012).

6.8.3 Electrical Tests

6.8.3.1 *Inclined Plane Test*

The tracking and erosion performance of the coatings when energized are tested with the inclined plane test as described in ASTM, D2303 (2004) or IEC 60587 (IEC, 2007). For this test the coated sample is installed at an angle of 45° with clamp on electrodes installed at the top and bottom of the sample, as seen in Fig. 6.94.

A mixture of salt water and a wetting agent is streamed across the sample between the electrodes. The contamination solution is pumped at a constant rate of 0.075 ml/min. The standards describe two energization options:

1. **Constant voltage method**, where the sample is energized at a predetermined, but constant voltage level during the whole test duration of 6 h
2. **Stepwise voltage method**, where the voltage is increased every hour by 250 V in a stepwise manner. The starting voltage is selected to represent an appropriate stress level. In this case, the test duration is 4 h.

Since this test has been developed for bulk insulation materials some adjustment to the test voltage used is necessary to accommodate the much smaller material thickness of these coatings. In general, this means that a lower test voltage is used for the tests on coatings.

The performance of the coating is measured by the time it takes for tracking or for erosion to extend across half the sample length (1 in.).

6.8.3.2 *Corona Exposure Test*

Corona exposure stress is considered as an important factor affecting the life of a coating. Important ageing factors are as follows:

Fig. 6.94 Inclined plane test (courtesy of EPRI)

- Ozone, which can be studied independently from corona. CIGRE WG D1.14 (2011-10) set up a project group dealing with ozone and combined ozone and corona evaluation. The subject is still under consideration.
- UV light associated with corona is in the UV-B range that has a shorter wavelength than that of sunlight. This UV exposure may directly age the material.
- There is also some evidence to suggest that corona in combination with water may result in an acid that may chemically attack the insulator or conductor coatings.

In this non-standard test, the coated test samples are continuously irradiated with a strong source of both positive and negative corona. A photo of the electrode set-up and corona is shown in Fig. 6.95. The gap between the energized electrode and the top of the coating surface is 2 mm. The test duration is 1 000 h.

The test samples are evaluated after each 250 h interval by visual examination and a hydrophobicity test.

6.8.3.3 *Tracking and Erosion Test*

Coatings on Insulators

The tracking and erosion resistance of insulating material is tested by the 1 000 h salt fog test specified by IEC 62217 (IEC, 2012) for composite insulators (IEC, 2008) which can be a good candidate for tracking and erosion test for coatings is a continuous stress test. Composite insulators are energized to a unified specific creepage distance (USCD) of 34.6 mm/kV and subjected to a standard salt fog of salinity between 1 and 10 kg/m^3 (depending on the geometry) for 1 000 h. The salinity is adjusted to avoid flashovers and to increase the erosion stress. This test is prescribed by IEC standard 62217 (IEC, 2012). The principles of this test can be

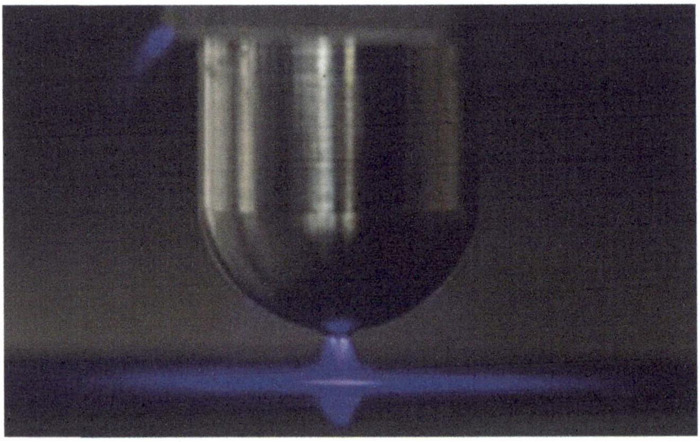

Fig. 6.95 Corona exposure test (courtesy of EPRI)

Fig. 6.96 General view of a 1 000 h salt fog chamber (courtesy of EPRI)

applied to coatings by adjusting electrical and salinity stresses for this specific application. A general view of a salt fog chamber is shown in Fig. 6.96.

Coatings for Conductors

The principles of the tracking and erosion tests of by IEC 62217 (IEC, 2012) have also been used as a basis for testing-coated conductor samples. The IEC test method is modified to include visual and hydrophobicity inspections (Gutman et al., 2013). The same salinity of 8 kg/m^3 is used. The visual inspections and wettability class (WC) measurements are done in accordance with IEC TS 62073 (IEC, 2003) before, during (at 500 h), and after the test. The test is carried out in a moisture-sealed corrosion-proof chamber with a volume of about 45 m^3.

A RTV- and a superhydrophobic-coated conductor were tested. The energized test conductors were suspended from the roof of the chamber by composite line insulators as shown in Fig. 6.97.

The applied voltage in Fig. 6.97 was selected to be sufficiently high enough to have initiated corona discharge on the surface of conductors (Gutman et al., 2013).

Both coatings passed this 1 000 h test without visible traces of deterioration; the adhesion of the coatings to the conductors was still good. However, it was found that the superhydrophobic coating lost its hydrophobicity within a few days of testing and the RTV one after 3 weeks. The recovery of the hydrophobicity was checked by performing a measurement at about 4 weeks after the end the test. By this time, the RTV coating showed almost complete recovery (WC 2-3), and the superhydrophobic conductor had only recovered (WC 2-3) it over part of its length, while other parts remain completely hydrophilic (WC 7).

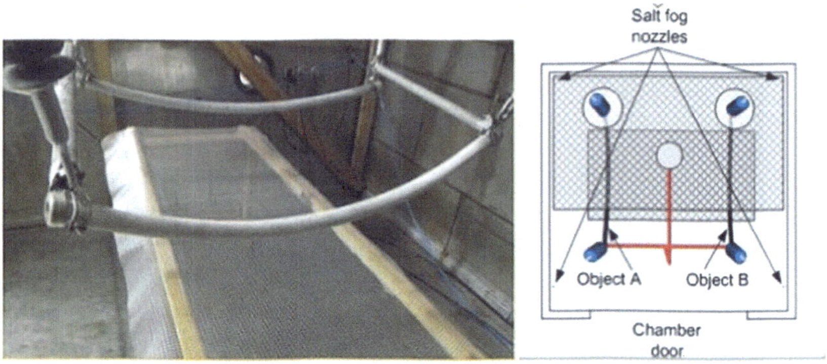

a) General picture of the test chamber b)schematic layout of the test chamber

Fig. 6.97 Set-up for 1000 h salt fog tracking and erosion test (Gutman et al., 2013)

6.8.4 Mechanical Wear and Tear Testing

6.8.4.1 *Scratch Test* (MAR *Testing*)

The susceptibility of the coating to scratching, marring, and similar physical damage due to handling is evaluated with the Taber Multi-Finger Scratch/Mar Tester according to ASTM, D5178 (2013). Figure 6.98 shows a picture of the test apparatus.

For the test, the specimen is secured to a pneumatically driven platform underneath five independent spline-shaft fingers, each loaded with a different weight. The interchangeable scratch tips exert a constant, vertical load on the

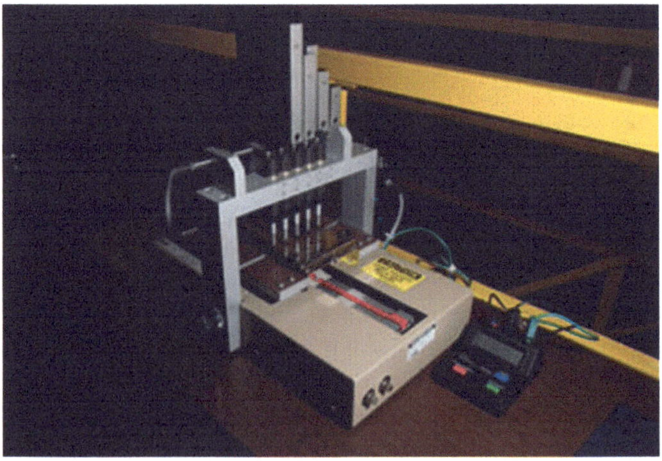

Fig. 6.98 Taber Multi-Finger Scratch/Mar Tester (courtesy of EPRI)

coating while the platform moves to produce scratches of varying depths in the coating surface.

The coating is evaluated by measuring the depths of the scratches left in the coating after the test.

6.8.4.2 *Abrasion Test*

The abrasion resistance of the coating is tested using the Taber Rotary Abraser as shown in Fig. 6.99.

During this test, which is described in ASTM, D4060 (2010), the coated test specimen is mounted onto a rotating turntable and then subjected to the rub wear action of two abrasive wheels. Driven by the test sample, the wheels produce abrasion marks that form an abrasion pattern in the shape a circular path. The number of turntable revolutions required to completely remove the coating is noted as the test result.

6.8.4.3 Gravelometer

The resistance of the coating to chipping caused by impact of gravel and other debris is tested with the gravelometer as per ASTM, D3170 (2012). The coated test specimen is mounted in the back of the gravelometer, and air pressure is used to eject 1 pint of gravel towards the sample. Then, the test object is removed and gently wiped with a clean cloth. Adhesive tape is then applied onto the entire tested surface and removed to capture any loose fragments of the coating. The appearance of the tested sample is then evaluated against reference pictures to determine the chipping ratings.

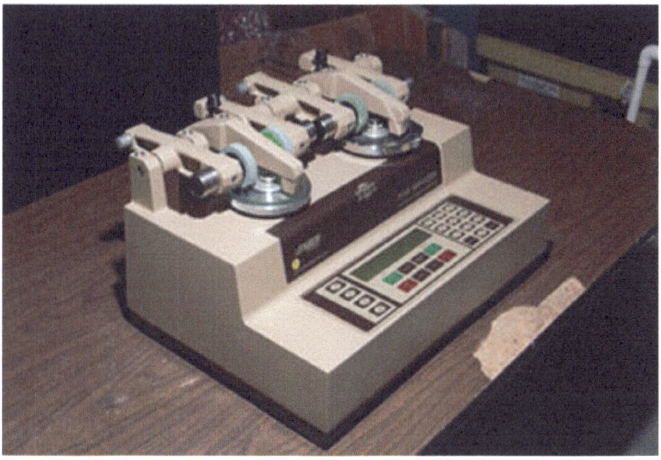

Fig. 6.99 Taber Rotary Abraser (courtesy of EPRI)

6.8.4.4 *Conductor* Pull Test

This test is intended to simulate mechanical stress during tension stringing. This test need not be applied to conductors where self-de-icing properties are obtained by a coating applied after installation of the conductor.

The test object for such tests is a conductor with a length of 12 m, where at least 100 cm of the mid-section has the self-de-icing properties intended to be tested (Fig. 6.100) (Petersson, 2015).

The test conductor shall be dragged five times back and forth over a wheel of diameter D lined with hard rubber, while loaded by a force F acting φ degrees relative to the pulling direction. The force F corresponds to 20% of the conductor's rated tensile strength. The diameter D is either 150 mm or is determined based on the diameter of the running block wheels intended to be used. The angle φ is 25° or determined based on the maximum bending angle expected over the running blocks in the first or last tower of the stringing section. Figure 6.101 shows a schematic of the set-up. Other arrangements that give the same stress may be used.

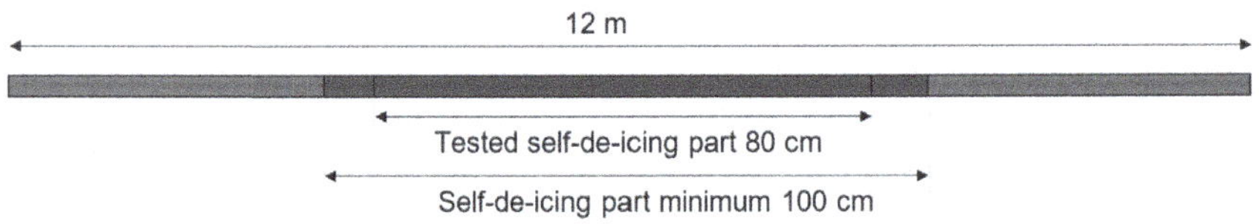

Fig. 6.100 Conductor sample for serviceability testing (Gutman et al., 2012)

Fig. 6.101 Principal set-up for mechanical test (Petersson, 2015)

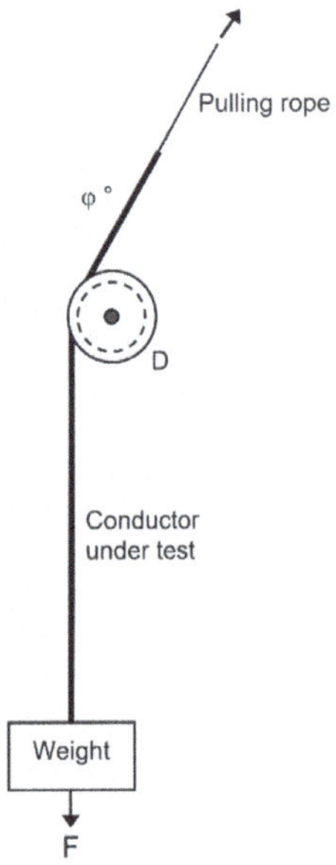

Surface hydrophobicity is visually confirmed after each test. Any visual damage or other degradation during the test is noted. The basic acceptance criterion is that the surface should maintain its hydrophobic properties after the final pull test.

6.8.5 Load Cycle Test

The conductor under test is tensioned by 20% of its rated breaking load at 20 °C. Wedge-type clamps may be used as grip ends. The conductor is heated by circulating a current through the conductor.

For this test the conductor is typically subjected to ten or more load cycles. The load cycle is shown in Fig. 6.102 and comprises the following five stages (Petersson, 2015):

1. The conductor is heated by current to 100 °C.
2. It is maintained at this temperature for 1 h.
3. The temperature is increased as fast as possible to 200 °C.
4. The load current is switched off to allow the conductor to cool down.
5. The next cycle is started when the conductor temperature is less than 10 K above the ambient.

The load cycle in Fig. 6.102 is repeated at least ten times.

6.8.6 Concluding Remarks

Durability is an important characteristic that needs to be verified to ensure that the coating will fulfil its intended purpose for its expected lifetime. This should be done by considering appropriate tests representative of the environmental, electrical, and mechanical stresses as recommended in this chapter. In some cases, these are already standardized test methods which are also considered appropriate for the coatings we have discussed. Other existing standards may need to be modified or adjusted to suit the specific characteristics of these coatings. In other cases, notably, for checking the electrical durability and ageing of the coatings there are no standard test methods available. A few possible tests methods are presented in this chapter to stimulate further development in view of standardization.

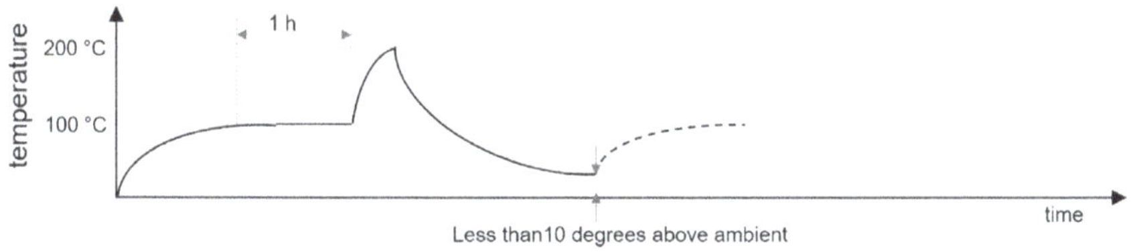

Fig. 6.102 Illustration of one load cycle

6.9 Recommendations for Coating Requirements and Testing

At present, there are no standards that directly apply to hydrophobic or icephobic coatings. Various parameters still need to be identified and investigated to develop test methods to qualify such coatings. In this respect, existing knowledge and ongoing work by laboratories and manufacturers constitute an important basis.

In the following sections, methods and existing testing techniques for these criteria are explained.

6.9.1 Requirements for Coatings

An approach to identify requirements and suitable test methods comprises the following steps:

- Identify suitable functional performance requirements
- Determine the primary (important) and secondary (less important) degradation modes
- Design and implement a set of tests to evaluate performance.

6.9.1.1 Functional Performance Requirement

The main functional performance requirement for coatings for protecting power network equipment in winter conditions is as follows:

- Functional requirement, i.e. icephobicity, hydrophobicity and self-cleaning
- Durability or longevity
- Cost effectiveness
- Ease of application

6.9.1.2 *Degradation Stresses*

The coatings discussed in this document are for application to the external—exposed—surfaces of power network equipment. These coatings should, therefore, be able to endure environmental stresses they are exposed to for their expected lifetime. These include exposure to

- Ultraviolet (UV) radiation from the sun
- Diurnal and seasonal temperature changes and the corresponding forces caused by thermal expansion and contraction
- Humidity and precipitation
- Aggressive chemical compounds that may be present, such as those in the contamination layer and acid rain.

Coatings should also be able to withstand the wear and tear associated with the handling, installation, and maintenance of the equipment involved. Examples are abrasion from tools, carts, and persons climbing along conductors and insulators when performing installation and maintenance tasks on components.

An additional complication of applying coatings to energized components is that it must be able to withstand the electrical stresses that it will be subjected to in combination with the environmental stresses. Coatings on conductors and insulators are continuously subjected to high electric fields that could result in an accelerated deterioration or even failure. It may also be expected that the coating will be subjected to some form of electrical discharge activity such as arcing, sparking, or corona.

Another principle that should be adhered to is that the application of the coating should not, in any way, negatively affect the performance of the system. This is specifically a concern if the coating should fail or lose its functionality. For example, a common problem with organic materials applied to insulating surfaces is the formation of tracking paths when the coating breaks down. Tracking is a conductive or partly conductive path which permanently degrades the electrical strength of the coating. In severe cases, the tracking on the insulating surface may have developed to such an extent that the apparatus can no longer withstand the operating voltage, leading to a long-term outage. In such cases, the power can only be restored after the affected equipment is removed or the coating from it. Such long-term outage may have large cost implications.

Based on these anticipated stresses a few key questions have been formulated to facilitate the introduction of new coating technologies to the power system.

6.9.1.3 *Identification of Suitable Tests*

To minimize risk and cost, a three-tier approach should be followed in the evaluation of coatings. This approach is illustrated in Fig. 6.103.

The tiers in Fig. 6.103 are defined by EPRI as follows:

- ***Tier 1*** Small-scale testing on coated material samples, e.g. glass or aluminium: These tests normally apply a single stress to small, standardized material samples. Examples of Tier 1 samples for insulator coatings are shown in Figs. 6.104a and 6.105a.
- ***Tier 2*** Laboratory testing on coated components, e.g. conductors or insulators: During these tests a limited number of stresses are applied to realistic samples in a controlled environment. An example of the small insulators used in Tier 2 tests is shown in Fig. 6.104b and Fig. 6.105b.
- ***Tier 3*** Field demonstrations at utility sites: A limited number of samples, applied to real equipment, are exposed to a typical service environment for example for one year. Normally, the services stresses are also monitored and recorded during the trials. In some cases, it may also be necessary to implement special protection measures to ensure that the power system is not adversely affected if a test sample fails.

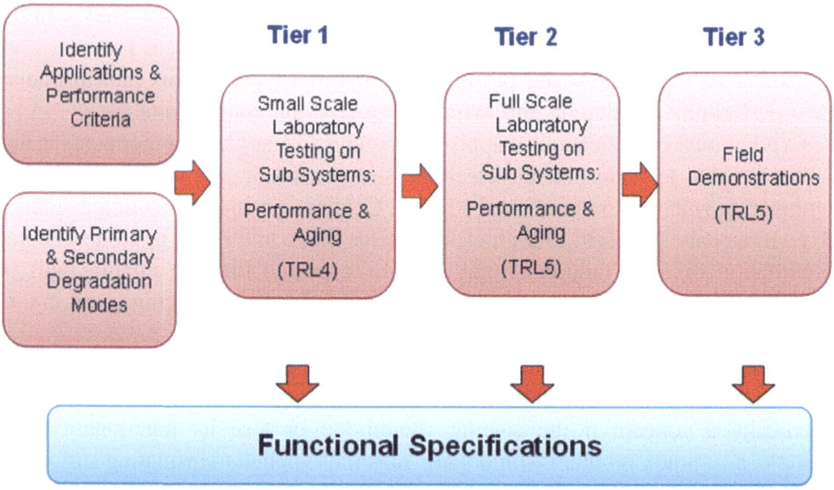

Fig. 6.103 Tiered approach to testing and development of functional specifications (EPRI, 2012)

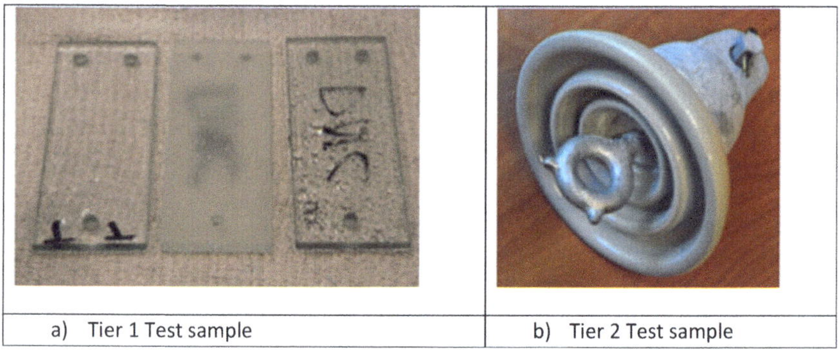

Fig. 6.104 Examples of samples used in the tiered approach for insulator coatings (EPRI, 2012)

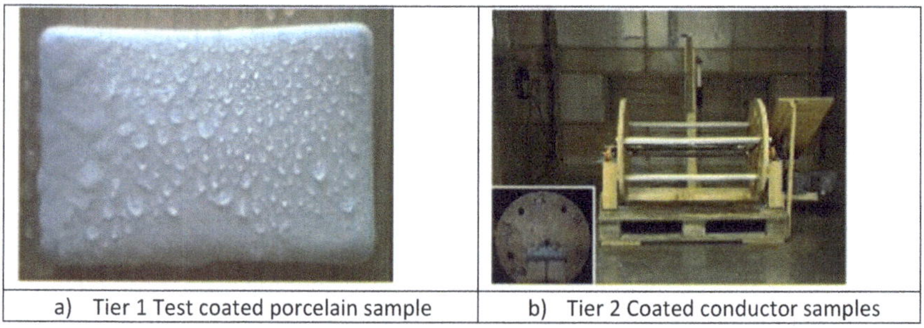

Fig. 6.105 Examples of samples used in the tests for insulator/conductor coatings (courtesy of STRI)

This approach not only evaluates potential test methods for the coating technologies to address transmission applications, but it also provides information about the performance of these coatings and will be of help in the future development of functional specifications that utilities can use to provide high-performance products. Basic requirements that test methods should meet to successfully qualify products include TB 142 (CIGRE TF 33.04.07, 1999).

- **Representative of service**: The conditions and stresses applied to the test subject should mimic those the material will be subjected to in service. This should be checked by verifying that failure modes observed during the tests are the same as those occurring under service conditions.
- **Repeatable**: It should be possible to define and control all stresses applied to the sample so that repeated testing would produce the results within an acceptable variance. In test set-ups with multiple samples, the outcome of the test should be influenced by the test position.
- **Reproducible**: The test methods should be defined and described in such a manner that comparable results are achieved if the test is performed by different laboratories. This can be achieved by ensuring that the test method is described in detail and that calibration procedures are in place to ensure the consistency of key parameters.

6.9.2 Concluding Remarks

The coatings discussed this book should be able to endure the environmental stresses they are exposed to for the duration of their expected design lifetime. This includes ultraviolet (UV) radiation from the sun, diurnal, and seasonal temperature changes and the coating stresses resulting from the thermal expansion and contraction of its substrate, humid environments, and exposure to various types of precipitation and chemical compounds that may be present. Coatings should also be able to withstand the wear and tear associated with the handling, installation, and maintenance of the equipment involved. It should also be taken into consideration that coatings on conductors and insulators are subjected to high electric fields, corona, sparking, and arcing that could result in an accelerated deterioration or even failure of the coating. Moreover, the application of the coating should not, in any way, degrade the performance of the system.

In this chapter, we introduced an approach to help to identify the testing methods to facilitate the evaluation of new coating before applying to the power network equipment. Based on the findings within, key questions and guidelines regarding the anticipated performance requirements of coatings for improved winter performance have been formulated.

7 Power System Reconfiguration Options for Anti- and De-Icing

This chapter reviews some reconfiguration options for protecting electric power systems before, during, and after adverse winter weather.

Loss of a single- or double-circuit line is common in electric power system operations and is considered a "first" or $(N - 1)$ contingency. For example, a double-circuit fault to ground from back flashover is expected on the phases whose instantaneous voltage opposes the polarity of the lightning flash. Double-circuit faults to ground on shielded transmission lines are detected quickly, and high-speed reclosing is successful in nearly all cases.

In contrast, ice and snow may lead to a sequence of several faults or may lead to a permanent fault that cannot (or should not) be reclosed if there is severe mechanical damage.

Both AC and DC can be used to heat line conductors for Joule effect anti-icing and de-icing, using many different power system reconfiguration options (Farzaneh et al., 2008). Using AC may involve lower equipment cost when the ice-melting current can be supplied directly from the existing network. However, to obtain the necessary value of current for effective melting in a limited time of 0.5–1 h, the melting voltage and corresponding total melting power must be sufficiently high, especially with long transmission lines. If the length of the line being heated and the required melting current and voltage are relatively small, AC can be successfully used. Melting of ice by DC is more advantageous for long, high power lines with conductors of large cross sections even when it requires purchase of rectifiers. For 220 kV lines with a single conductor rather than bundles of subconductors, the capacity of DC power supply for ice management is about 20% of the required AC power supply.

The power system reconfiguration options for de-icing of AC transmission lines have three important classifiers:

M. Farzaneh and W. A. Chisholm, *Techniques for Protecting Overhead Lines in Winter Conditions*, Compact Studies, https://doi.org/10.1007/978-3-030-87455-1_7

- Those that use DC rectifiers rather than AC sources
- Those using earth return, rather than return of de-icing current through other phase conductors
- Those that maintain normal AC system operation, using series capacitors or one-phase-open configurations, compared to those that take the AC line out of service and isolate it from the network.

7.1 Generation and Load Rejection Schemes

One disadvantage of Joule effect methods for anti- and de-icing is that they usually require reconfigurations of the power system that represent single or multiple contingencies. Any line removed from normal service sets a single contingency in reliability assessment, and in icing conditions, multiple, simultaneous contingencies may also be anticipated. The isolated circuit is energized from a temporary connection to a source of current for generating I^2R heating in the conductors. Despite the complexity, power system reconfiguration has been accepted as a necessary process in many icing regions.

Under deliberate $(N - 1)$ contingency conditions, such as isolation of a line for de-icing, many utilities plan for additional contingencies with some form of "load and generation rejection" (LGR). When significant load loss is detected, there is a simultaneous adjustment of generation to minimize over-frequency swing, increase stability across critical interfaces and avoid islanding. Arming of LGR schemes can mitigate some of the reliability concerns with Joule anti- and de-icing methods.

In addition to loss of individual lines or components, some Joule de-icing methods make use of forced loading, phase shifting, and short-circuiting or open-phase operations that are not part of normal operation. Faults under these conditions may be difficult to detect. For example, a paper Hogan and Pebler (1955-01) presents the problem of line protection under conditions of sleet melting (forced loading, phase shifting, or short-circuiting) and proposes a means of detecting faults on lines being ice-melted by the short-circuit method. The source of power for the short-circuiting method may be an isolated generator with a step-up transformer that may be adjusted to supply the proper current. Some lines because of their length may lend themselves to ice melting by direct connection to a station bus. Under these conditions, the normal relay settings cannot generally be set to give adequate fault protection. In the instance where several lines are connected in series for sleet melting, the relays at the sectionalizing points between extreme terminals must be tagged out of service, to be restored in a documented process after the de-icing is complete.

7.2 Reconfigurations for Joule Effect Anti- and De-Icing with AC

7.2.1 Sleet Bus Methods

In the period after construction of transmission lines and before adoption of line design criteria such as 50-year wind and ice loads, many electric utilities operated transmission lines with marginal ice load capability with the use of so-called sleet bus arrangements. Under icing conditions, threatened circuits were removed from normal service and connected to an independent electric supply. Current derived from the sleet bus was circulated, either to reduce the level of accretion during storms or to remove it afterwards. The sleet bus approach was used for both copper and aluminium phase conductors.

Icing precipitation was a concern to Pennsylvania Water and Power Company since 1910 (Shealy et al., 1952-08). The evolution in combating sleet on line conductors from simply heating a line was described as: (i) forecasting the appearance of ice formations; (ii) detecting accumulations on conductors; and (iii) applying heating current to the conductors to remove the ice.

Another early experience involving ice melting occurred on the New England Company power network during a storm in 1920 (Oliver, 1925-11). The line conductor size was 1/0 AWG copper (8.25 mm diameter) with a cross section of 53.5 mm^2 (106 kcmil). After six days of ineffective work trying to clear ice accretion from their lines, this utility organized a temporary configuration of 2.3 kV/220 V transformers in series and parallel to provide 6–8 kV and 300 A, for a current density of about 2.8 A/kcmil. This is equivalent to a current density of about 1.7 A/kcmil in an aluminium conductor with higher resistance per km than copper of the same size.

Much to the surprise of all, Joule de-icing with 2.8 A/kcmil current density worked amazingly well, removing the ice but not damaging the conductor by overheating. Improvements to the de-icing system were made during remainder of the winter. The total apparatus tied up for sleet thawing consisted of one 60-cycle, 2.3 kV generator, and two three-phase transformers. One of many conclusions of Oliver (1925-11) is that "wherever possible it is advisable to install a thawing device if it can be done at a reasonable figure…It will pay for itself many times on transmission lines thus protected from the sleet storm".

7.2.2 Load Shifting Method with Generation and Load Rejection

Experience with accreted ice conditions on the Niagara Mohawk System indicated that adequate planning is necessary if satisfactory results are to be secured from any ice melting program (Smith & Wilder, 1952-08). The planning extends from system design to operating procedures, to training. All must be coordinated to weather a severe storm successfully.

The Niagara Mohawk System recognized three principal methods of icing prevention or ice melting:

- **Load current method**: The heating effect of load currents is used to prevent or remove ice on the conductor. Where necessary, normal operating conditions are modified to force more load current through a particular circuit. The use of this procedure is emphasized since switching is performed at manually operated stations, no markups are necessary, and a minimum of time is required to make the procedure operative.
- **Short-on-a-generator method**: One or more generators are isolated from the rest of the system and connected to a circuit, which has also been isolated. The three phases of this isolated circuit are short-circuited at the far end. The generator field is gradually increased (from a subnormal value) to a value sufficient to produce the desired melting current.
- **Short-on-a-bus method**: One or more transmission circuits are isolated from the rest of the system. The total length of these transmission circuits is predetermined so that, with normal operating voltage applied at one end and a three-phase short circuit established at the far end, the desired ice melting current will flow. Then, having established the three-phase fault, the near end is connected to its regular bus (at normal voltage) through a circuit breaker.

The following reconfiguration procedures were formalized by Niagara Mohawk:

- The system connections were changed as indicated by the AC network analyser studies. In this instance, a complete auxiliary bus was added in a station to facilitate ice melting. The final decision as to design of two new 115 kV generating station buses was influenced by their adaptability for ice melting purposes.
- Step-by-step instructions and detailed drawings were developed for the various ice melting procedures.
- Ice melting tables were provided for the operating personnel. In those tables, the following information is given for the system operators:
  - **For ice prevention** Conductors maintained at a temperature slightly above freezing would not accumulate ice. The current necessary to maintain this temperature depends on air temperature, wind velocity, conductor cross section, and conductivity.
  - **For ice melting** Once ice has formed on the conductors, it is necessary to go to sleet-melting currents, which, obviously, will be greater than sleet-preventing currents. The required current depends on thickness of ice formation, air temperature, wind velocity, amount of time available to remove the ice, conductor cross-sectional area, and conductivity.

The load shifting method requires no additional equipment on the system and consists of using the heating effect of load currents to prevent conductor icing or to remove ice from conductors. However, the current carried by HV lines is generally not sufficient to produce enough heat to prevent or melt ice. Normal operating conditions must be modified in order to force more load current through a particular circuit by transferring or shifting loads from other circuits linking the same two substations (Polhman & Landers, 1982; Prud'Homme et al., 2005b). Hence, if the load is high enough, depending on the ambient temperature and wind speed, the current in the remaining circuit will induce ice melting. This method is suitable for single conductor lines as bundled conductors require too much current. One problem with this method resides in the difficulty of controlling the current flow during the de-icing period which is mainly determined by the power demand of customers. Loss of power load could lead to de-icing failure, whereas too much power load could lead to the overheating of conductors. Moreover, the power load available must be in accordance with the climatic conditions in the area where de-icing must take place. For these reasons and considering the large number of parameters to consider, to be efficient, the load shifting method requires a well-defined de-icing strategy and decision tools (Huneault et al., 2005-04b; Prud'Homme et al., 2005a).

7.2.3 Short-Circuit Methods

In both the former Soviet Union and North America, ice melting on all three phases using short-circuit methods offers the advantage of being the most expedient. Five reconfiguration schemes in Fig. 7.1 have been used in the former USSR for ice-melting.

Melting by method of short circuit in Fig. 7.1a short-circuits the line at one end. The other end is connected to the voltage source. This applies the required melting current, possibly working through busbars, transformers, etc.

Melting by the method of phase opposition (Fig. 7.1b, c) reconfigures the conductors of the heated line to different phases of supply sources at each substation. Each phase of the line will be then at $U_L = \sqrt{3}U_{\mathrm{ph}}$, i.e. phase voltage will be 73% higher, which results in higher melting currents. This approach can de-ice longer lines but must be used with care to ensure that any equipment and arrester ratings are adequate.

Melting by method of two-phase short-circuit (Fig. 7.1d) and series connection of line phases (Fig. 7.1e) can be suitable for de-icing shorter lines.

In large EHV systems, for example, 500 kV networks, melting by AC at the rated voltage can be applied. The line being heated is short-circuited at the end and is connected directly to 500 kV bus as shown in Fig. 7.1a. In this configuration, the line operates as a high-power reactor. Any shunt reactors must be disconnected. The inductive melting current is compensated by the capacitive network current, and the resulting load of the supply source can be relatively small. In some cases, the existing sources can supply only high melting currents, which can result in

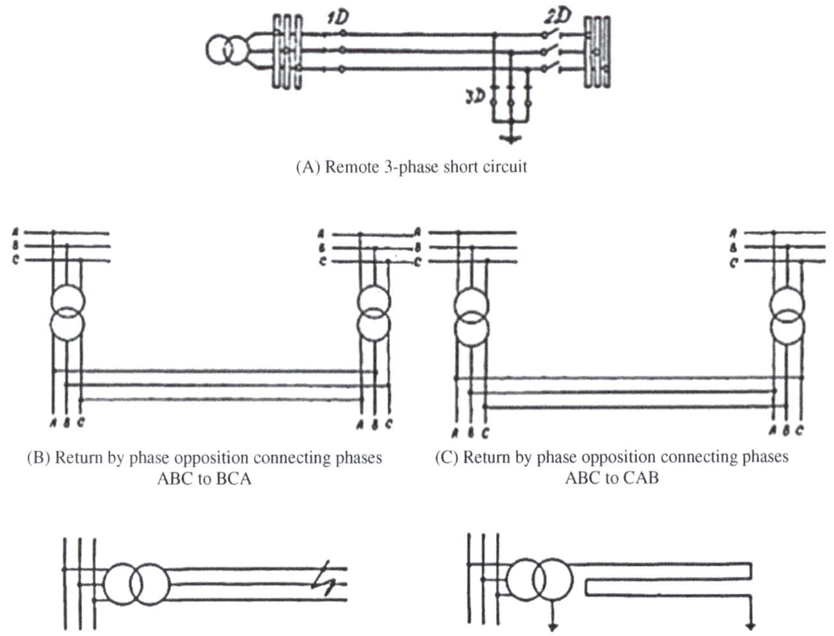

(A) Remote 3-phase short circuit

(B) Return by phase opposition connecting phases ABC to BCA

(C) Return by phase opposition connecting phases ABC to CAB

(D) Remote 2-phase short circuit

(E) "Snake" scheme with remote 1-phase short circuit

Fig. 7.1 Schemes for off-line AC ice melting using short-circuit methods in the former Soviet Union (Timoshova et al., 2003)

conductor overheating if applied for a long time. In such cases, the method of heavy intermittent currents is used for melting. Heavy current is supplied to the line during a certain time, well below the thermal time constant of the conductors (typically 10 min). An appropriate duty cycle, such as 3–5 min on, 3–10 min off, gives heating and cooling cycles that are repeated several times until the ice falls off.

7.2.4 Reduced Voltage Short-Circuit Methods

Many electric power utilities in the world have some experience with short-circuit heating. For instance, in the early 1970s, Manitoba Hydro began using three-phase short circuits to melt ice as an experimental procedure (Adolphe, 1992-11). Today, they have the capability to de-ice several thousands of kilometres of distribution and subtransmission lines with conductors ranging in size up to 218 mm^2 (336.4 kcmil) ACSR. Currently, 90 stations, at 33, 66, and 115 kV, are equipped for short-circuiting. Ice melting is routinely carried out by Manitoba Hydro, not only during severe widespread ice storms, but also during less severe weather conditions, as a preventive measure against the slow build-up of ice on conductors. The achieved short-circuit current is a function of the applied voltage, circuit length, and conductor electrical characteristics. As an approximate rule, for applied voltages of 12, 25, or 69 kV, it is possible to get the required ice-melting current for circuit lengths of 12, 25, or 69 km, respectively, on Hydro-Québec's single conductor lines, within a margin of 15% (Gingras et al., 2000-08). The short-circuit de-icing method requires some equipment to be added in advance. Switches on the short-circuit side may be needed to produce a three-phase fault, which may be isolated from ground if there are no pre-existing ground switches. On the source side, switches and connections are required to power the lines to be de-iced. Also, the overload capability of existing equipment may need to be increased, and system protection devices reconfigured to support the flow of de-icing current without interruption.

7.2.5 De-Icing with Phase Shifting Transformers

Most of the Joule effect de-icing methods mentioned require disconnecting the sections to be de-iced from the network. To overcome this problem, phase shifting transformers can maintain normal operation while increasing the current in a selected circuit.

A paper by Oehlwein (1953-12) presents the phase shift method with emphasis on 60° shift that can be obtained conveniently by reconfiguration of transformer connections. The system operator must have a clear picture of the problems involved to correctly apply the phase shift method of ice melting. Considerations include provision adequate active- and reactive-power capacity as well as thermal overload limitations on both line and station equipment. The system operation must be arranged for varying conditions due to storm location and line availability, being always mindful of a primary responsibility to keep the power flowing.

This de-icing approach was used in the USA and described by Ekstrom (1958-04). Transformer windings were reconnected to provide 60° phase shift to mitigate ice formation on 34.5 kV lines in a limited 40–46 km range of line length.

Fixed angle (60°) phase shifting provided limited benefit, however, as it could only be used on 34.5 kV lines of only 25 miles (40–56 km) in length, depending upon design impedance and conductor size. In general, the greater the length, the greater the impedance will be, eventually reducing the circulating current to a value too low to produce sufficient current to melt the ice at reasonable length of time. Adding many of short lines in series like in the more densely populated areas to achieve the minimum length of 25 miles (40 km) involved a greater liability in the amount of load and number of customers affected, hence the need for different phase shift angles.

Figure 6.40a–i from Ekstrom (1958-04) shows how phase shifts of 19°, ±41°, 79°, and 101° can be achieved by reconfiguring a three-winding closed-tertiary transformer bank or polyphase transformer. Advantages of lesser phase-angle shifts include application to lines of shorter length or lower voltage; less deviation from normal service voltage at mid-point of the line from which ice is being melted; anti-icing on longer lines; reduction of relaying problems due to fewer lines being connected in series and reduced hazard to customer service. Introducing the 120° roll has an added advantage in that the direction of a 60° phase shift can thereby be reversed, that is from lag to lead. This provides flexibility to meet specific conditions of in-service ice melting such as location in the line of the larger conductors, service to primary networks, and capacity of facilities serving the remote end of various lines. Greater phase-angle shifts provide another range of non-standard short-circuit voltages which enable a greater range of line length in Joule effect de-icing with the line out of service.

A new concept, the load network de-icer (ONDI) was developed by CITEQ and Hydro-Québec (Brochu et al., 2005). The ONDI concept is based on the use of a variable phase shifting transformer (PST) which is a special three-phase transformer employed for controlling power flows on transmission lines. This offers greater flexibility than the fixed configurations shown in Fig. 7.2. During de-icing, the PST is connected in series with one of the two lines to create an AC current loop in the lines, as illustrated in Fig. 7.3. By correctly adjusting the phase shift of the PST, it is possible to increase the current flow by a factor of four in the line opposite to the PST line while maintaining the voltage in the network during the de-icing period.

From the network simulations conducted by CITEQ, the ONDI system should be able to de-ice over 900 km of 230 kV or 315 kV lines and this by adequately switching existing circuit breakers to configure the network for de-icing. The variable phase shifting transformer method can also be used to prevent icing of conductors. However, this approach is not yet suitable for de-icing lines made of bundles of three or four conductors which would require additional rating for the ONDI.

7.2.6 Contactor Load Transfer for Bundle Conductors

This de-icing method was specially developed for bundled phases (Couture, 2004-07). Conventional metallic bundle spacer/dampers are replaced by new spacers equipped with contactor devices that can control the current flow in the bundle. During a de-icing sequence, three contactors open, which forces all the phase current through a single subconductor. The Joule input increases by a factor of 16, which under sufficient load can melt accreted ice. The process is repeated for each conductor of the bundle until complete de-icing is obtained. The system can be automated and easily controlled remotely. However, this method is presently at a conceptual stage, and further studies are necessary to estimate the real need and the cost of its development and implementation.

7.2.7 Open-Phase Configuration with Two Phases in Service

In most cases, the line to be heated must be taken out of service during melting. If this is not desirable, ice melting can be carried out using the phase-by-phase method where one phase is taken out of service, while power is transmitted over the two other phases.

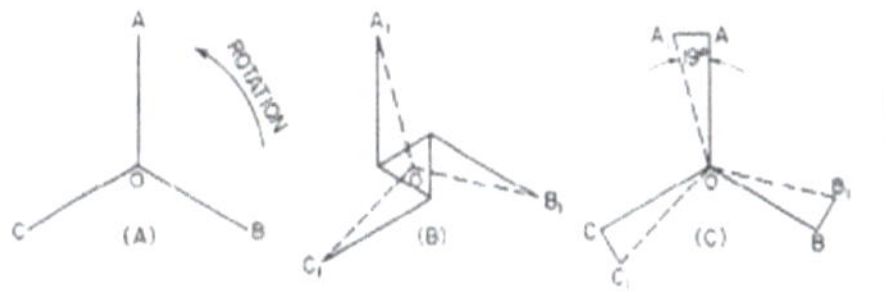

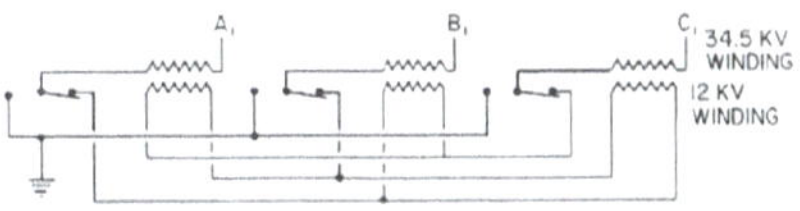

(A) Development of the 19° phase shift:

a) Voltage vector of normal 34.5-kV system

b) Neutral ends of 34.5-kV windings reconnected to 12 kV tertiary

c) Phase displacement between phase-shifted transformer and normal system

(B) Switching arrangement for the 19° phase shift

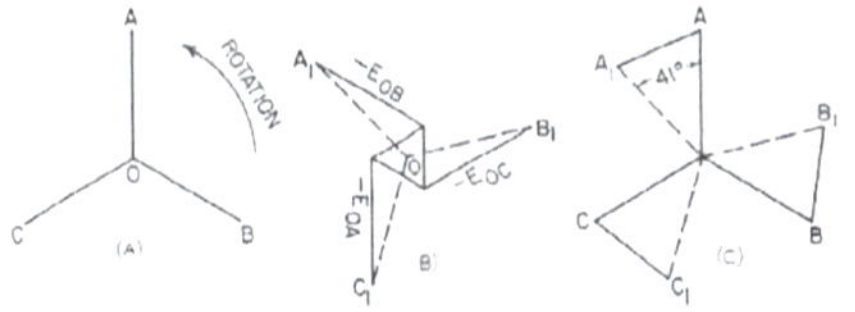

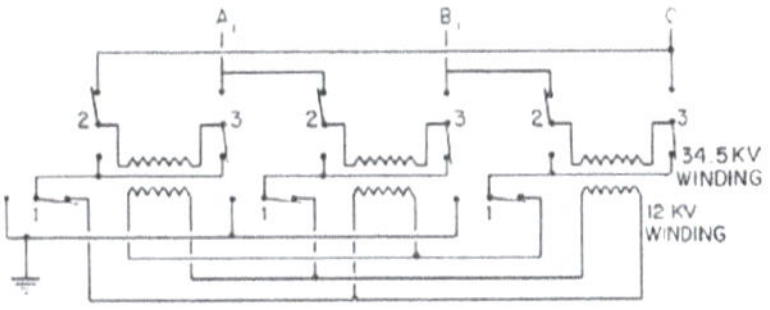

(C) Development of the 41° phase shift:

a) Voltage vector of normal 34.5-kV system

b) 34.5-kV windings reconnected to 12-kV tertiary

c) Phase displacement between phase-shifted transformer and normal system.

(D) Switching arrangement for the 41° phase shift

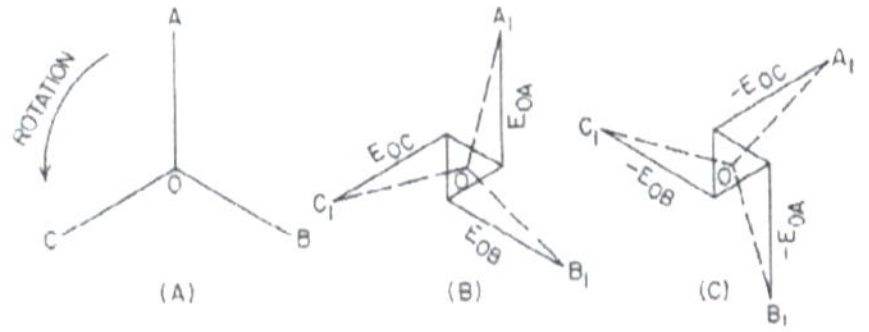

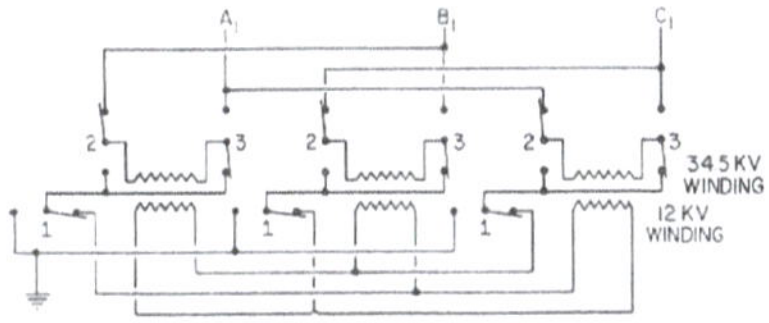

(E) Illustration of how reversing the delta tertiary reverses the direction of the phase shift

a) Voltage vector of normal 34.5-kV system

b) Neutral ends of 34.5-kv windings reconnected to reversed 12-kV tertiary for 19° lag

c) Neutral ends of 34.5-kv windings reconnected to reversed 12-kV tertiary for 41° lag.

(F) Switching arrangement for the -41° (lag) phase shift

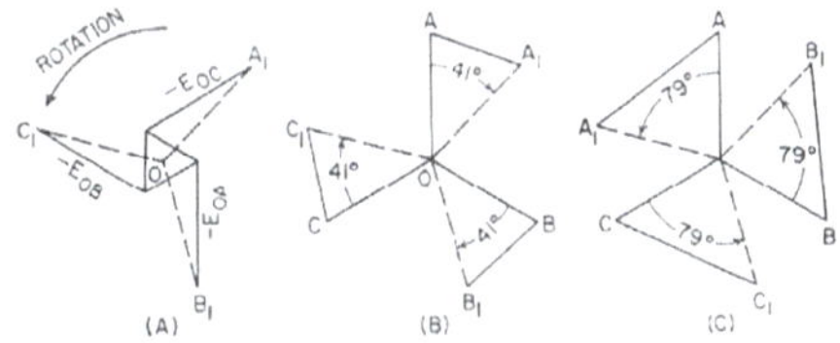

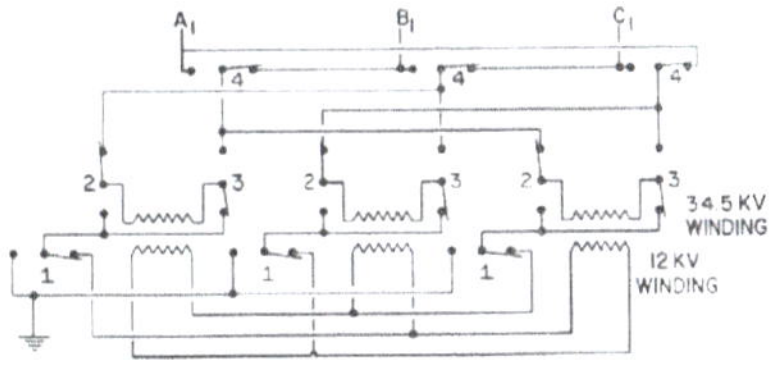

(G) Development of the 79° phase shift:

a) Development of 41° lag phase shift

b) 41° lag phase displacement

c) 79° lead phase displacement

(H) Switching arrangement for the 79° phase shift

Fig. 7.2 Configurations of phase shifting transformers for improved ice melting (Ekstrom, 1958-04)

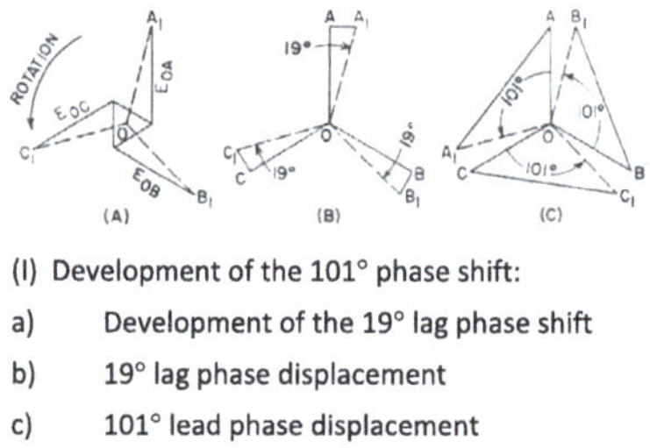

Fig. 7.2 (continued)

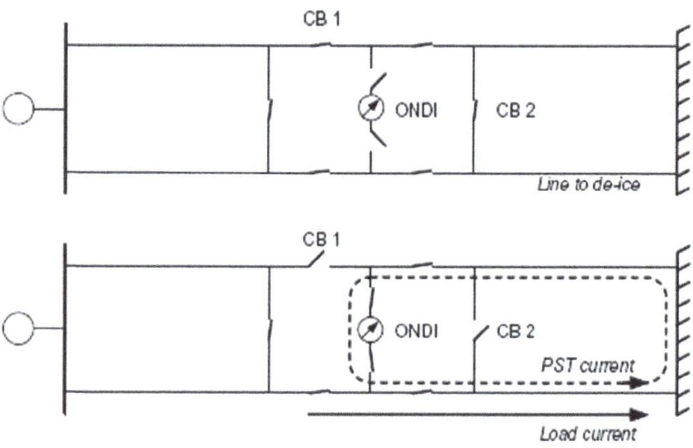

Fig. 7.3 On-line de-icing (ONDI) energization and network configuration for de-icing (Brochu et al., 2005)

When the switches marked “4” and “5” in Fig. 7.4 are open, two of the three phases can transfer power and the third phase is open, isolated from the power system. An appropriate AC source can then inject current that, schematically, returns through the remote substation earth.

Depending on power system configuration, it may be feasible to operate the AC system with one phase open for sufficient duration to de-ice each conductor. For example, sustained operation of wye-delta (Y-Δ) transformers, Y-Y with three-leg core, or Y-Δ-Y with buried Δ, with less than 10% drop in positive sequence voltage can support light to moderate system loads. Figure 7.5 exploits this possibility. Two phases of the AC line remain connected to the network and provide AC power flow. The third phase is disconnected from the network and subjected to DC injection. Ground is used as the DC return path. This configuration could also be considered with an AC source.

When the transformer neutrals are grounded, opening one phase causes zero-sequence and negative-sequence currents. The latter condition, together with the melting current will influence line communication (if power-line carriers, PLC,

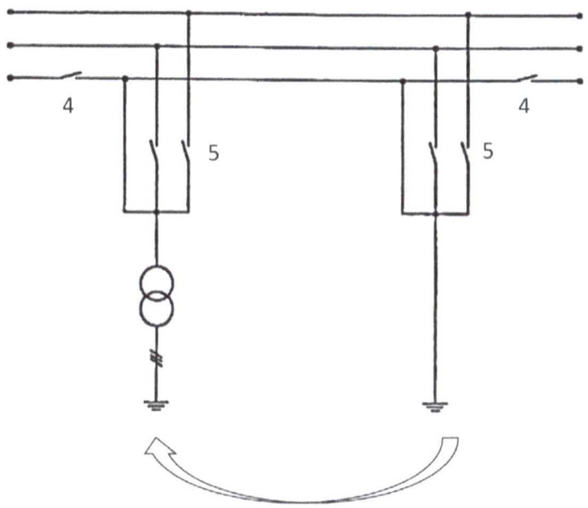

Fig. 7.4 On-line AC ice melting (one phase at a time) with earth return and one-phase-open configuration

are still used). Permissibility of such influence should be checked. Negative-sequence currents sometimes can exceed the admissible values for generators and motors. All this can limit application of the above schemes if special measures improving the quality of electric power at consumer buses are not taken.

The solution of the problem may lie in the artificial balancing of unbalanced line conditions. One simple and efficient method is compensating line inductive reactance of zero sequence by installing capacitor banks in the neutrals of the power transformers.

7.3 Reconfigurations for Joule Effect Anti- and De-Icing with DC

To obtain the necessary value of anti-icing or de-icing current, the applied voltage and corresponding total melting power must be sufficiently high, especially with long transmission lines. If the required melting current and voltage are relatively small, AC can be used successfully. DC is more advantageous for long high-voltage power lines with large cross-sectional conductors. DC has a significant advantage over AC current in heating conductors because the supply voltage can be much lower. The DC source does not need to overcome the inductive reactance of the circuit.

The DC current ice-melting technology has been developed, implemented, and successfully used on a large scale in the former USSR to melt ice along long 500 kV lines with bundled phase conductors (Motlis, 2002).

7.3.1 Off-Line Ice Melting with DC

Five different methods for reconfiguring a three-phase AC transmission line for of DC ice melting were evaluated for feasibility by Iravani (1999).

Figure 7.5a shows a schematic diagram of a three-phase AC line with a DC ice-melting system. During an ice-melting process involving a three-phase AC line with

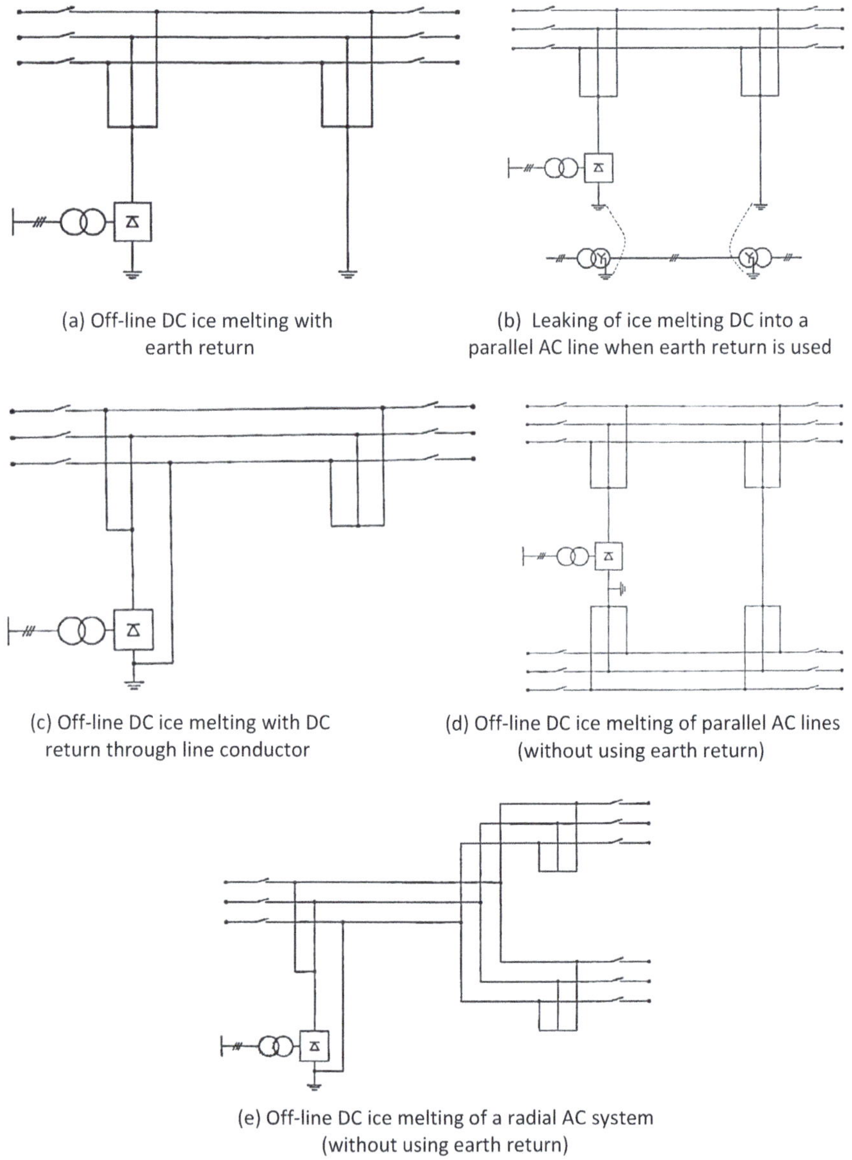

(a) Off-line DC ice melting with earth return

(b) Leaking of ice melting DC into a parallel AC line when earth return is used

(c) Off-line DC ice melting with DC return through line conductor

(d) Off-line DC ice melting of parallel AC lines (without using earth return)

(e) Off-line DC ice melting of a radial AC system (without using earth return)

Fig. 7.5 Schemes for off-line ice melting by DC (Iravani, 1999)

a DC ice-melting system, the line is disconnected from the network. The injected DC is supplied by an AC-DC converter. The converter is supplied from the network through a transformer, and the return path for DC is through the ground. Depending on the rating of the converter, 1, 2, or 3 phases can be injected at a time.

In Fig. 7.5b, parallel AC lines with *Y* transformers having grounded neutrals provide a fraction of the DC return path. Any DC current flow through the AC lines may cause transformer saturation. This happens as well in areas of high soil resistivity during monopole operation of HVDC systems and in response to geomagnetic-induced currents.

Figure 7.5c illustrates a configuration where the return path for the ice-melting DC is through the line itself and not through the ground. The return path can use 1 or 2 phases depending on the ice-melting cycle preferred by the user.

Where the AC-DC converter can supply the necessary power the configuration as Fig. 7.5d allows two parallel lines (in the same corridor) to be simultaneously de-iced.

If the AC transmission system has a radial structure, ice melting can be extended for a wider portion of the network as shown in Fig. 7.5e.

Following the January 1998 ice storm, Hydro-Québec carried out digital simulations to determine the minimum network backbone needed to ensure the supply of a minimum load to guarantee electricity for essential services. This established that the system could operate with several lines removed from service for periods long enough to allow for de-icing. On the basis of these results, Hydro-Québec commissioned a system incorporating a 250 MW rectifier at the Lévis substation near Québec City to de-ice four lines at 735 kV and one double-circuit line at 315 kV (Gingras et al., 2000-08; Granger et al., 2005-06). One of the 735 kV lines in this program has a length of 242 km, and all have a history of icing accretion. The evaluations considered short-circuit de-icing methods with AC sources but rejected these as requiring too much reactive power and excessive voltage magnitude. The DC rectifier power rating has been determined based on its capacity to de-ice a 242 km long 735 kV transmission line consisting of 1354 kcmil (686 mm^2) quad bundles. The Lévis system can supply a 7200 A current at a voltage level of ±20.8 kV for 90 min of de-icing (Horwill et al., 2006-05).

Figure 7.6 shows the reconfiguration of the 735 kV transmission line to allow DC de-icing. All three phases of the treated line are isolated from the AC network. Then, bridging connections are installed at a remote substation, and the DC source is applied locally. This configuration matches Fig. 7.5c.

The longest 735 kV lines are de-iced one phase at a time, in three stages. The other two phases are operated in parallel for the DC current return path to minimize the rectifier's power consumption. This is shown in Fig. 7.7 by the direction arrows, one pointing to the left and two pointing to the right to illustrate the direction of current flow.

Shorter 735 kV lines can be de-iced two phases at a time, so that the de-icing process has only two steps. For a 315 kV double-circuit line that is 183 km long, two circuits can de-iced simultaneously in a single step, as illustrated in Fig. 7.7. These lines have small-diameter twin-bundled conductors.

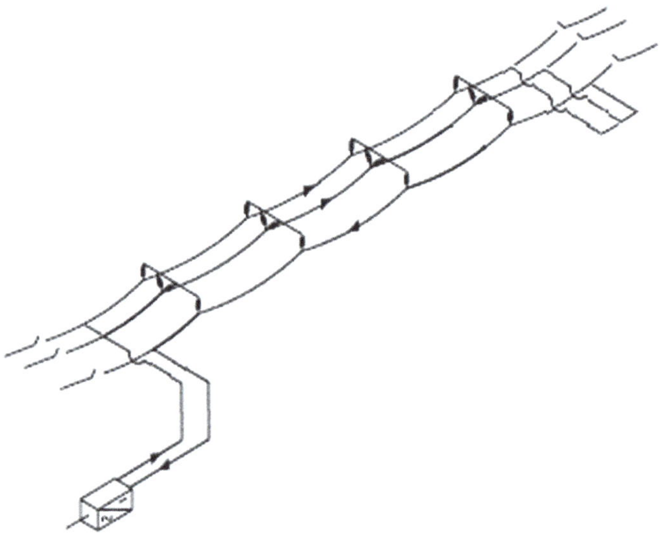

Fig. 7.6 One of the three-step sequences used to de-ice 735 kV line phase conductors

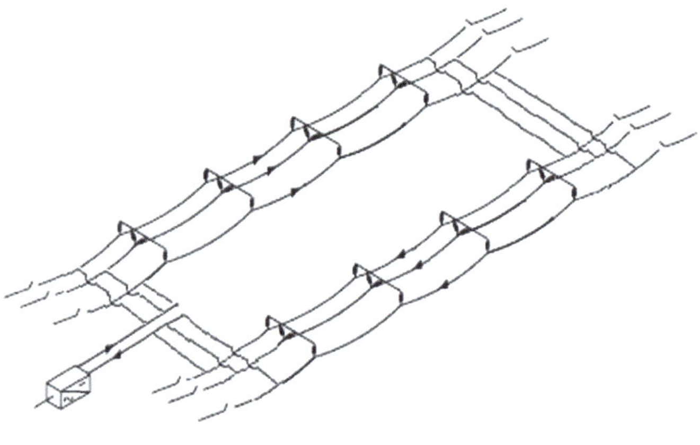

Fig. 7.7 Configuration used to de-ice double-circuit 315 kV line phase conductors

Each de-icing step takes about an hour to complete. The optical ground wire (OPGW) can also be de-iced at a lower current intensity using the same system because it is insulated from the tower. Installing a DC rectifier is relatively expensive for de-icing duty only, so the Hydro-Québec rectifier circuit was designed to be reconfigurable as a Static Var Compensator under normal operating conditions (Déry & Gingras, 2005) to provide added value.

Initiation of de-icing is based on the decision of a team of specialists who study weather conditions and forecasting. Their analysis considers the weather conditions near the line, as well as the weight of the ice measured on the line to be de-iced. Based on the analysis, the de-icing parameters are quantified. For example, the 7200 A current would be sufficient to melt 12 mm of ice in 30 min at −10 °C and wind speed of 11 km/h. In the absence of any ice or wind, the conductor temperature should be about 95 °C, which is the maximum allowable temperature. In other cases, however, different lower temperatures may be allowed.

Development and testing of an appropriate disconnect switch was also required to make sure it could operate safely considering the high current level with ice accumulation (Granger et al., 2005-06).

In the former USSR, ice melting with DC was also found to be advantageous for long transmission lines (Timoshova et al., 2003). In this case, at a constant voltage level, the current value was observed to reduce by 30–40% due to the heating of conductor sections not covered by ice. To maintain a constant current level, special circuits of converter supply are used. These schemes enable to arrange ice melting both on very short 35–220 kV lines (less than 1 km) and on lines of medium length (100–150 km). The melting current is proportional to the load of autotransformer or shunt reactor and is regulated by changing the transformation ratio of the booster transformer, with the voltage being self-regulated within certain limits.

7.3.2 On-Line Ice Melting with DC

On-line DC ice-melting process consists in the injection of DC in an AC line with no interruption of the AC power flow. The main challenges of on-line DC ice melting process are as follows:

- to contain the injected DC within the intended AC line and prevent it from being distributed in the remainder of the AC network
- to prevent the flow of system AC current in the DC source
- to prevent magnetic saturation of power transformers on the energized line due to DC

There are two approaches to deal with the above challenges. The first is based on DC injection through the neutral of a wye-connected power transformer, as Fig. 7.8a. This transformer must be fitted with extra windings to prevent saturation of the core. The second circuit configuration for DC injection is seen in Fig. 7.8b. Reactors (1) provide a high impedance against the AC flow in the converter, and DC is contained within the ice-coated line by means of series capacitor banks (3).

There are several options for providing a return path for the impressed DC current. Figure 7.9 illustrates the first option where three phases of the AC line remain connected to the AC network through capacitors (3). The DC current from (2) returns to the source via a remote connection to ground through reactors (1).

Thus, with respect to AC operation, the three phases contribute to AC power flow. This configuration has the same disadvantages as monopole operation of a DC system with earth return.

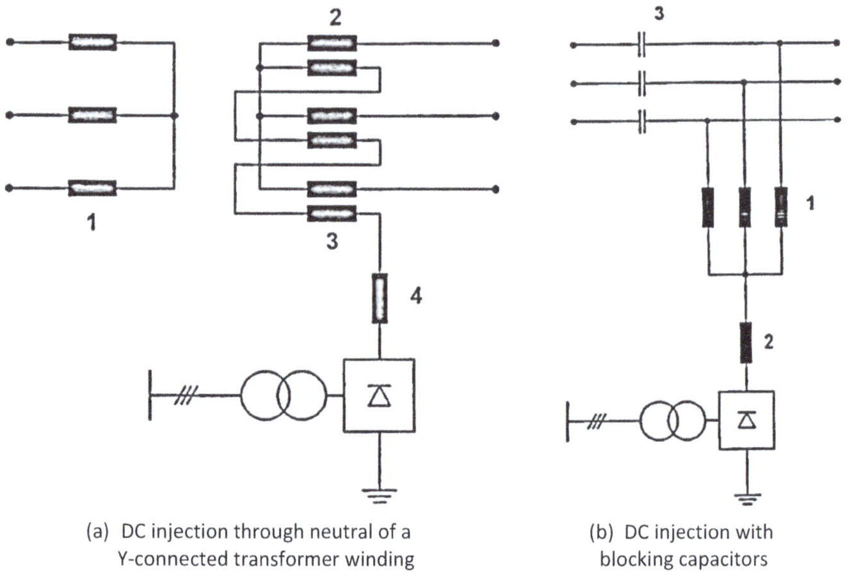

Fig. 7.8 DC injection into energized AC line

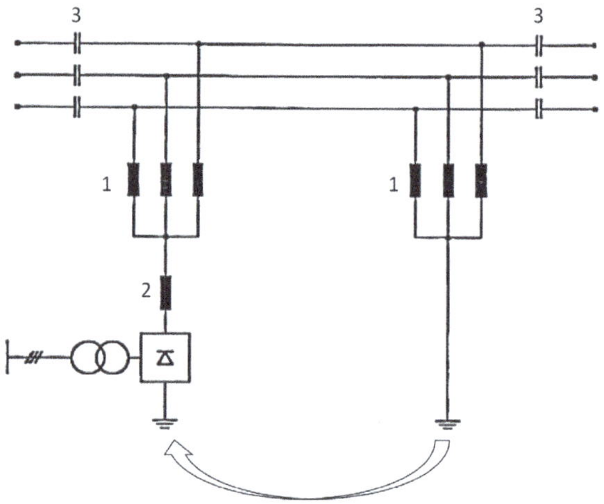

Fig. 7.9 On-line DC ice melting with earth return

If two parallel lines are to be simultaneously de-iced, then the configuration of Fig. 7.10 can be adopted for on-line ice melting to mitigate the effects of DC current in ac transformers connected through neutral or wye connections to the station ground. This is especially important in areas of high soil resistivity, where

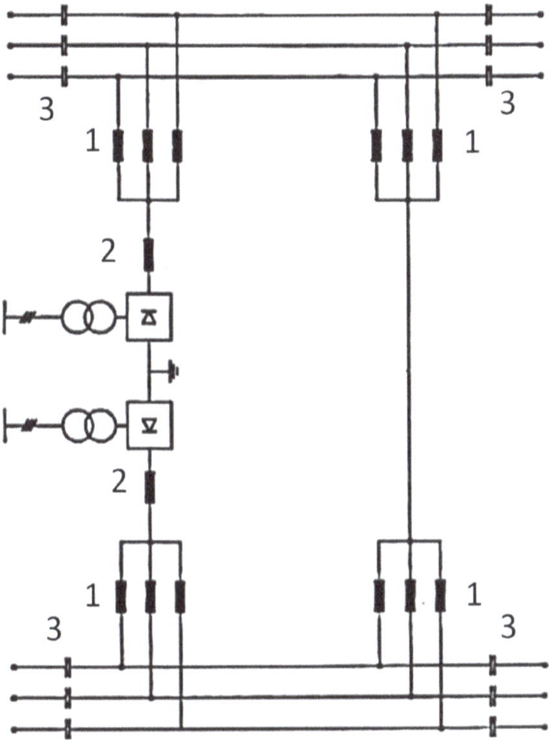

Fig. 7.10 On-line DC ice melting of two circuits without earth return

the AC system then carries a large fraction of DC return current, despite extensive DC earth electrode systems.

Figure 7.11 provides an alternative circuit configuration to that of Fig. 7.9. In common with Fig. 7.6, one phase of the AC line is used as the return path for DC, doubling its current and quadrupling its Joule loss. Earth return is not utilized for the DC current.

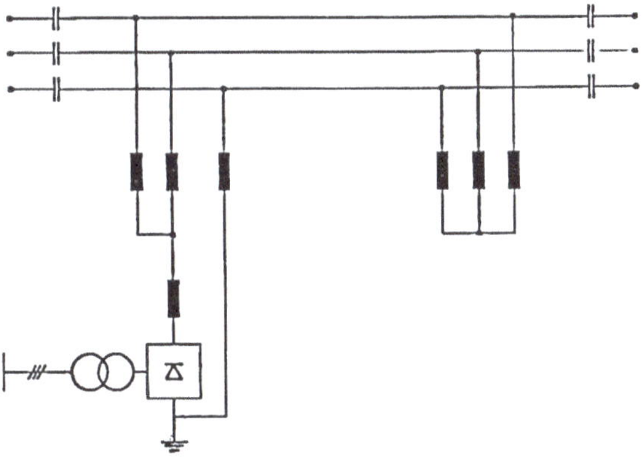

Fig. 7.11 On-line DC ice melting of single phase without earth return

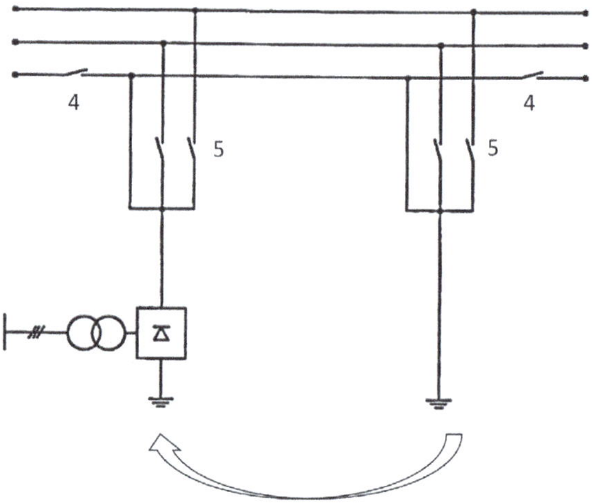

Fig. 7.12 On-line DC ice melting (one phase at a time) with earth return

When the switches marked "4" and "5" in Fig. 7.12 are open, two of the three phases can transfer power and the third phase is open.

Depending on power system configuration, it may be feasible to operate the AC system with one phase open. For example, sustained operation of wye-delta (Y-Δ) transformers, Y-Y with three-leg core, or Y-Δ-Y with buried Δ, with less than 10% drop in positive sequence voltage can support light loads. Figure 7.12 shows how tolerance to single-open-phase condition can be exploited to de-ice one phase at a time. Two phases of the AC line remain connected to the network (without the need for series capacitors) and provide AC power flow. The third phase is disconnected from the network and subjected to DC injection. Ground is used as the DC return path. This configuration could also be considered with an AC source.

Figure 7.13 is an alternative configuration for on-line DC ice melting without using ground as the return path. DC is injected in two phases of the AC line through reactors. DC is confined to the line by means of series capacitor banks at both ends

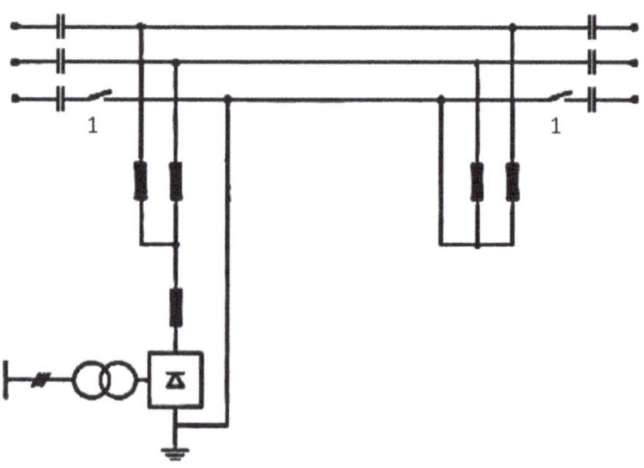

Fig. 7.13 On-line DC ice melting with one phase open and no earth return

of the line. The third phase is disconnected from the AC network and used as the return path.

In Figs. 7.8, 7.9, 7.10, 7.11, 7.12, and 7.13, an AC-DC converter provides injected DC for ice melting. There are two types of converters which can be used for AC-DC conversion for ice-melting applications. The first type is a line-commutated converter which utilizes conventional thyristor valves, with ratings up to 4 kA DC. For the Hydro-Québec Lévis de-icer project, two thyristor valves were connected in parallel to provide 8 kA DC. The second type of AC-DC converter uses forced-commutated semiconductors such as gate turn-off (GTO), insulated gate bipolar transistors (IGBT), or integrated gate-commutated thyristor (IGCT) devices with continuous ratings of around 1 kA.

7.4 Reconfigurations for Passive De-Icing

Depending on the seasonal system load patterns, it may be feasible to reconfigure double-circuit transmission lines for summer and winter operation. In summer, meeting peak air conditioning load and allowing for loss of a single circuit from a lightning outage may dominate planning requirements. Winter configurations where one of the two circuits is removed from service may provide sufficient ampacity and/or where overhead ground wires are removed.

7.4.1 Phase Rearrangement for Double-Circuit Lines

Another successful de-icing method, combining features of passive and Joule effects, limits the effects of conductor icing and sleet jump by phase rearrangement. Generally, the circuit arrangement of double-circuit lines is a complete circuit on each side of the tower. The phases can be rearranged to place the middle phase in each circuit, on the opposite side of the tower to the other two phases in the circuit Fig. 7.14. This again allows the top phase to unevenly sag under ice and not clash

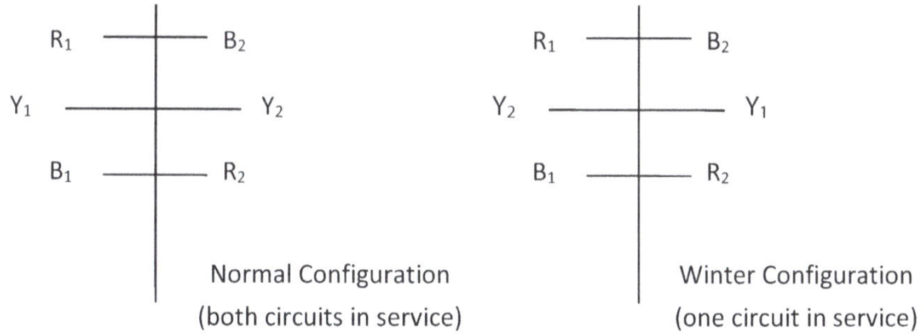

Fig. 7.14 Alternative phasing configuration for anti-icing

with its adjacent phase, although it may still come close to the off-potential phase of the other circuit. A contact may be tolerated in any event if the air-break disconnect switches at each remain open to isolate the unused conductors if these can be operated individually. One limitation to this phase rearrangement is that the system must accommodate the single ($N - 1$) contingency as one of the two circuits will be out of service during the icing season.

This phase rearrangement method is put in place just before winter by rearranging the phase connections at the substations at each end or special towers at either side of the higher altitude line section. If an icing event is forecast, one of the circuits is switched out of service, and the situation monitored. Normal phasing is returned during the summer months to allow for safe operation and maintenance opportunity on the individual circuits. Care and suitable communication is needed during these temporary phase rearrangement periods to ensure safety of operational and maintenance staff, as normal phase markings on affected structures could not be readily changed or modified.

7.4.2 Lines Without Overhead Ground Wires

In areas with severe winter weather and limited lightning activity, such as British Columbia and Newfoundland and Labrador in Canada, transmission lines at 230 and 500 kV are constructed without overhead ground wires. In Newfoundland, the lightning performance deficit is mitigated using single-pole reclosing, so that most lightning flashes to the phase conductor do not cause system instability associated with three-phase reclosing operations. The reliability of single-pole reclosing protection can be improved by fitting transmission line surge arresters with sufficient 60 mm ZnO block diameter to absorb the typical charge associated with the full lightning flash.

A similar approach was adopted after overhead ground wire failures on the 400 kV Gutinas–Braslov transmission line in Romania. Florea (2019) noted that shield wires were removed from two separate line sections, seven and sixteen spans, in 1985. Non-gapped line arresters were installed in stages, improving the mechanical performance of the arrester attachment and flexible lead. Florea concluded that application of surge arresters on the 400 kV line was an effective solution in a mountainous area where shield wires cannot be installed because of high ice accretion levels on conductors.

7.5 Reconfigurations for Mechanical Shock-Wave De-Icing

Studies on the Hydro-Québec power system indicated that EIDI method with multiple reclosing sequences could be applied on 315 kV lines with twin-bundle conductors, but only for emergencies during severe ice storms. For 735 kV lines, however, the required short-circuit currents to move the four-conductor bundle and reclosing sequences to excite movement at a subspan resonant frequency, posed unacceptable threats to network stability, and therefore, the method was not suitable for their EHV system.

7.6 Systems with Redundant Stations

With typical extent of a severe ice storm of approximately 100 km, power systems can be planned and operated effectively on considering the loss of an entire station as a contingency. The cost of this redundancy for a 500 kV network is on the order of USD 100 M but, when balanced against the societal costs of a two-day loss of electrical service, usually has a favourable business case.

7.7 Renewable Generation and Storage Performance in Adverse Winter Weather

The IEEE and CIGRE technical literature remain relatively weak in discussion of a search of terms (<wind turbine> and (<adverse weather> or <icing>)). Ice accretion on wind turbine blades, and the consequences of ice being thrown from the blades, dominates the research interest. This partly relates to the modest impairment in energy yield, even considering operations that shut down the turbine when icing events start. The WMO COST 727 (WMO, 2009-12) studies identified some issues in the effects of icing on wind energy production:

1. The accretion of ice on the wind turbine blades may disrupt aerodynamics and cause a reduction of power
2. The accretion of ice on blades may lead to a complete loss of production due to the shutdown of the turbine by operating policy.
3. Ice accretion may cause turbine overloads due to delayed stall, where the turbine speeds up suddenly at a wind speed much higher than the normal stall speed (around 1 m/s).
4. Non-uniform ice formation or shedding from blades can increase the fatigue of turbine components due to rotor imbalance.
5. Ice throws from the blades when production is resumed represent a specific danger around the operating wind turbine with throws of impressive ice blocks tens of metres around the vicinity.

Generally, wind turbines are now operated in icing conditions until the accretion brings the machine to a halt. Large wind turbines may lose 20% of the available energy in the winter season, which is used in evaluating the business case for de-icing systems.

The performance of wind generation in icing conditions may be sufficiently reliable to support black-start capability. Normally, this is provided by combustion turbine generators or hydraulic generation.

When solar panels are aligned to maximize winter output, they are often placed at vertical angles that shed snow cover rather well. Solar panels set to maximize energy during summer peak load periods may be more vulnerable, and for most large installations, clearing is not feasible. While the solar cells operate with higher efficiency, the same is not true of batteries. The charging rates and ageing effects for lithium-ion batteries in cold winter temperatures are particularly deficient.

It was suggested in the 1980s that renewable energy had extra value to offer power system operators, in terms of improved power quality, reliability, and black-start capability. It has proved difficult to include these factors in pricing renewable energy compared to conventional sources, some (such as combustion turbines) that provide the same benefits. Generally, operation of renewable resources in winter conditions is an area where cooperation among stakeholders is encouraged.

7.8 Integrated Resource, Reliability, and Network Planning Example: USSR

One indication of the degree of cooperation within a utility for successful ice melting is found in the power systems of the former USSR. These developed and standardized the resources and planning for large-scale ice melting technologies (Timoshova et al., 2003) on networks of all voltages up to and including 500 kV. The operational experience with ice melting systems has been described in many publications. The principles and philosophy adopted by Russia and other Republics of the former USSR are briefly given below.

7.8.1 Technical Constraints and Parameters

Analysis of design and operation experience has shown that there are several technical constraints, which determine the choice of parameters of ice melting devices (IMD). Among these are the following:

- Load capacity of supply for ice melting
- Maintaining an acceptable level of network voltage during ice melting procedure
- Maintaining system frequency during IMD switching operation
- Limitation of harmonic levels and voltage unbalance at consumers busbars, based on the existing standards
- Possible conductor overheating by melting current at ice-free line sections
- Ensuring the rate of ice melting on a group of lines, exceeding the rate of ice formation.

Development of each ice melting system requires the following technical parameters to be considered:

- line voltage
- type and size of conductors
- scheme how the conductors will be connected
- capability and other parameters of the feeding transformer(s)
- special requirements for protection and control (P&C)
- requirements to balance reactive power
- requirements to the level of harmonics on the AC and DC side
- requirements to combine the function of ice melting with the function of a static compensator
- special requirements for the system specific for a project.

7.8.2 Choice of AC or DC Voltage and Current Levels

Ice melting devices/technologies in the former Soviet Union were custom-made to meet the requirements for successful ice melting depending on the parameters of the line where IMD should be used. Based on their experiences, when designing a new IMD scheme, one should keep in mind that it is determined mainly by the supply scheme, by the scheme of ice melting, and by the type of melting current, AC or DC. The scheme of the IMD supply depends on the system configuration and on the types and parameters of equipment installed at the substation, supplying ice-melting facilities. The type of melting current is chosen based on the length of the network lines.

Capacities and voltages are compared with the parameters required for ice melting on the lines, connected to the node. Melting of ice by AC is envisaged for the lines when the available parameters meet the requirements. If the capacity and voltage required for AC melting exceed the available parameters at substation, melting by DC should be considered.

7.8.3 Consideration of Substation Configurations

Selection of optimal IMD supply schemes is based on the features of the main substation scheme:

- Substations with voltage control by tapped switches of transformers (autotransformers)
- Substations with voltage control by line regulating transformers
- Substations with voltage control by booster transformers.

In case the required AC voltage for ice melting on certain lines is lower than the available voltage, the following measures have been considered by reconfiguration of elements within substations:

- Regulation of melting current by means of substation transformers with tapped switches
- Use of line regulating transformers
- Use of booster transformers
- Increasing the length of the heated loop.

7.8.4 Observation Posts, Monitoring and Warning Systems

In the former Soviet Union, ice-melting projects included provisions for special observation posts and warning technologies. In some cases, preheating of conductors was done based on forecast of icing as an anti-icing measure.

7.8.5 Network Planning Aspects

In Russia, typical ice-melting applications for conductor size up to 300 mm^2 are characterized using practical combinations of line length and supply voltage as follows:

- 10–15 km (6 kV)
- 20–30 km (10 kV)
- 50–70 km (35 kV)
- 150–200 km (110 kV).

The organization of the IMD includes the following planning components:

- Infrastructure to coordinate all activities and services in preparation to winter season and operational performance during the icing periods
- Ice-melting devices/technologie
- Database/look-up tables of “ice melting current—melting time”, with examples in Fig. 5.17, for each line considering weather conditions and terrain
- Options of power utility’s power capacity balance during ice-melting event
- Sequence of lines to melt ice on
- Warning system(s) for the utility’s personnel about ice accretion and wind speeds
- P&C systems to ensure safe operation during the ice-melting event for IMD, substation equipment, and service personnel
- Telecommunication systems e.g. warning about the dangerous ice accretion, communication during ice melting event, post-event system restoration.

Conclusions and Recommendations

8

Conclusions and some recommendations for additional work are organized by chapter.

8.1 Winter Conditions: Ice and Snow Effects and Climatology

Heavy ice or wet snow accretion on ground wires and phase conductors of overhead lines may lead to major service outages. Winter conditions also lead to accelerated corrosion of components located near salted roads. Direct impact of icing events was found to be catastrophic in the "expected" countries that have severe winter weather—Canada, Iceland, and Norway—as well as in countries with milder climates in the usual classification methods. Ice and snow accretion continue to cause failures of overhead lines, despite a century of design and operation experience leading to well-accepted design standards.

There has been an improved appreciation of the climate and weather factors that lead to corrosion and electrical faults, as well as progress in monitoring and preventing cascade failures. Many anti-icing (AI) and de-icing (DI) techniques require reliable monitoring and predicting of icing events. This is increasingly accurate with modern meteorological forecasting systems, which serve both power transmission system and wind turbine generation stakeholders. The described resources, some based on new monitoring methods, can assist engineers in establishing representative levels of ice and snow accretion, insulator pollution, and other parameters needed for adequate overhead line dimensioning. Monitoring can also improve the understanding of repeated ice accretion on the thermal rating of the overhead line, important both for de-icing strategies and for useful operating margin in summer conditions. The influence of global climate change is predicted to affect the icing climatology, but estimates suggest it will not eliminate the risk of future problems in winter and may exacerbate thermal rating problems in warmer summer conditions.

M. Farzaneh and W. A. Chisholm, *Techniques for Protecting Overhead Lines in Winter Conditions*, Compact Studies, https://doi.org/10.1007/978-3-030-87455-1_8

8.2 Dimensioning Overhead Lines for Winter Conditions

Electrical clearances to prevent conductor clash under light ice accretion and steady, high wind speed are established using galloping ellipses. In some cases, midspan clearance issues can be mitigated with application of phase spacers. CIGRE literature treats this problem in detail, based on hundreds of observed galloping episodes around the world.

In contrast, standard treatment of the thermal rating of conductors considers an ambient temperature that exceeds 60 °F (+16 °C). Any mandated adoption of "Ambient Adjusted Rating" in winter conditions should probably rely on the observations of heat transfer on dry conductors in icing wind tunnels—perhaps at −10 °C rather than +16 °C. This is an area of research where additional supporting data—at low temperature and low wind speed and on conductors with a wider range of rugosity—should be accumulated and summarized.

Overhead line insulators are exposed to long-term accumulation of pollution—months of winter without rain—that affects the selection of leakage distance. Also, ice accumulation can bridge the leakage distance, leading to flashovers that are related to the line voltage gradient along the dry arc distance. Generally, good performance in areas of modest pollution can be achieved with HV suspension insulators, but the scaling to EHV configurations argues for the use of V-strings, which are less affected by ice and snow bridging than I-strings.

8.3 Monitoring and Predicting Ice Accretion and Shedding

From the point of view of power line maintenance, it remains difficult to develop a common strategy to protect all overhead power lines against the damage caused by severe ice and snow accretion. The initial approach is to consider expected ice loads from available data bases. Several methods then exist to reduce or prevent icing on conductors and ground wires. Many utilities in cold climate regions have developed and used methods and strategies to reduce ice loads using anti-icing (AI) and/or de-icing (DI) methods. Generally, AI methods are employed before or early during ice accumulation, whereas DI methods are activated during and sometimes after ice has accumulated.

Opportunities for improved DI and AI initiatives are presented by increasing utility applications of dynamic line rating systems, intended for meeting summer peak load. Instrument-based DLR systems provide estimates of the mechanical state of critical lines. Depending on sampling rate, these systems can also detect or localize failures caused by severe winter conditions. As ice and snow accretion and wind-on-ice loads affect the input observations as well as the calculated outputs of DLR systems, it is prudent to monitor their process variables closely during forecast or observed ice accretion. Weather sensors are common components in DLR systems, and these indicate the duration of instrumented icing periods, making resumption of normal operation timelier.

The possibility of having detailed photographs of ice accretion on structures and conductors has increased exponentially with the wide adoption of telephones with high-resolution cameras and adequate memory. This shifts the problems of icing observation to managing the data well. New technologies such as 3D digital modelling of icing from multiple photographs have not yet been exploited, but the ability to produce a realistic replica through 3D printing is interesting for improved wind tunnel studies of aerodynamic parameters, complementing 3D modelling of airflow in software.

There are so many problems occurring during severe winter weather that the complaints about excessive audible noise from corona and partial discharge, only under these adverse conditions, seem trivial. However, in many countries, low levels of AN must be maintained as part of authorization procedures and service. Even if the predicted values of AN are violated only in winter storms, this may lead to revised line routing, adapted conductor types, and perhaps more subconductors. Given the low cost of monitoring audible nose, understanding its characteristics on existing lines may lead to improved early-warning systems of ice and snow accretion as well as improved design guidance for new projects.

8.4 Preventing Accretion of Ice on Overhead Conductors

The most promising methods to combat ice accumulation on overhead line ground wires (GW) and phase conductors, insulators and towers have been discussed in this Green Book, summarizing many years of work by various CIGRE technical committees, working groups, and task forces. To fully understand these methods, the text of Chap. 4 also covered the fundamental aspects and processes of ice adhesion and accretion, as well as the mechanical and thermodynamic behaviour of ice and snow, including the shedding process. Conclusions were formed from decades of shared experience in developing CIGRE Technical Brochures 438 (CIGRE WG B2.29, 2010), 631 (CIGRE WG B2.44, 2015) and many other resources.

One of the most practical solutions to combat icing problems is a procedure for monitoring ice and snow accumulation before its load reaches a dangerous level, allowing time to remove the accreted ice or snow or adjust system operations. In this regard, dynamic line rating systems normally supporting peak summer loads should be used to provide advance warning of problems. In the best case, DLR systems should be specified to work well in winter conditions, even when coated with ice or covered in snow.

8.5 Removing Accretion of Ice from Overhead Conductors

This volume organized and presented many operational or potential DI and AI systems. However, for a better understanding and appropriate application of the AI/DI techniques, some fundamental aspects of ice and snow shedding must be understood. Moreover, study of the promising potential of new materials as AI coatings has been supported by a thorough set of references.

Traditionally, Joule methods based on an expanded DLR conductor thermal model have been used to disrupt the adhesion of an ice layer to a conductor. In the first fifty years of power system operation, some standards and processes did not define adequate ice design loads. Deficient transmission lines based on these norms did not satisfy today's understanding of climatic loads that are considered by design engineers on a case-by-case basis for each project. Despite design limitations, some of these lines were successfully operated for many winters by isolating them individually and reconfiguring the electric supply to provide current that usually exceeded summer-limit rating (1 A/kcmil) by a factor of two or three.

Anti- and de-icing methods using Joule heating of ice-covered line conductors by electrical currents have been accepted worldwide as an efficient engineering approach to minimize the sometimes-catastrophic consequences of severe ice events, thus increasing power system reliability. Either AC or DC has been used for ice melting. The technologies for both types of current are available, the methodology has been developed, and operational experience with ice-melting systems has been acquired for several decades. For overhead lines up to 110–230 kV, AC is recommended, but for higher voltages (including UHV lines), DC should be used to provide the required large power to melt ice on long lines with large conductor bundles, which is typical for lines of 345 kV and higher voltages. Also, successful deployment of DI and AI techniques depends on reliable monitoring and predicting icing events, reinforcing the important role of modern meteorological forecasting as well as remote real-time observation.

A review of the existing de-icing (DI) and anti-icing (AI) technology showed interesting non-thermal methods, including passive methods, active coatings and devices, and mechanical methods, but their scope is generally limited to local intervention. For example, a helicopter, a long wood pole, an insulating cable, and an experienced electrical utility staff pilot are needed for shock wave methods, which continue to be used because they can de-ice hundreds of spans in a few days.

8.6 Protective Coating Considerations for Winter Conditions

Surface engineering and the application of coatings could offer several opportunities to improve the performance of overhead power network equipment and their visibility impact in different environments.

- **Corrosion of metal parts of power lines**: In power networks, many economic losses are linked to the corrosion of line sections located in coastal or industrial areas. In fact, marine and industrial pollutants cause corrosion on metal components, such as conductors, fittings, insulator electrodes, and support structures. Corrosion rates in areas with low average temperature are considered.
- **Icing of conductors, ground wires and insulators**: Accretion of ice and snow on overhead line conductors and ground wires, especially in combination with the effects of wind, can pose a real threat to the mechanical integrity of the line. In addition, the electrical performance of insulators under these conditions is of concern.

Advances in materials science and surface engineering in recent years have resulted in the development of a variety of advanced coatings with great potential for application to overhead power network equipment to improve their reliability and quality of service.

Winter conditions in urban areas often lead to marine-like exposure to road salt on structures, conductors, and insulators. The role of protective coatings for protecting power system components is well understood and relevant in the maintenance programs at many utilities, not just those facing cold weather. Special considerations for selecting anti-corrosion coatings that will survive cold temperature were identified.

Anti-corrosion characteristics are especially desirable for overhead lines that are exposed to long periods of accumulation of pollution, as well as those that experience many freeze–thaw cycles each winter. Proximity to roads that are kept free of ice with salt or sand in the winter can provide an exceptionally severe exposure, similar in many respects to substations located at the edge of the sea.

There are several practical techniques for evaluating the natural corrosion process of metallic components of transmission lines in an electrical network. The key point is the correct and timely diagnosis to find and use the most effective way to protect each part against corrosion. The choice of a coating and its application on a transmission line component strongly depends on the environmental parameters and the characteristics of the component to be protected. Taking this into account, it can be decided to apply an appropriate type of coating to the element to be protected. Testing the performance of a coating in the field is the best way for its evaluation even though the process usually takes several years. The use of certain anticorrosive materials may be suitable in some countries but not in others since factors such as rain, sand, salinity, and temperature play a critical role in choosing the right product.

The concept of an anti-corrosion coating that also provides some degree of icephobicity is intriguing. Many new anti-icing coatings were discussed, but at this time, the two desirable features are not related. This approach is promising as a long-term goal as it improves the business case for future applications.

This book has described potential applications and benefits of self-cleaning, superhydrophobic and icephobic coatings for protection of overhead power networks in winter conditions, using the accomplishments of CIGRE WG B2.44

(2015). Methods of testing and characterizing various properties of these coatings were also reviewed. Promising benefits are anticipated from the use of coatings for future development and applications to power networks under adverse winter weather and aggressive environmental conditions. At the same time, the book outlines some deficiencies in existing standard methods to assess the functional properties and durability of advanced coatings for their profitable use.

Understanding the concepts of icephobicity starts with understanding the fundamental aspects of wetting but does not end there. Static contact angle and dynamic contact angles are two important concepts for icephobic characterization. It is shown that contact angle hysteresis is one of the most important factors determining the self-cleaning effect and anti-icing behaviour of a coating.

Examination of existing passive coatings has shown that although many anti-icing coatings have been developed and tested, none of them are effective and durable enough for application on conductors and ground wires. The few active coatings, such as those based on the electrolysis of ice and the injection of a pulsed current into a conductive coating on a wire are impractical or inefficient. One of the active coatings based on the use of a ferromagnetic coating appears to be effective in melting wet snow accumulated on conductors. However, its large-scale application may not meet a business case. In addition, further R&D efforts are needed for its application to melting ice. Regarding insulators, some coatings such as RTV silicone have already been widely used to effectively reduce pollution flashovers. However, applying RTV to improve insulator performance under icing conditions is ineffective and therefore not recommended. Likewise, several anti-icing coatings intended to prevent icing insulator flashover have been found to be ineffective. In addition, several methods developed in different laboratories to test and characterize the performance and durability of anti-icing coatings in the absence of appropriate standards were presented and discussed. Although a number of these methods can be applied to test these coatings after making some modifications, there is still a need for their refinement as well as the development of new and better suited techniques. Finally, additional R&D efforts are needed to develop effective and durable anti-icing coatings.

A variety of techniques have been developed for applying coatings to protect surfaces from corrosion and/or impart icephobic behaviour under winter conditions. Successful application of any coating is preceded by a suitable surface pre-treatment including cleaning, removal of loose material and physicochemical modifications. Protective coatings for overhead power network equipment such as conductors and insulators are classified as either active or passive categories depending on whether they need electrical energy to be activated. Some of these techniques, such as spray coating are already implemented in some industrial applications while others, like plasma sputtering, are still at development stage.

Extensive testing is required to ensure that a coating has the appropriate properties to fulfil its intended function. Different institutions have developed various test methods for assessing self-cleaning and anti-corrosion properties of coatings as well as their visual impact when applied to conductors or insulators. Fortunately, there are many standards and already available methods, some of which, as discussed in this volume, may be applied to these new coatings with some modifications. The adaptations include the evaluation of coating wetting properties, intrinsic material properties, as well as electrical and mechanical characteristics. However, further refinement of existing tests and development of new tests is essential, as new strengths and weaknesses of these new coatings are discovered.

Among the testing of various coating parameters, durability is an important characteristic that needs to be verified to ensure that the coating will fulfil its intended purpose for its expected lifetime. This should be done by considering appropriate tests which are representative of the environmental, electrical, and mechanical stresses as recommended in this document. In some cases, these are already standardized test methods which are also considered appropriate for the coatings. Other existing standards may need to be modified or adjusted to suit the specific characteristics of these coatings. In other cases, notably, for checking the electrical durability and ageing of the coatings there are no standard test methods available. An approach was introduced to help identify the appropriate test methods that can be used to facilitate the evaluation of new coating before applying them to power network equipment. A few possible tests methods are presented to stimulate further development of standardization.

8.7 Reconfiguring Power Systems for Anti- or De-Icing

Power systems must generally be reconfigured to deploy AI or DI technologies based on Joule heating. Methods for the prevention of ice accretion should be a part of system planning specifications for new projects, standards, and operational procedures in regions that are often affected by atmospheric icing events. Many countries and utilities have already integrated methods to de-ice lines that significantly reduce the risk of catastrophic damage. One challenge going forward is to continue to use and refine these methods, often with single contingency consequences, while maintaining the reliability of the grid using modern analysis and substation automation.

Reconfiguration of lines for winter service is a passive measure that may find suitable applications. Lately, reconfiguration of lines may remove problematic overhead ground wires, and make up the performance deficit using transmission line surge arresters. This solves one set of problems (sleet jump and high overturning moment) but introduces a new set of problems (mechanical response of arresters in wind and ice conditions) that may also be resolved after some experience is gained.

Increasing penetration of renewable resources into the electricity supply introduces new challenges and opportunities in managing overhead lines in winter conditions. Weather monitoring at wind turbines is robust, and strategies for dealing with ice accretion, including some acceptance of output impairment, are well organized.

There are several practical combinations of system voltage, conductor cross section, and line length that can be considered when planning a power system for de-icing. In particular, Russian success with de-icing 150–200 km of 110 kV transmission line suggests an appropriate spacing of substations for other systems exposed to winter conditions.

Definitions and Acronyms

General terms

Table A.1 Definition of general terms for organizations

Acronym	Phrase	Definition
IEC	International Electrotechnical Commission	
IEEE	Institute of Electrical and Electronics Engineers	
ISO	International Organization for Standardization	
TB	Technical Brochure	A publication produced by CIGRÉ representing the state-of-the-art guidelines and recommendations produced by an SC WG. Hardcopy TBs can be purchased, or individual members or staff of a collective member can download the PDF for free using their login credentials (copyright restrictions for use within their own CIGRE membership only)
SC	Study Committee	One of the 16 technical domain groups of CIGRE
TSO	Transmission System Operator	
WG	Working Group	A group formed by a SC to develop a TB on a subject of interest

M. Farzaneh and W. A. Chisholm, *Techniques for Protecting Overhead Lines in Winter Conditions*, Compact Studies, https://doi.org/10.1007/978-3-030-87455-1

Specific terms

Table A.2 Definition of technical terms used in this Green Book

Acronym	Phrase	Definition
–	Abrasive blast cleaning	Operation of forcibly propelling a stream of abrasive material against a surface under high pressure to smooth a rough surface, roughen a smooth surface, shape a surface, or remove surface contaminants. A pressurized fluid, typically compressed air, or a centrifugal wheel is used to propel the blasting material (often called the media)
ASA	Acrylonitrile styrene acrylate	Acrylonitrile styrene acrylate, also called acrylic styrene acrylonitrile, is an amorphous thermoplastic developed as an alternative to acrylonitrile butadiene styrene, but with improved weather resistance, and is widely used in the automotive industry
--	Active coatings de-icing methods	De-icing techniques which require electrical energy to activate heating effects of the coating material, but much less so than thermal methods. For example, some of these methods use the heat losses from dielectric materials covering a conductor or are based on the use of ferromagnetic coatings to heat the conductor surface
–	Active coatings for conductors	De-icing techniques which require electrical energy to activate heating effects of the coating material, but much less so than for thermal methods. For example, some of these methods may use heat losses from dielectric materials covering a conductor or be based on the use of ferromagnetic coatings to heat the conductor surface
–	Active coatings for insulators	De-icing techniques which require electrical energy to activate heating effects of the coating material, but much less so than thermal methods. For example, the use of semiconducting glaze or conductive silicone rubber coatings on insulator surfaces which are used to reduce ice adhesion and weight of accumulated ice
–	Anti-icing techniques	Techniques used to prevent or reduce ice and snow from accumulating on phase conductors or ground wires by weakening ice adhesion strength or by preventing freezing of supercooled water droplets on impact, particularly by the use of icephobic coating techniques
Ra	Arithmetical mean roughness	Arithmetic mean of the absolute ordinate values of the surface profile within a sampling length
–	Atmospheric icing	The expression "atmospheric icing" comprises all processes where drifting or falling water droplets, rain, drizzle, or wet snow in the atmosphere freeze or stick to any object exposed to the weather

(continued)

Table A.2 (continued)

Acronym	Phrase	Definition
AN	Audible noise	Any sound a human is capable of hearing
–	B-H curve	Magnetization curve of a material
BPA	Bonneville Power Administration	The Bonneville Power Administration is a non-profit federal power marketing administration based in the Pacific Northwest USA. Although BPA is part of the U.S. Department of Energy, it is self-funding and covers its costs by selling its products and services. BPA markets wholesale electrical power from 31 federal hydroelectric projects in the Northwest, one non-federal nuclear plant and several small non-federal power plants
CAPEX	Capital expenditure	A capital expenditure is incurred when a business spends money, uses collateral, or takes on debt to either buy a new asset or add to the value of an existing asset with the expectation of receiving benefits for longer than a single tax year. Essentially, a capital expenditure represents an investment in the business. Capital expenses are recorded as assets on a company's balance sheet rather than as expenses on the income statement. The asset is then depreciated over the total life of the asset, with a period depreciation expense charged to the company's income statement, normally monthly. Accumulated depreciation is recorded on the company's balance sheet as the summation of all depreciation expenses, and it reduces the value of the asset over the life of that asset
–	Coat	Continuous layer of metal material or a continuous film of paint resulting from a single application
–	Coating	A complete or partial covering that is applied onto the surface of an object via various chemical or physical methods for decorative and/or functional purposes
CA	Contact angle	The value of the angle a drop of water makes with the surface
–	Coating system	Sum of the coats of metal material and/or paints or related products which are to be applied or which have been applied to a substrate to provide corrosion protection, required surface colour, or other surface properties
	Corona, overhead power lines	Corona occurring at the surfaces of electrodes during the positive or negative polarity of the power-line voltage
	Corona extinction gradient	The gradient on that part of an electrode surface at which continuous corona last persists as the applied voltage is gradually decreased

(continued)

Table A.2 (continued)

Acronym	Phrase	Definition
	Corona extinction voltage	The voltage applied to the electrode to produce the corona extinction gradient
	Corona inception gradient	The gradient on that part of an electrode surface at which continuous corona first occurs as the applied voltage is gradually increased
	Corona inception voltage	The voltage applied to the electrode to produce the corona inception gradient
	Corona noise	Noise created by a corona (discharge)
–	De-icing techniques	Techniques used to remove or reduce ice accretion on phase conductors and ground wires. Contrarily to anti-icing techniques, they are activated late during or sometimes after the ice storm. They comprise passive methods, active coating methods, mechanical methods, and thermal methods
DFT	Dry film thickness	Thickness of coating remaining on the surface when the coating has hardened/cured
–	Dielectric material	A material that acts as an electrical insulator that may be polarized by the action of an applied electric field
–	Dielectric permittivity	A measure of how an electric field affects and is affected by a dielectric medium
	Dry band arcing	A partial arc occurring across a dry band formed on a polluted insulator
–	Dry snow	A type of precipitation icing that accretes at subfreezing temperatures. This type of accretion has low density and low adhesion and appears rarely and when the wind speed is very low, i.e. below 2 m/s
	Electric field	Physical field that acts on electric charges through the Coulomb force
	Electric field strength	Describes the strength and direction of an electric field, i.e. the ability of this field to exert force on charges
EDS	Everyday stress	Everyday stress of the conductor in the overhead line
–	Etching	In this TB, process of using a chemical to increase the surface roughness
FFT	Fast Fourier Transform	A fast Fourier transform (FFT) is an algorithm that computes the discrete Fourier transform (DFT) of a sequence, or its inverse (IDFT). Fourier analysis converts a signal from its original domain (often time or space) to a representation in the frequency domain and vice versa. The DFT is obtained by decomposing a sequence of values into components of different frequencies
–	Flux	A chemical, often zinc ammonium chloride, used in the hot-dip galvanization process. Applied to the steel, after cleaning and prior to galvanizing, to inhibit oxidation of the cleaned surface upon exposure to air

(continued)

Table A.2 (continued)

Acronym	Phrase	Definition
–	Glaze ice	A type of precipitation icing resulting in transparent ice accretion of density 700 kg/m^3 to 900 kg/m^3, sometimes with the presence of icicles underneath the wires. It very strongly adheres to objects and is difficult to knock off
–	Ground wire	A conductor having grounding connections at intervals and that is suspended usually above but not necessarily above the line conductor to provide protection against lightning discharges
HV	High voltage	–
HVAC	High voltage alternating current	–
HVDC	High voltage direct current	–
	Hydrophilic, hydrophilicity	A substance interacts strongly with water
–	Icephobic coatings	Anti-icing method used to reduce or eliminate ice adhesion force on coated equipment such as phase conductors or ground wires
–	Interferometry	Measurement methods that use the superposition or interference of waves to determine the quantities to be measured
–	Icephobic surface	A surface on which ice adhesion is low
–	In-cloud icing	A porous, opaque ice deposit which is formed by the impaction and freezing of supercooled water droplets on a substrate
–	Ionization	The process or the result of any process by which a neutral atom or molecule acquires either a positive or a negative charge
–	Joule effect	A physical phenomenon by which heat is generated by the electrical current flowing through a conductor
LWC	Liquid water content	The mass of water per unit volume of air, typically expressed in g/m^3
–	Magnetic field	Physical field generated by a permanent magnet or by moving electric charge
	Magnetostriction effect	Deformation of magnetic materials or substances due to an applied magnetic field
–	Mechanical de-icing methods	Methods involving ice breaking in order to accelerate ice shedding. Generally, most of the mechanical methods are based on two strategies. One strategy consists in breaking the ice by scraping methods and the second by shock waves methods
–	Natural ice or snow shedding	The process of reduction of the accumulated ice and snow on ground wires and phase conductors as it occurs naturally, without human intervention

(continued)

Table A.2 (continued)

Acronym	Phrase	Definition
–	Optical fibre	Element for optical data transmission
–	Passive de-icing methods	De-icing techniques which do not require any external energy input other than from natural forces: wind, gravity, incident radiation, and temperature variations
–	Phase conductor	Any conductor other than the neutral conductor
–	Photocatalytic	A chemical reaction triggered by light
–	Pinholes	Pinholes are small holes in the coating layer, which can vary in size from small to larger pores
–	Powder coating	Powder coating is a type of coating that is applied as a free-flowing, dry powder. Unlike conventional liquid paint, which is delivered via an evaporating solvent, powder coating is typically applied electrostatically and then cured under heat or with Ultraviolet light
PDMS	Polydimethylsiloxane	Polydimethylsiloxane, also known as dimethylpolysiloxane or dimethicone, belongs to a group of polymeric organosilicon compounds that are commonly referred to as silicones
–	Precipitation icing	A type of atmospheric icing which is caused by rain droplets or snowflakes that freeze or stick to the icing body
OPGW	Optical ground wire	Type of ground wires that is used in overhead power lines containing optical fibres. Such a wire combines the functions of grounding and communications
OPEX	Operating expense	Expenses incurred by a business in the course of its normal business activities. Often abbreviated as OPEX, operating expense includes rent, equipment, inventory costs, maintenance costs, payroll, insurance, milestone costs, and more. General repairs and maintenance of existing fixed assets such as buildings and equipment are also considered operating expenses unless the improvements will increase the useful life of the asset
R_f	Surface roughness	In thermal rating, the ratio of strand diameter d to (2 (D-d)) where D is the outer diameter of the stranded conductor
RH	Relative humidity	Relative humidity (RH) is the ratio of the partial pressure of water vapour to the equilibrium vapour pressure of water at a given temperature. Relative humidity depends on the temperature and pressure of the system of interest. The same amount of water vapour results in higher relative humidity in cold air than warm air. A related parameter is the dew point
RTS	Rated tensile strength	Rated tensile strength of the conductor or shield wire
RTV	Room temperature vulcanizing	RTV silicone (room temperature vulcanizing silicone) is a type of silicone rubber which cures at room temperature

(continued)

Table A.2 (continued)

Acronym	Phrase	Definition
–	Scraping methods	A mechanical de-icing technique consisting in removing accreted ice by using scrapers, rollers, or cutters attached to a rope which is pulled by line crews
–	Shock wave methods	A mechanical de-icing technique consisting in creating a shock wave which is propagated along the conductor to induce ice shedding
SiR	Silicone rubber	
SLIPS	Slippery lubricant-infused porous surface	Slippery lubricant-infused porous surfaces (SLIPS), which typically rely on surface structure and chemical properties to store and maintain the lubricant, usually lack fast self-replenishing ability due to the limited storage capacity of the lubricant or structure obstruction to fluid flow
–	Sound (level) meter	Portable dynamic signal analyser for acoustic measurements
–	Streamer	A repetitive corona discharge characterized by luminous filaments extending into the low electric field intensity region near either a positive or a negative electrode, but not completely bridging the gap
–	Sublimation	The release of vapour molecules to the ambient air from the ice surface
–	Supercooled water droplets	Water droplets whose temperature is below 0 °C, but which are still in the liquid state
–	Superhydrophobic surface	A surface whose contact angle is greater than 150°. Such a surface has a repelling effect on water droplets
–	Surface profile	Micro-roughness of a surface or profile that results from the intersection of the real surface by a specified plane
–	Surface roughness	Surface roughness often shortened to roughness is a component of surface texture. It is quantified by the deviations in the direction of the normal vector of a real surface from its ideal form. If these deviations are large, the surface is rough; if they are small, the surface is smooth
	Surface voltage gradient	The voltage gradient on the surface of a component, e.g. conductor, insulator, and fitting
–	Tensile strength	Ratio of the maximum load a material can support without fracture when being stretched to the original area of a cross section of the material
–	Thermal de-icing methods	De-icing techniques based on Joule effect by which line conductors are heated by the passage of electrical current, AC or DC, in order to melt the ice deposits and hence cause ice shedding
UHV	Ultra-high voltage	–
UHVDC	Ultra-high voltage direct current	–

(continued)

Table A.2 (continued)

Acronym	Phrase	Definition
	Voltage gradient	A vector *E* equal to and in the direction of the maximum space rate of change of the voltage at the point specified
–	Wet snow	A type of precipitation icing which is observed when the air temperature is just above freezing point, usually between 0.5 and 2 °C. The density of wet snow varies in a wide range but is normally significantly higher than that of snow on the ground
–	Run	A thicker part of a hot-dup galvanizing coating where a droplet of molten zinc solidified while running/rolling off the surface

References

Acatay, K., Simsek, E., Ow-Yang, C., & Menceloglu, Y. Z. (2004-09). Tunable, superhydrophobically stable polymeric surfaces by electrospinning. *Angewandte Chemie International Edition, 43*(39), 5210–5213.

Admirat, P., Maccagnan, M., & Goncourt, B. (1988). Influence of Joule effect and of climatic conditions on liquid water content of snow accreted on conductors. In: *International Workshop on Atmospheric Icing of Structures (IWAIS), Paris* (pp. 129–133, 367–371).

Admirat, P., Sakamoto, Y., Lapeyre J. L., & Maccagnan, M. (1986). Quantitative results and proposed mechanisms on wet snow accretion in the Ishiuchi wind-tunnel facilities. In: *3rd International Workshop on Atmospheric Icing of Structures (IWAIS), Vancouver* (pp. 155–160).

Adolphe, W. (1992-11). Ice melting of sub-transmission lines. In: *Proceedings of CEA Annual Conference, Toronto, CA.*

Adomou, M. (2011-08). *Étude expérimentale de l'utilisation des revêtements glaciophobes et/ou hydrophobes sur les pales d'éolienne* (UQAC Masters thesis). Chicoutimi, QC, Canada.

Alizadeh, A., et al. (2012). Dynamics of ice nucleation on water repellent surfaces. *Langmuir, 28* (6), 3180–3186.

Allaire, M. A., & Laforte, J. L. (2003). Method and apparatus for breaking ice accretions on an aerial wire. *USA, Patent No. 6518497*

Anderson, D. N., & Reich, A. D. (1997-01-06). Tests of the performance of coatings for low ice adhesion. In: *AIAA 35th Aerospace Sciences Meeting & Exhibit, Reno NV.*

Anderson, L. O., Golander, C. G., & Persson, S. (1994). Ice adhesion to rubber materials. *Journal of Adhesion Science and Technology, 8*, 117–132.

Andersson, P., Jelle, B. P., & Zhang, Z. (2017). Passive snow repulsion: A state of the art review illuminating research gaps and possibilities. *Energy Procedia, 132*, 423–428.

ANSI, C29.2. (1992). *Insulators—Wet process porcelain and toughened glass—Suspension type.* American National Standards Institute.

Antonini, C., et al. (2011). Understanding the effect of superhydrophobic coatings on energy reduction in anti-icing systems. *Cold Regions Science and Technology, 67*(1–2), 58–67.

Asadollahi, S., Farzaneh, M., & Stafford, L. (2019). On the icephobic behavior of organosilicon-based surface structures developed through atmospheric pressure plasma deposition in nitrogen plasma. *Coatings, 9*(10), 679.

Ashby, M. F., & Hallam, S. D. (1986). The failure of brittle solids containing small cracks under compressive stress states. *Acta Metallaurgy, 16*, 107–144.

ASTM SC D01.46. (2016). *ASTM D6386, standard practice for preparation of zinc (hot-dip galvanized) coated iron and steel product and hardware surfaces for painting.* ASTM International.

ASTM SC G01.05. (2019). *ASTM B117-19, standard practice for operating salt spray (fog) apparatus.* ASTM International.

M. Farzaneh and W. A. Chisholm, *Techniques for Protecting Overhead Lines in Winter Conditions*, Compact Studies, https://doi.org/10.1007/978-3-030-87455-1

ASTM, B504. (2011). *Standard test method for measurement of thickness of metallic coatings by the coulometric method.* ASTM International.

ASTM, D1005. (2013). *Standard test method for measurement of dry-film thickness of organic coatings using micrometers.* ASTM International.

ASTM, D1475. (2013). *Standard test method for density of liquid coatings, inks, and related products.* ASTM International.

ASTM, D2303. (2004). *Standard test methods for liquid-contaminant, inclined-plane tracking and erosion of insulating materials.* ASTM International.

ASTM, D2584. (2011). *Standard test method for ignition loss of cured reinforced resins.* ASTM International.

ASTM, D2794. (2010). *Standard test method for resistance of organic coatings to the effects of rapid deformation (impact).* ASTM International.

ASTM, D3170. (2012). *Standard test method for chipping resistance of coatings.* ASTM International.

ASTM, D3276. (2015). *Standard guide for painting inspectors (metal substrates).* ASTM International.

ASTM, D3359. (2008). Standard test method for measuring adhesion by tape test. In: *ASTM International*, West Conshhocken, PA, USA.

ASTM, D3363. (2011). *Standard test method for film hardness by pencil test.* ASTM International.

ASTM, D4060. (2010). *Standard test method for abrasion resistance of organic coatings by the taber abraser.* ASTM International.

ASTM, D4414. (2013). *Standard practice for measurement of wet film thickness by Notch Gages.* ASTM International.

ASTM, D4541. (2002). *Standard test method for pull-off strength of coatings using portable adhesion tester.* ASTM International.

ASTM, D4587. (2011). *Standard practice for fluorescent UV-condensation exposures of paint and related coatings.* ASTM International.

ASTM, D5178. (2013). *Standard test method for mar resistance of organic coatings.* ASTM International.

ASTM, D5324. (2010). *Standard guide for testing water-borne architectural coatings.* ASTM International.

ASTM, D5796. (2010). *Standard test method for measurement of dry film thickness of thin-film coil-coated systems by destructive means using a boring device.* ASTM International.

ASTM, E1131. (2014). *Standard test method for compositional analysis by thermogravimetry.* ASTM International.

ASTM, E1545. (2011). *Standard test method for assignment of the glass transition temperature by thermomechanical analysis.* ASTM International.

ASTM, E831. (2014). *Standard test method for linear thermal expansion of solid materials by thermomechanical analysis.* ASTM International.

ASTM, G106. (2010). *Standard practice for verification of algorithm and equipment for electrochemical impedance measurements.* ASTM International.

ASTM, G151. (2010). *Standard practice for exposing nonmetallic materials in accelerated test devices that use laboratory light sources.* ASTM International.

ASTM, G154. (2012). *Standard practice for operating fluorescent ultraviolet (UV) lamp apparatus for exposure of nonmetallic materials.* ASTM International.

ASTM, G155. (2013). *Standard practice for operating xenon arc light apparatus for exposure of non-metallic materials.* ASTM International.

ASTM, G59. (2014). *Standard test method for conducting potentiodynamic polarization resistance measurements.* ASTM International.

Asto Soto, S., & Silva, D. A. D. (2019). Performance of RTV silicone rubber pre-coated toughened glass insulators in zones of high salt contamination—Peruvian experience. In: *INMR World Congress, Tucson, Arizona, USA.*

Atmospheric Composition Analysis Group. (2020). *Surface PM2.5.* [Online]. Available at: http://fizz.phys.dal.ca/~atmos/martin/?page_id=140. Accessed March 29, 2021.

Baalmann, A., Hennemann, O., Liebermann, J., & Vissing, K. (2002). Method for producing an electrical insulator. *USA, Patent No. 6497923*

Baalmann, A., Hennemann, O., Liebermann, J., & Vissing, K. (2003). Insulator having a porcelain body and a hydrophobic coating. *USA, Patent No. 6541118*

Bahadur, V., et al. (2011). Predictive model for ice formation on superhydrophobic surfaces. *Langmuir, 27*(23), 14143–14150.

Bakic, K., Jakl, F., & Debenjak, B. (2009). The first experiences with ice-monitoring system for transmission network in Slovenia. In: *International Workshop on Atmospheric Icing of Structures (IWAIS), Andermatt.*

Balordi, M., Cammi, A., & Chemelli, C. (2019a). Effect of micro-roughness of aluminum alloy surface on anti-ice and durability properties of hierarchical structured coatings. In: *International Workshop on Atmospheric Icing of Structures (IWAIS), Reykjavik, Iceland.*

Balordi, M., Cammi, A., Magistris, G. S. d., & Chemelli, C. (2019b). Role of micrometric roughness on anti-ice properties and durability of hierarchical super-hydrophobic aluminium surfaces. *Surface & Coatings Technology, 374*, 549–556.

Balordi, M., et al. (2019-06). A comparison of anti-ice and anti-snow coatings performances: laboratory and field testing. In: *International Workshop on Atmospheric Icing of Structures (IWAIS), Reykjavík.*

Barthlott, W., & Neinhuis, C. (1997-04). Purity of the sacred lotus, or escape from contamination in biological surfaces. *Planta, 202*(1), 1–8.

Bartlett, S. C., Imburgia, C. A., & McDaniel, G. H. (1952-08). Sleet melting on the American gas and electric system. *Transactions of the American Institute of Electrical Engineers. Part III: Power Apparatus and Systems, 8, 71*(3), 704–709.

Bascom, W., Cottington, R., & Singleterry, C. (1969). Ice adhesion to hydrophilic and hydrophobic surfaces. *Journal of Adhesion, 1*, 246–263.

Bayliss, D., & Deacon, D. H. (2004). *Steelwork Corrosion Control.* Spon Press.

Beaunier, I., Epelboin, J., Lestrade, C., & Takenouti, H. (1976). Étude électrochimique par microscopie électronique à balayage, du fer recouvert de peinture. *Surface Technology, 4*, 237–254.

Beck, H., et al. (2018-10). Present and future Koppen-Geiger climate classification maps at 1-km resolution. *Nature, 5*(180214), 12.

Berry, W. B., Sachs, J. L., & Kleinman, R. L. (1993-04). *Radio frequency (RF) third rail deicing—A comparison with heated rail.* In: *Proceedings of IEEE/ASME Joint Railroad Conference, Pittsburgh, PA, USA* (pp. 41–45).

Beshkar, F., Khojasteh, H., & Salavati-Niasari, M. (2017-06). Flower-like CuO/ZnO hybrid hierarchical nanostructures grown on copper substrate: Glycothermal synthesis, characterization, hydrophobic and anticorrosion properties. *Materials, 10*(7), 697.

Bico, J., Thiele, U., & Quére, D. (2002). Wetting of textured surfaces. *Colloids and Surfaces A, 206*, 41–46.

Blackburn, C., Laforte, J. L., Gagnon, D., & St-Louis, M. (2002). Comparative study of the quantity of rime and glaze accreted on free-rotating and fixed overhead ground cables: Effect of anti-torsion device. In: *Proceedings of the 10th International Workshop on Atmospheric Icing of Structures (IWAIS 2002), Brno, Czech Republic.*

Blackett-Voltshield, J. (2009). Anti-polluant treatment for glass and glazed porcelain insulators. In: *20th International Conference on Electricity Distribution, Paper 1055, Prague, Czech Republic.*

Boinovich, L. B., Emelyanenko, A. M., Ivanov, V. K., & Pashinin, A. S. (2013). Durable icephobic coating for stainless steel. *ACS Applied Material Interfaces, 5*(7), 2549–2554.

Boluk, Y. (1996). *Adhesion of freezing precipitates to aircraft surfaces.* Transport Canada No. TP 12860E, Ottawa, ON.

Bormashenko, E., & Grynyov, R. (2012). Plasma treatment induced wetting transitions on biological tissue (pigeon feathers). *Colloids Surfaces B Biointerfaces, 92*, 367–371.

Bourdages, M. (2000-05). Ground wire de-icing by joule effect. In: *Proceedings of International OPGW Symposium, Montréal, CA.*

Brassard, J. D., Laforte, C., Guerin, F., & Blackburn, C. (2018). Icephobicity: Definition and measurement regarding atmospheric icing. In C. H. Cham (Ed.), *Contamination mitigating polymeric coatings for extreme environments* (pp. 123–143). Springer International Publishing AG.

Brochu, J., Cloutier, R., & Bergeron, A. (2005). On-load network de-icer for HV transmission lines. In: *11th International Workshop on Atmospheric Icing of Structures (IWAIS 2005), Montréal, CA.*

Byrd, N. (2004). Polysiloxane(amide-ureide) anti-ice coating. *USA, Patent No. 6797795*

Byrkjedal, Ø., et al. (2016-09). New Advances in icing measurements and icing predictions. In: *WindEurope Summit, Hamburg.*

Byrkjedal, Ø., & Nygaard, B. (2019). *FRonTLINES—Development of a toolbox for assessing frost and rime ice impact on overhead transmission lines* (Report no. KVT/ØB/2018/R053). Kjeller, Norway: Kjeller Vindteknikk.

Byrkjedal, Ø., Nygaard, B., & al., e. (2019). *FRonTLINES—Development of a toolbox for assessing frost and rime ice impact on overhead transmission lines* (Report no. KVT/ØB/2018/R053). Kjeller, Norway: Kjeller Vindteknikk.

Cassie, A., & Baxter, S. (1944). Wettability of porous surfaces. *Transactions of the Faraday Society, 40*, 546.

CEATI. (2002-04). *De-icing techniques before, during, and following ice storms.* CEATI No. T003700-3303 (two volumes), Montreal.

CEATI. (2008). *Performance evaluation of an automatic ice control (AIC) sensor on an energized 230kV.* CEATI International Report T023700 3309B, Montréal, CA.

Cheng, D., Jiang, X., & Wang, L. (2009). Anti-icing effect of PRTV coating on DC transmission line. *High Voltage Apparatus (in Chinese), 45*, 154–156.

China Electricity Council. (2019). *T/CEC 184–2018, test of ice-resistant coating for insulators.* Beijing: National Standard Network Era 2020-07-14.

Chisholm, W. A. (2005). Ten years of application experience with RTV silicone coatings in Canada. In: *INMR World Congress, Hong Kong* (no 17, pp. 1–20).

Chisholm, W., Ahlholm, C., & Gutman, I. (2019 10 23). Dust & sea salt pollution mapping: Satellite versus ground truth results. In: *INMR World Congress, Tucson, AZ.*

Chisholm, W. A., Li, Z., & Mogilevsky, A. (2016-05). *Calibration of satellite resources for applications of EHV and UHV polymer insulators.* CIGRE.

Chongqing Daily. (2019). *Dream chaser on snow peak mountain* [Online]. Available at: https://www.cqrb.cn/contents/2019-02/09/content_184582.htm. Accessed September 9, 2021.

Cicala, G., et al. (2003-10). Morphological and structural study of plasma deposited fluorocarbon films at different thicknesses. *Diamond and Related Materials, 12*(10–11), 2020–2025.

CIGRE TF 22.06.01. (2001). *TB 179: Guidelines for field measurement of ice loadings on overhead line conductors.* CIGRE.

CIGRE TF 33.04.07. (1999). *TB 142: Natural and artificial ageing and pollution testing of polymeric insulators.* CIGRE.

CIGRE TF 33.04.09. (1999). Influence of ice and snow on the flashover performance of outdoor insulators—Part I: Effects of ice. *Electra, 187*, 91–111.

CIGRE TF 33.04.09. (2000). Influence of ice and snow on the flashover performance of outdoor insulators—Part II: Effects of snow. *Electra, 188*, 55–61.

CIGRE TF B2.11.06. (2007-06). *TB 322: State of the art of conductor galloping.* CIGRE.

CIGRE TF B2.16.03. (2006). *TB 291: Guidelines for meteorological icing models, statistical methods and topographical effects.* CIGRE.

CIGRE WG 22.06. (1995-06). *TB 48: Tower top geometry.* CIGRE.

CIGRE WG 22.12. (2002). *TB 207: Thermal behaviour of overhead conductors.* CIGRE.

CIGRE WG B2.06. (2008). *TB 344: Big storm events: What we have learnt.* CIGRE.

CIGRE WG B2.16. (2004-10). *TB 256: Report on current practices regarding frequencies and magnitude of high intensity winds.* CIGRE.

CIGRE WG B2.16. (2006). *TB 291, Guidelines for meteorological icing models, statistical methods and topographical effects.* CIGRE.

CIGRE WG B2.22. (2012-10). *TB 515: Mechanical security of overhead lines containing cascading failures and mitigating their effects.* CIGRE.

CIGRE WG B2.29. (2010). *TB 438: Systems for prediction and monitoring of ice shedding, anti-icing and de-icing for overhead power line conductors and ground wires.* CIGRE.

CIGRE WG B2.44. (2015). *TB 631: Coatings for protecting overhead power network equipment in winter conditions.* CIGRE.

CIGRE WG B2.55. (2019). *TB 763: Conductors for the uprating of existing overead lines.* CIGRE.

CIGRE WG B2.69. (2021-06). *TB 837: Coating for improvement of electrical performance of outdoor insulators under pollution conditions.* CIGRE.

CIGRE WG B2.69. (2021-06). *TB 838: Coatings for protecting overhead power networks against icing, corona noise, corrosion and reducing their visual impact.* CIGRE.

CIGRE WG C4.303. (2008-06). *TB 361: Outdoor insulation in polluted conditions: Guidelines for selection and dimensioning, part 1: General principles and the AC case.* CIGRE.

CIGRE WG C4.303. (2012-12). *TB 518: outdoor insulation in polluted conditions: Guidelines for selection and dimensioning part 2: The DC case.* CIGRE.

CIGRE WG C4.303. (2013). *TB 555: Artificial pollution test for polymer insulators.* CIGRE.

CIGRE WG C4.409. (2014-06). *TB 578: Lightning protection of wind turbine blades.* CIGRE.

CIGRE WG D1.14. (2011-10). *TB 478: RTV silicone rubber insulator coatings.* CIGRE.

CIGRE Working Group 22.12. (1992). Mathematical model for evaluation of conductor temperature in the steady state and the application thereof. *Electra, 144*.

CIGRE Working Group 22.12. (1997). Mathematical model for evaluation of conductor temperature in the unsteady state. *Electra, 174*.

Clem, J. E. (1930-12-06). Currents required to remove conductor sleet. *Electrical World, 12, 96*, p. 1053–1057.

Colbeck, S., & Ackley, S. F. (1982). Mechanisms for ice bonding in wet snow accretions on power lines. In: *International Workshop on Atmospheric Icing of Structures (IWAIS), Hanover* (pp. 25–30).

Cong, Q., Chen, G.-H., Fang, Y., & Ren, L. (2004). Superhydrophobic characteristics of butterfly wing surface. *Journal of Bionics Engineering, 1*(4), 249–255.

Corey, C., Selfridge, H., & Tomlinson, H. (1952-08). Sleet-thawing practices of the New England electric system. *Transactions of the American Institute of Electrical Engineers. Part III: Power Apparatus and Systems, 71*(3), 649–657.

Couture, P. (2004-07). Switching modules for the extraction/injection of power (without ground or phase reference) from a bundled HV line. *IEEE Transactions on Power Delivery, 7, 19*(3), 1259–1266.

Cretchley, D. (2001). *Interruption to KwaZulu natal.* ESKOM Internal Report, Sunninghill, ZA.

Crouch, V., & Hartley, R. (1992). Adhesion of ice to coatings and the performance of ice release coatings. *Journal of Coatings Technology, 64*, 41–53.

Darque-Ceretti, E., & Felder, E. (2003). *Adhésion et adhérence.* CNRS Editions.

Dash, J. G., Rempel, A. W., & Wettlaufer, J. S. (2006). The physics of premelted ice and its geophysical consequences. *Reviews of Modern Physics, 78*, 115–167.

DELO. (2014). Newspaper DELO, Ljubljana. *DELO*, 03-03 02-20.

Déry, A., & Gingras, J. P. (2005). Hydro-Québec TransÉnergie's Lévis substation de-icing project. In: *11th International Workshop on Atmospheric Icing of Structures (IWAIS 2005), Montréal, CA.*

Di Mundo, R., & Palumbo, F. (2011). Comments regarding "an essay on contact angle measurements." *Plasma Processes and Polymers, 8*(1), 14–18.

Doan, Y., Ma, Z., & Yang, Q. (2017). Characteristics of consecutive dry days variation in China. *Theoretical and Applied Climatology, 130*, 701–709.

Druez, J., Louchez, S., & Bouchard, G. (1990). Study of ice shedding phenomenon on cables. In: *9th International Conference on Offshore Mechanics and Arctic Engineering, Houston* (pp. 143–148).

Druez, J., Louchez, S. P., & McComber, P. (1995). Ice shedding from cables. *Cold Regions Science and Technology, 23*, 377–388.

Druez, J., Phan, C. L., Laforte, J. L., & Nguyen, D. D. (1978). Adhesion of glaze and rime on aluminium electrical conductors. *Transactions of the Canadian Society for Mechanical Engineering, 5*, 215–220.

Egbert, R., et al. (1989-07). An investigation of power line de-icing by electro-impulse methods. *IEEE Transactions on Power Delivery, 4*(3), 1855–1861.

Ekstrom, I. R. (1958-04). More angles of phase shift added to previously known ice-melting methods. *Transactions of the American Institute of Electrical Engineers. Part III: Power Apparatus and Systems, 77*(3), 1449–1452.

Elíasson, Á., Gunnlaugsson, P., Thorsteins, E. (2009). *Ice accumulation at measuring site Hallormsstadahals.* In: *Proceedings of 13th International Workshop on Atmospheric Icing of Structures (IWAIS 2005), Andermatt CH.*

Elíasson, Á., Hannesson, G., & Thorsteinson, E. (2017). Wet snow icing—Analysis of field measurements 1999–2016. In: *Proceedings of the 17th International Workshop on Atmospheric Icing of Structures, IWAIS 2017, Chongqing, China.*

Elíasson, Á., Ísaksson, S., Ágústsson, H., & Thorsteins, E. (2015). Wet snow icing—Comparing simulated accretion with observational experience. In: *Proceeding of the 16th International Workshop on Atmospheric Icing of Structures, IWAIS 2015, Uppsala, Sweden.*

Elíasson, Á., Ísaksson, S., & Þorsteins, E. (2019-06). *Registration of observed icing on overhead lines in Iceland.* In: *Proceedings of the 18th IWAIS 2019, Reykjavik.*

Elíasson, Á., & Thorsteinson, E. (2007). Ice load measurements in test spans for 30 years. In: *Proceedings of the 12th International Workshop on Atmospheric Icing of Structures, IWAIS 2007, Yokohama.*

EPRI. (2012). *Application of nano technology coatings to transmission components* (EPRI Report 1025805). Palo Alto, CA.

Eskandarian, M. (2005). *Ice shedding from overhead electrical lines by mechanical breaking* (Ph. D. thesis). Université du Québec à Chicoutimi, Chicoutimi, QC, Canada.

Farhadi, S., Farzaneh, M., & Kulinich, S. (2011). Anti-icing performance of superhydrophobic surfaces. *Applied Surface Science, 257*, 6264–6269.

Farias, A. R. (1999). De-icing technologies—Introduction to manitoba hydro's use of NIce melting and NIce rolling technologies to remove ice from subtransmission and 115 kV transmission lines. In: *Proceedings of CEA Electricity, Vancouver, CA* (p 99).

Farid, H., Saeidi, A., & Farzaneh, M. (2018). An atmospheric ice empirical failure criterion. *Cold Regions Science and Technology, 146*, 81–86.

Farzaneh, M. (Ed.). (2008). *Atmospheric icing of power networks*. Springer.

Farzaneh, M. (2014-10). Insulator flashover under icing conditions, whitehead memorial lecture. *IEEE Transactions on Dielectrics and Electrical Insulation, 21*(5), 1997–1201.

Farzaneh, M. (2019). Overview of impact and mitigation of icing on power network equipment. In: *INMR World Congress, Tucson, USA* (pp. 1–12).

Farzaneh, M., et al. (2003). Insulator icing test methods and procedures: A position paper prepared by the IEEE task force on insulator icing test methods. *IEEE Transactions on Power Delivery, 18*(4), 1503–1515.

Farzaneh, M., & Chisholm, W. A. (2009). *Insulators for Icing and Polluted Environments* Hoboken, New Jersey: Wiley / IEEE Press Series on Power Engineering.

Farzaneh, M., & Chisholm, W. A. (2014). 50 years in Icing Performance of Outdoor Insulators. *IEEE Electrical Insulation Magazine, 30*(1), 41–24.

Farzaneh, M., Farokhi, S., & Chisholm, W. (2012). *Electrical design of power transmission lines*. McGraw Hill.

Farzaneh, M., & Volat, C. (2008). Anti-icing and de-icing techniques for overhead lines (Chap. 6). In: *Atmospheric icing of power networks* (p. 320). Springer.

Farzaneh, M., Volat, C., & Leblond, A. (2008). Anti-icing and de-icing techniques for overhead lines. In M. Farzaneh (Ed.), *Atmospheric icing of power networks* (pp. 229–268). Springer.

Feng, L., et al. (2003-02). Creation of a Superhydrophobic Surface from an Amphiphilic Polymer. *Angewandte Chemie International Edition, 42*, 800–802.

Ferrick, M. G., et al. (2012). Minimization of ice adhesion to space shuttle component surfaces. *Adhesives Science and Technology, 26*(4–5), 473–503.

Fikke, S. (2005). Modern meteorology and atmospheric icing. In: *International Workshop on Atmospheric Icing of Structures (IWAIS), Montreal, Canada*.

Fikke, S. (2006-12). *Measurements and data collection on icing: State of the Art. Report of COST Action 727, WG1*. MeteoSchweiz, Bundesamt für Meteorologie und Klimatologie.

Fikke, S., Hansen, J., & Rolfseng, L. (1982-06). Long range transported pollutants and conductivity of atmospehric ice on insulators. *IEEE Transactions on Power Apparatus and Systems, 101*(6), 1497–1503.

Fikke, S., Kristjansson, J. E., & Nygaard, B. E. (2008). Modern meteorology and atmospheric icing. In: M. Farzaneh (Ed.), *Atmospheric icing of power networks*. Springer.

Fikke, S., et al. (1994-06). Effect of long range airborne pollution on outdoor insulation. In: *Proceedings of Nordic Insulation Symposium NORD-IS 94, Vaasa, Finland*.

Fikke, S., et al. (2007). Atmospheric icing on structures, measurements and data collection on icing: State of the art. *s.l.:COST-727, Publication of MeteoSwiss, 75*, 110.

Florea, M. (2019). Application experience with transmission line arresters. *INMR, 19*(01), 1–19.

Forrest, J. (1942). The characteristics and performance in service of HV porcelain insulators. *Journal of the IEE, 89*, 60–92.

Fortin, G., Beisswenger, A., & Perron, J. (2010-08). Centrifuge adhesion test to evaluate icephobic coatings. In: *AIAA Atmospheric and Space Environments Conference, Toronto*.

Frankenstein, S., & Tuthill, A. M. (2002). Ice adhesion to locks and dams: Past work; future directions? *Journal of Cold Regions Engineering, 16*(2), 83–96.

Gagnon, D. (2012). *Hydro-Quebec DAC de-icer actuated by cartridge* [Online]. Available at: http://www.hydroquebec.com/innovation/en/pdf/2012G069-DAC.pdf. Accessed June 15, 2021

Gingras, J. P., et al. (2000-08). Stratégie de renforcement du réseau d'Hydro-Québec à la suite du verglas exceptionnel de janvier 1998: une démarche pour sécuriser davantage l'alimentation électrique. In: *CIGRÉ Session Paper 37-101, Paris*.

Granger, M., Dutil, A., Nantel, A., & Laurin, S. (2005-06). Performance aspects of lévis substation DC de-icing project. In: *Proceedings of 11th International Workshop on Atmospheric Icing of Structures (IWAIS 2005), Montréal, CA*.

Gu, I. Y., Berlijn, S., Gutman, I., & Bollen, M. (2013-06). Practical applications of automatic image analysis for overhead lines. In: *22nd International Conference on Electricity Distribution CIRED-2013, Paper 0560, Stockholm, Sweden*.

Guo, P., et al. (2012). Icephobic/anti-icing properties of micro/nanostructured surfaces. *Advanced Materials, 24*(19), 2642–2648.

Guo, Z., Liu, W., & Su, B.-L. (2011). Superhydrophobic surfaces: From natural to biomimetic to functional. *Journal of Colloid and Interface Science, 353*(2), 335–355.

Gutman, I., et al. (2008-08). Field experience and test verification of HV composite apparatus insulators. In: *CIGRE Session Paper A3-302, Paris*.

Gutman, I., et al. (2011-08). Influence of air-borne volcano ashes on outdoor insulation. In: *17th ISH, Paper C-014, Hanover, Germany*.

Gutman, I., Radojcic, M., & Carlshem, L. (2013). Reducing ice accretion using superhydrophobic coatings on conductors & insulators. *INMR*, Q2, *21*(2), 139–144.

Gutman, I., Seifert, J., & Greiling, C. (2012). Recovery of hydrophobicity of a new nano-composite coating for ceramic insulators. In: *IEEE International Symposium on Electrical Insulation-2012 Paper 12, San Juan, PR, USA* (pp. 48–52).

Hacker, D., White, B., & Ferrill, R. (2000-10). Ice release coatings for helicopter rotor blades. In: *78th Annual Meeting of the International Coatings Exposition, Chicago, IL, USA*.

Hammer, M., et al. (2020). Global estimates and long-term trends of fine particulate matter concentrations (1998–2008). *Environmental Science and Technology, 54*, 7879–7890.

Hansen, R. (2004). Means and method for removing extraneous matter like ice/snow an overhead line. *USA, Patent No. 2004065458*

Hassan, M., Lee, H. P., & Lim, S. P. (2010). The variation of ice adhesion strength with substrate surface roughness. *Measurement Science and Technology, 21*(7).

Hayashi, T., et al. (2013-08). Anti-icing performance of insulators by special surface treatments. In: *18th International Symposium on High Voltage Engineering, Seoul, South Korea*.

He, B., Patankar, N. A., & Lee, J. (2003-05). Multiple equilibrium droplet shapes and design criterion for rough hydrophobic surfaces. *Langmuir, 19*(12), 4999–5003.

He, G., et al. (2019a). Impact of icing severity on positive corona generated audible noise characteristic of rime ice-covered conductor. *IEEE Transactions on Industry Applications, 55* (5), 5269–5276.

He, G., et al. (2019b). Influence of freezing water conductivity on the positive corona performance of soft rime ice-covered conductor. *International Journal of Electrical Power and Energy Systems, 112*, 137–143.

He, Y., et al. (2015). Reducing ice accumulation and adhesion by using a flexible micro-rod film. *Cold Regions Science and Technology, 118*, 57–63.

Hefny, R., Kollar, L., & Farzaneh, M. (2012). Simulation of snow adhesion on power transmission cables. *International Journal of Mechanical Engineering and Mechatronics, 1*(2), 102–108.

Henson, W., Stewart, R., & Kochtubajda, B. (2007). On the precipitation and related features of the 1998 Ice Storm in the Montréal area. *Atmospheic Research, 83*, 36–54.

Hesse, K. H. (1988). The management of devastating ice storms. In: *Proceedings of the 4th International Workshos of Atmospheric Icing of Structure (IWAIS 1988), Paris* (pp. 330–334).

Higuchi, N. (1972). *Snow accumulation prevention methods*. Hokkaido Electric Power.

Hogan, J. C., & Pebler, C. G. (1955-01). Relay protection for lines being sleet-melted by the short-circuit method. *Transactions of the American Institute of Electrical Engineers. Part III: Power Apparatus and Systems, 12, 74*(3), 1067–1072.

Horwill, C., Davidson, C., Granger, M., & Déry, A. (2006-05). An application of HVDC to the de-icing of transmission LInes. In: *IEEE/PES Transmission and Distribution Conference and Exhibition, Dallas, TX*.

Huneault, M., Langheit, C., & Caron, J. (2005-04a). Combined models for glaze ice accretion and de-icing of current-carrying electrical conductors. *IEEE Transactions on Power Delivery, 4*, 20 (2), 1611–1616.

Huneault, M., et al. (2005-04b). A dynamic programming methodology to develop de-icing strategies during ice storms by channeling load currents in transmission networks. *IEEE Transactions on Power Delivery, 1 4, 20*(2), 1604–1610.

Hydro-Quebec. (1988). *Janvier de cristal - Le Quebec dans le coeur d'hydro*. Hydro-Quebec.

IEC Standard 60826. (2003). *Designing criteria for overhead transmission lines*. International Electrotechnical Commission.

IEC Standard 60865. (1993). *Short-circuit currents—Calculation of effects—Part 1: Definitions and calculation methods*. International Electrotechnical Commission.

IEC. (1991). *IEC 60507 "Technical specification artificial pollution tests on high-voltage insulators to be used on A.C. systems"*. International Electrotechnical Commission.

IEC. (1997). *IEC 61284 "Overhead lines—Requirements and tests for fittings*. International Electrotechnical Commission.

IEC. (2003). *IEC TS 62073 "Guidance on the measurement of wettability of insulator surfaces"*. International Electrotechnical Commission.

IEC. (2007). *IEC 60587 "Electrical insulating materials used under severe ambient conditions—Test methods for evaluating resistance to tracking and erosion"* (3rd ed.). International Electrotechncial Commission

IEC. (2008). *IEC 60815–1, Selection and dimensioning of high-voltage insulators intended for use in polluted conditions—Part 1: Definitions, information and general principles.* International Electrotechnical Commission.

IEC. (2012). *IEC 62217 "Polymeric HV insulators for indoor and outdoor use - General definitions, test methods and acceptance criteria".* International Electrotechnical Commission.

IEC. (2017). *IEC 60826 "Design criteria of overhead transmission lines".* International Electrotechnical Commission.

IEEE DEIS. (2009-10). *IEEE standard 1783, guide for test methods and procedures to evaluate the electrical performance of insulators in freezing conditions.* IEEE.

IEEE DEIS. (2018-10). *Std 1523-2018, IEEE guide for the application, maintenance and evaluation of RTV silicone rubber coatings for outdoor ceramic insulators.* IEEE Standards Association.

IEEE PES. (2005-06). *Standard 643-2004, IEEE guide for power-line carrier applications.* IEEE.

IEEE PES. (2013). *Standard 738-2012, IEEE standard for calculating the current-temperature relationship of bare overhead conductors.* IEEE.

Igoshin, V. A., & Berdnikov, A. G. (1989). Study of adhesion and friction of fluoroplast coating in contact with ice. *Soviet Journal of Friction and Wear (english Translation of Trenie i Iznos), 10*, 83–87.

Ingvaldsen, K. (2019-02-15). *Hardingnuten, tinn municipality, Norway, site and data description* (Report no. KVT/KI/2018/R067). Kjeller, Norway: Kjeller Vindteknikk.

Iravani, M. R. (1999). *Feasibility of DC ice-melting for high-voltage power lines* (Ph.D. thesis). University of Toronto, Toronto, CA.

ISO, 16773-1, 16773-2, 16773-3,16773-4. (2007). *16773-1, 16773-2, 16773-3,16773-4, Paints and varnishes—Electrochemical impedance spectroscopy (EIS) on high-impedance coated specimens Part 1: Terms and definitions, Part 2: Collection of data, Part 3: Processing and analysis of data from dummy cells, Part 4: Examples of spectra of polymercoated specimens.* International Organization for Standardization.

ISO. (1993). *ISO 3231, "Paints and varnishes—Determination of resistance to humid atmospheres containing sulfur dioxide."* International Organization for Standardization.

ISO. (1995). *ISO 527–3, "Plastics—Determination of tensile properties—Part 3: Test conditions for films and sheets."* International Organization for Standardization.

ISO. (1999). *ISO 12944–4 "Paints and varnishes—Corrosion protection of steel structures by protective paint systems—Part 4: Types and surface preparation."* International Organization for Standardization.

ISO. (2005). *ISO 11997–1, "Paints and varnishes—Determination of resistance to cyclic corrosion conditions—Part 1: Wet (salt fog)/dry/humidity)."* International Organization for Standardization.

ISO. (2006a). *ISO 1520, "Paints and varnishes—Cupping test."* International Organization for Standardization.

ISO. (2006b). *ISO 8502–6 "Preparation of steel substrates before application of paints and related products—Tests for the assessment of surface cleanliness—Part 6: Extraction of soluble contaminants for analysis—The Bresle method."* International Organization for Standards.

ISO. (2007). *ISO 2808 "Paints and varnishes—Determination of film thickness."* International Organization for Standardization.

ISO. (2009). *ISO 1461 "Hot dip galvanised coatings on fabricated iron and steel articles—Specification and test methods."* International Organization for Standardization.

ISO. (2011a). *ISO 1518-1, 1518-2, Paints and varnishes—Determination of scratch resistance: Part 1: Constant loading method, Part 2: Variable-loading method.* International Organization for Standardization.

ISO. (2011b). *ISO 1519, "Paints and varnishes—Bend test (cylindrical mandrel)."* International Organization for Standardization.

ISO. (2011c). *ISO 2881 "Paints and varnishes—Determination of density: Part 1: Pyknometer method, Part 2: Immersed body (plummet) method, Part 3: Oscillation method, Part 4: Pressure cup method."* International Organization for Standardization.

ISO. (2011d). *ISO 4628–6 "Paints and varnishes—Evaluation of degradation of coatings—Designation of quantity and size of defects, and of intensity of uniform changes in appearance—Part 6: Assessment of degree of chalking by tape method."* International Organization for Standardization.

ISO. (2011e). *ISO 6272-1, 6272-2, Paints and varnishes—Rapid-deformation (impact resistance) tests Part 1: Falling weight test, large-area indenter, Part 2: Falling-weight test, small-area indenter.* International Organization for Standardization.

ISO. (2012a). *ISO 1431-1: "Rubber, vulcanized or thermoplastic—Resistance to ozone cracking—Part 1: Static and dynamic strain testing."* International Organization for Standardization.

ISO. (2012b). *ISO 15184, "Paints and varnishes—Determination of film hardness by pencil test."* International Organization for Standardization.

ISO. (2012c). *ISO 19840 "Paints and varnishes—Corrosion protection of steel structures by protective paint systems—Measurement of, and acceptance criteria for, the thickness of dry films on rough surfaces."* International Organization for Standardization.

ISO. (2012d). *ISO 4628-8 "Paints and varnishes—Evaluation of degradation of coatings—Designation of quantity and size of defects, and of intensity of uniform changes in appearance—Part 8: Assessment of degree of delamination and corrosion around a scribe or ...."* International Organization for Standards.

ISO. (2012e). *ISO 9223 "Corrosion of metals and alloys—Corrosivity of atmospheres—Classification, determination and estimation."* International Organization for Standardization.

ISO. (2012f). *ISO 9227, "Corrosion tests in artificial atmospheres—Salt spray tests."* International Organization for Standardization.

ISO. (2013). *ISO 2409, Paints and varnishes—Cross-cut test.* International Organization for Standardization.

ISO. (2016a). *ISO 4624 "Paints and varnishes—Pull-off test for adhesion."* International Organization for Standardization.

ISO. (2016b). *ISO 4628-1 "Paints and varnishes—Evaluation of degradation of coatings—Designation of quantity and size of defects, and of intensity of uniform changes in appearance—Part 1: General introduction and designation system."* International Organization for Standardization.

ISO. (2016c). *ISO 4628-10 "Paints and varnishes—Evaluation of degradation of coatings—Designation of quantity and size of defects, and of intensity of uniform changes in appearance—Part 10: Assessment of degree of filiform corrosion."* International Organization for Standardization.

ISO. (2016d). *ISO 4628–2 "Paints and varnishes — Evaluation of degradation of coatings — Designation of quantity and size of defects, and of intensity of uniform changes in appearance — Part 2: Assessment of degree of blistering."* International Organization for Standardization.

ISO. (2016e). *ISO 4628-3 "Paints and varnishes—Evaluation of degradation of coatings—Designation of quantity and size of defects, and of intensity of uniform changes in appearance—Part 3: Assessment of degree of rusting."* International Organization for Standardization.

ISO. (2016f). *ISO 4628-4 "Paints and varnishes—Evaluation of degradation of coatings—Designation of quantity and size of defects, and of intensity of uniform changes in appearance—Part 4: Assessment of degree of cracking."* International Organization for Standardization.

ISO. (2016g). *ISO 4628-5 "Paints and varnishes—Evaluation of degradation of coatings—Designation of quantity and size of defects, and of intensity of uniform changes in appearance—Part 5: Assessment of degree of flaking."* International Organization for Standardization.

ISO. (2016h). *ISO 4628-7 "Paints and varnishes—Evaluation of degradation of coatings—Designation of quantity and size of defects, and of intensity of uniform changes in appearance—Part 7: Assessment of degree of chalking by velvet method."* International Organization for Standards.

ISO. (2016i). *ISO 7784-1, 7784-2, 7784-3: "Paints and varnishes—Determination of resistance to abrasion—Part 1: Rotating abrasive-paper-covered wheel method, Part 2: Rotating abrasive rubber wheel method, Part 3: Reciprocating test panel method."* International Organization for Standardization.

ISO. (2017a). *ISO 12494 "Atmospheric icing of structures."* International Organization for Standardization.

ISO. (2017b). *ISO 12944-2 "Paints and varnishes—Corrosion protection of steel structures by protective paint systems—Part 2: Classification of environments."* International Organization for Standardization.

ISO. (2019). *ISO 12944-5 "Paints and varnishes—Corrosion protection of steel structures by protective paint systems—Part 5: Protective paint systems."* International Organization for Standardization.

Jafari, R., Menini, R., & Farzaneh, M. (2010). Superhydrophobic and icephobic surfaces prepared by RF-sputtered polytetrafluoroethylene coatings. *Applied Surface Science, 257*(5), 1540–1543.

Jakl, F. (2013). Icing events and their impacts on overhead lines 400, 220 and 110 kV in Slovenia on January 1997 and 2009. In: *CIGRE WG B2.44 Presentation, Vienna.*

Jakl, F., Mirosevic, G., & Tumpa, G. (2003). Wind storm Bora in Croatia. In: *CIGRE WG B2.16 Presentation, Reykjavik.*

Jellinek, H. (1962). Ice adhesion. *Canadian Journal of Physics, 40*(10), 1294–1309.

Jellinek, H., et al. (1978). Ice releasing block-copolymer coatings. *Colloid and Polymer Science, 256*, 544–551.

Jenkner, P., Just, E., & Liebermann, J. (2002). Surface-modified insulator and method of modifying the surface of an insulator. *USA, Patent No. 6426150*

Ji, H., et al. (2002-02). Hydrophobic fluorinated carbon coatings on silicate glaze and aluminium. *J. Thin Solid Films, 405*(1–2), 104–108.

Johnson, R., & Dettre, R. (1964). Contact angle hysteresis. I. Study of an idealized rough surface. *Advances in Chemistry, 43*(7), 112–135.

Jones, K. (2003-01). *Ice storms in the St. Lawrence valley region.* Springfield.: K. F. Jones, Ice Storms in the St. Lawrence Vall. U.S. Army Engineer Research and Development Center, ERDC/CRREL TR-03-1.

Jorcin, J.-B. (2007-03). *Spectroscopie d'impédance électrochimique locale :caractérisation de la délamination des peintures et de la corrosion des alliages de Al-Cu, Institut national de polytechnique de Toulouse* (Doctorat thesis). Institut national de polytechnique de Toulouse, Tolouse.

Kako, T., et al. (2004). Adhesion and sliding of wet snow on a super-hydrophobic surface with hydrophilic channels. *Journal of Materials Science, 39*(2), 547–555.

Kasaai, M. R., & Farzaneh, M. (2004). A critical evaluation of the existing methods to determine ice adhesion strength. In: *Proceedings of the 23rd International Conference on Offshore Mechanics and Arctic Engineering (OMAE04), Vancouver.*

Kelly, R. G., Scully, J. R., Shoesmith, D., & Buchheit, R. G. (2002). *Electrochemical techniques in corrosion science and engineering*. CRC Press.

Kemell, M., et al. (2005). Superhydrophobic surfaces by self-assembly of hydrophobic monolayers on nanostructured surfaces. In: *Proceedings of the TNT 2005, Oviedo, ES.*

Kermani, M., & Farzaneh, M. (2008). Fracture toughness of atmospheric ice. In: *19th IAHR International Symposium on Ice, Vancouver.*

Kermani, M., Farzaneh, M., & Gagnon, R. E. (2007). Compressive strength of atmospheric ice. *Cold Regions Science and Technology, 49*, 195–205.

Kermani, M., Farzaneh, M., & Gagnon, R. E. (2008). Bending strength and effective modulus of atmospheric ice. *Cold Regions Science and Technology, 53*, 162–169.

Kern, J., & Zadnik, B. (1987). Icing in the Slovenian electric power industry. *Ujma (in Slovenian), 1*, 31–34.

Kim, P., et al. (2012). Liquid-infused nanostructured surfaces with extreme anti-ice and anti-frost performance. *ACS Nano, 6*(8), 6569–6577.

Kitamura, T., et al. (2003). Snow-melting magnetic material wire. *Fujikura Technical Review.*

Kiuru, M. (2004-08-20). *Experimental studies on diamond-like carbon and novel diamond-like carbon—Polymer—Hybrid coatings.* Faculty of Science of the University of Helsinki.

Koleske, J. V. (1995). *Paint and coating testing manual.* ASTM Publications.

Koshel, D., et al. (2003). Characterization of CFx films plasma chemically deposited from C3F8yC2H2 precursors. *Surface and Coatings Technology, 173*, 161–171.

Kulinich, S. A., Farhadi, S., & Nose, K. (2011). Superhydrophobic surfaces: Are they really ice-repellent. *Du Langmuir, 27*, 25–29.

Kulinich, S. A., & Farzaneh, M. (2004). Alkylsilane self-assembled monolayers: Modeling their wetting characteristics. *Applied Surface Science, 230*, 232–240.

Kulinich, S. A., & Farzaneh, M. (2005-08). On wetting behaviour of fluorocarbon coatings with various chemical and roughness characteristics. *Vacuum, 79*(3–4), 255–264.

Kulinich, S. A., & Farzaneh, M. (2009). How wetting hysteresis influences ice adhesion strength on superhydrophobic surfaces. *Langmuir, 8, 25*, 8854–8856.

Kumar, P., Brisson, P., Weinwurm, G., & Clark, J. (1999). The harsh-environment initiative—Meeting the challenge of space-technology transfer. *ESA Bulletin, 99*, 1–9.

Laakso, T., et al. (2003-04). *State-of-the-art of wind energy in cold climates.* International Energy Agency (IEA) R&D Wind Programme.

Lacroix, A., & Manwell, J. F. (2000-06). *Wind energy: Cold weather issues.* University of Massachusetts at Amherst.

Laforte, C. (2001-09). *Étude de l'adhérence de la glace sur des solides à caractère glaciophobe* (UQAC Masters thesis). Chicoutimi, QC, Canada.

Laforte, C., & Beisswenger, A. (2005-06). Icephobic material centrifuge adhesion test. In: *International Workshop on Atmospheric Icing of Structures (IWAIS), Montréal.*

Laforte, C., Carrière, J. C., & Laforte, J. L. (2002). How a solid coating can reduce the adhesion of ice on a structure. In: *Proceedings of the 10th International Workshop on Atmospheric Icing of Structures (IWAIS 2002), Brno, CZ.*

Laforte, J., & Allaire, M. (1998). Apparatus for deicing overhead wires and cables. *Canada, Patent No. 2207152.*

Laforte, J. L., Allaire, M. A., & Asselin, D. (1994-12). *État de l'art du déglaçage des conducteurs et des câbles aériens,* Montreal, CA: Hydro-Québec Rapport HQ-94–01.

Laforte, J. L., Allaire, M. A., & Gagnon, D. (2005). Ice shedding of 200 m-long artificially iced overhead cables at an outdoor test site. In: *Proceedings of 11th International Workshop on Atmospheric Icing of Structures (IWAIS 2005), Montréal, CA.*

Laforte, J. L., Allaire, M. A., & Laflamme, J. (1998). State-of-the-art on power line de-icing. *Atmospheric Research, 46*, 143–158.

Laforte, J. L., Allaire, M. A., & Laforte, C. (2005-06). Demonstration of the feasibility of a new mechanical method of cable de-icing. In: *Proceedings of 11th International Workshop on Atmospheric Icing of Structures (IWAIS 2005), Montréal.*

Lahti, K., Lahtinen, M., & Nousiainen, K. (1997). Transmission line corona losses under hoar frost conditions. *IEEE Transactions on Power Delivery, 12*(2), 928–933.

Lambrinos, G., Aguirre-Puente, J., & Sakly, M. (1987). Experimental research on the sublimation of ice samples. *Annales Geophysicae, 5B*, 589–594.

Landry, M., Beauchemin, R., & Venne, A. (2001). De-icing EHV overhead transmission lines using electromagnetic forces generated by moderate short-circuit currents. In: IEEE ESMO—2000 IEEE Proceedings of 9th International Conference on Transmission and Distribution Construction, Operation and Live-Line Maintenance, Montréal, CA (pp. 10–14).

Landy, M., & Freiberger, A. (1967). Studies of ice adhesion. *Journal of Colloid and Interface Science, 25*, 231–244.

Langmuir, I., & Blodgett, K. B. (1945). Mathematical investigation of water droplet trajectories. *Collected works of Irving Langmuir* (pp. 335–393). Oxford University Press.

Lanoie, R., Bouchard, D., & Lessard, M. (2005). Using steam to de-ice energized substation disconnect switch. In: *Proceedings of 11th International Workshop on Atmospheric Icing of Structures (IWAIS 2005), Montréal, CA*.

Lasia, A. (1999). *Electrochemical impedance spectroscopy and its applications*. Kluwer Academic/Plenum Publishers.

Lau, K., et al. (2003-10). Superhydrophobic carbon nanotube forests. *Nano Letters, 3*(12), 1701–1705.

Leblond, A., Asano, Y., & Isozaki, M. (2009). *Performance study of LC-spiral rods under icing conditions*. In: *Proceedings of of 13th International Workshop on Atmospheric Icing of Structures (IWAIS 2009), Andermatt*.

Leblond, A., & Côté, A. (2004-05-03). Atmospheric icing of OPGW: Monitoring and de-icing. In: Proceedings of Int'l OPGW Symposium, Montréal, CA (pp. 3–5).

Leblond, A., et al. (2005-06). Development of a portable de-icing device for overhead ground wires. In: *Proceedings of 11th International Workshop on Atmospheric Icing of Structures (IWAIS 2005), Montréal, CA*.

Leng, C., et al. (2015). Effects of temperature, pH, and ionic strength on the Henry's law constant of triethylamine. *Geophysical Research Letters, 42*, 3569–3575.

Li, J., et al. (2014-08). Anti-icing performance of a superhydrophobic PDMS/modified nanosilica hybrid coating for insulators. *Journal of Adhesion Science and Technology, 26*, 37–41.

Liu, H. W., & Miller, K. J. (1979). Fracture toughness of fresh water ice. *Journal of Glaciology, 22*, 135–143.

Liu, Q., et al. (2015). Durability of a lubricant-infused Electrospray Silicone Rubber surface as an anti-icing coating. *Applied Surface Science, 346*, 68–76.

Liu, X., et al. (2019). Slippery liquid-infused porous electric heating coating for anti-icing and de-icing applications. *Surface and Coatings Technology, 374*, 889–896.

Li, W., Zhang, X., Yang, J., & Miao, F. (2013). In situ growth of superhydrophobic and icephobic films with micro/nanoscale hierarchical structures on the aluminum substrate. *Journal of Colloid Interface Science, 410*, 165–171.

Lugschitz, H., Ernst, A., & Gros, T. (2004-08). Corrosion protection of steel towers and camoflage of lines using the DUPLEX-system. In: *Proceedings of CIGRE Session 2004, Paper B2-313, Paris*.

Maissan J. F. (2001-03-19). Wind power development in sub-artic conditions with severe rime icing. In: *Circumpolar Climate Change Summit and Exposition, Whitehorse, YK, CA*.

Makkonen, L. (2000). Models for the growth of rime, glaze, icicles and wet snow on structures. *Philosophical Transactions of the Royal Society, London, 358*, 2913–2939.

Ma, M., et al. (2005-04). Electrospun poly(styrene-block-dimethylsiloxane) block copolymer fibers exhibiting superhydrophobicity. *Langmuir, 21*(12), 5549–5554.

Ma, M., et al. (2005-10). Superhydrophobic fabrics produced by electrospinning and chemical vapor deposition. *Macromolecules, 38*(23), 9742–9748.

Marais, P. (2004). *Ice loading on lines in RSA*. ESKOM Internal Reseach Report RES/FR/03/22761, Sunninghill, ZA.

Marcacci, P., Lacavalla, M., & Pirovano, G. (2019). Monitoring, measurements and mitigation for wet snow accretion on overhead conductors. In: *International Workshop on Atmospheric Icing of Structures (IWAIS), Reykjavík*.

Marmur, A. (2003). Wetting on hydrophobic rough surfaces: To be heterogeneous or not to be? *Langmuir, 19*, 8343–8348.

McComber, P. (2000). A non-circular accretion shape freezing rain model for transmission line icing. In: *Proceedings of the 9th International Workshop on Atmospheric Icing of Structures (IWAIS 2000), Chester, UK*.

McComber, P., Druez, J., & Felin, B. (1990). Cable rime accretion measurements on a Laurentian mountain. *Canadian Journal of Civil Engineering, 17*, 1022–1032.

Menini, R., & Farzaneh, M. (2009). Elaboration of Al_2O_3/PTFE icephobic coatings for protecting aluminum surfaces. *Surface Coating Technologies, 203*(14), 1941–1946.

Menini, R., Ghalmi, Z., & Farzaneh, M. (2011). Highly resistant icephobic coatings on aluminum alloys. *Cold Region Science and Technology, 65*(1), 65–69.

Meteorological Service Center of China. (2008). *Rare rain, snow and freezing rain disaster in the south of China* [Online]. Available at: http://www.weather.com.cn/zt/kpzt/229278.shtml. Accessed May 23, 2021.

Ming der Ger, K. H., Wang, L. M., & Hwang, B. J. (2002-01). The fwww.weather.govriction and wear of Ni-P-PTFE composite deposits under water lubrication. *Materials Chemistry and Physics, 77*(3), 755–764.

Mishchenko, L., et al. (2010). Design of ice-free nanostructured impacting water droplets. *ACS Nano, 4*(12), 7699–7707.

Mizoe, H., Ohno, K., & Nakayama, I. (2007-10). Effect of snow accretion resistance measures on overhead transmission lines. In: *International Workshop on Atmospheric Icing of Structures (IWAIS), Yokohama.*

Mizuno, S., et al. (2003). Water-repellent coating and coating film. *USA, Patent No. 6579620*

Mobarakeh, L. F., Jafari, R., & Farzaneh, M. (2013). The ice repellency of plasma polymerized hexamethyldisiloxane coating. *Applied Surface Science, 284*, 459–463.

Mock, U., Forster, R., Menz, W., & Ruhe, J. (2005-02). Towards ultrahydrophobic surfaces: A biomimetic approach. *Journal of Physics of Condensed Matter, 17*(9), 639–648.

Momen, G., & Farzaneh, M. (2011-06). A simple process to fabricate a superhydrophobic silicone rubber coating. *Micro & Nano Letters, 6*(6), 405–407.

Momen, G., & Farzaneh, M. (2012-05). A ZnO-based nanocomposite coating with ultra-water repellent properties. *Applied Surface Science, 258*(15), 5723–5728.

Momen, G., & Farzaneh, M. (2014). Facile approach in the development of icephobic hierarchically textured coatings as corrosion barrier. *Applied Surface Science, 299*, 41–46.

Montambault, S., Côté, J., & St-Louis, M. (2000). Preliminary results on the development of a teleoperated compact trolley for live-line working. In: *Proceedings of 9th IEEE Conference onTransmission and Distribution Construction, Operation and Live-Line Maintenance (ESMO 2000), Montréal, CA* (pp. 21–27).

Montanyà, J., et al. (2016). Global distribution of winter lightning: A threat to wind turbines and aircraft. *Natural Hazards and Earth System Sciences, 16*, 1465–1472.

Morgan, V. T. (1991). *Thermal behaviour of electrical conductors-steady, dynamic and fault-current ratings*. Research Studies Press & Wiley.

Motlis, Y. (2002). *Melting ice on overhead-line conductors by electrical current.* CIGRE Internal Working Document SC22/WG12 Draft 4.

Mulherin, N. D., & Haehnel, R. B. (2003). *Ice engineering: Progress in evaluating surface coatings for icing control at corps hydraulic structures.* US Army Engineer Research and Development Center.

Müller, M., & Oehr, C. (2011). Comments on "an essay on contact angle measurements." *Plasma Processes and Polymers, 8*(1), 19–24.

Murase, H., & Fujibayashi, T. (1997). Characterization of molecular interfaces in hydrophobic systems. *Progress in Organic Coatings, 31*(1–2), 97–104.

Murase, H., et al. (1994-12). Interactions between heterogeneous surfaces of polymers and water. *Journal of Applied Polymer Science, 54*(13), 2051–2062.

Murase, H., et al. (1994). Interactions between heterogeneous surfaces of polymers and water. *Journal of Applied Polymer Science, 54*, 2051–2062.

Myers, D. (1999). *Surfaces, interfaces, and colloids: Principles and applications* (2nd ed.). Wiley.

Nakagami, M. (2007-10). Verification of the effects of PTFE tapes on reducing snow accretion on overhead power transmission lines. In: *International Workshop on Atmospheric Icing of Structures (IWAIS), Yokohama.*

Nakajima, A., Hashimoto, K., & Watanabe, T. (2001-01). Recent studies on super-hydrophobic films. *Monatshefte fur Chemie, 132*, 31–41.

Neinhuis, C., & Barthlott, W. (1997-06). Characterization and distribution of water-repellent, self-cleaning plant surfaces. *Annals of Botany, 79*(6), 667–677.

Newell, H. H., Liao, T. W., & Warburton, F. W. (1968-04). Corona and RI caused by particles on or near EHV conductors: II-Foul weather. *IEEE Transactions on Power Apparatus and Systems, PAS-87*(4), 911–927.

Nexans. (2016). *Les conducteurs aériens quasi invisibles de Nexans se fondent dans le paysage d'une réserve naturelle en Norvège pour un projet de Statnett* [Online]. Available at: www.nexans.fr/eservice/France-fr_FR/navigatepub_333666_-35195/Les_conducteurs_aeriens_quasi_invisibles_de_Nexans.html. Accessed April 26, 2021.

Nguyen, T. B., Park, S., Jung, Y., & Lim, H. (2019). Effects of hydrophobicity and lubricant characteristics on anti-icing performance of slippery lubricant-infused porous surfaces. *Journal of Industrial and Engineering Chemistry, 69*, 99–105.

Nguyen, T. B., Park, S., & Lim, H. (2018). Effects of morphology parameters on anti-icing performance in superhydrophobic surfaces. *Applied Surface Science, 435*, 585–591.

Nigol, O., & Buchan, P. (1981-02). Conductor galloping part 1—den Hartog mechanism. *IEEE Transactions on Power Apparatus and Systems, PAS-100*(2), 699–707.

NOAA National Weather Service. (2008). *10th Anniversary of the Devastating 1998 Ice Storm in the Northeast* [Online]. Available at: https://www.weather.gov/media/btv/events/IceStorm1998.pdf. Accessed October 1, 2021.

Noormohammed, S. (2009). *Nanostructured thin film for icephobic applications* (Ph.D. thesis). Université du Québec à Chicoutimi, Canada.

Nourai, A., & Hayes, R. M. (2003). Power line ice-shedder. *USA, Patent No. 6660934*

NTT. (2015). *HIREC case study: NTT DoCoMo* [Online]. Available at: www.ntt-at.com/product/hirec/pdf/CaseStudy.pdf. Accessed August 02, 2021.

Nygaard, B. (2009). Evaluations of icing simulations for the COST727 icing test sites in Europe. In: *Proceedings of 13th International Workshop on Atmospheric icing of Structures (IWAIS 2009), Andermatt, CH.*

Nygaard, B., Fikke, S. L. E., & Harstveit, K. (2007-10). Modeling icing in exposed mountain terrain. In: *Proceedings of 12th International Workshop on Atmospheric icing of Structures (IWAIS 2007), Yokohama, JP.*

Nygaard, B., & Honkasaari, M. (2017). *Ålvikfjellet, Kvam kommune, Hordaland, Isingsmålinger for perioden 2014–2017* (Report no. KVT/BEN/2017/R097). Kjeller, Norway: Kjeller Vindteknikk.

Oberst, G. (1998). Lattice tower ground line corrosion & mitigation; a case study. In: *8th IEEE International Transmission & Distribution Construction, Operation & Live-Line Maintenance, Orlando, FL, USA.*

Oehlwein, O. L. (1953-12). A system operator's view of ice melting on a power line while in service [includes discussion]. *Transactions of the American Institute of Electrical Engineers. Part III: Power Apparatus and Systems, 72*(6), 1200–1207.

Oliver, C. R. (1925-11). Sleet and ice troubles on transmission lines in New England. *Journal of the AIEE, 5, 44*(11), 1171–1176.

Onodera, T., Inukai, H., & Odashima, T. (2007 10). Overview of power outage in the Niigata Kaetsu area caused by a snowstorm. In: *International Workshop on Atmospheric Icing of Structures (IWAIS), Yokohama, JP.*

Parent, O., & Ilinca, A. (2011). Anti-icing and de-icing techniques for wind turbines: Critical review. *Cold Regions Science and Technology, 65*, 88–96.

Parker, A. R., & Lawrence, C. R. (2001-11). Water capture by a desert beetle. *Nature, 414*, 33–34.

Penn, L. S., & Meyerson, A. (1992). *Ice-pavement bond prevention: Fundamental study*. Chemical Engineering Department, Polytechnic University.

Petersson, E. (2015). *Conductor pull test.* Private Communication.

Péter, Z. (2006). *Modelling and simulation of the ice melting process on a current-carrying conductor* (PhD thesis). UQAC, Chicoutimi, QC.

Péter, Z., Farzaneh, M., & Kiss, L. I. (2007). Assessment of the current intensity for preventing ice accretion on overhead conductors. *IEEE Transactions on Power Delivery, 22*(1), 565–574.

Petrenko, V. (2000). Systems and methods for modifying ice adhesion strength. *USA, Patent No. 6027075*

Petrenko, V. F., & Peng, S. (2003). Reduction of ice adhesion to metal by using self-assembling monolayers (SAMs). *Canadian Journal of Physics, 81*, 387–393.

Petrenko, V. F., & Ryzhkin, I. (1997-08). Surface states of charge carriers and electrical properties of the surface layer of ice. *The Journal of Physical Chemistry B, 101*(32), 6285–6289.

Petrenko, V. F., & Ryzhkin, I. A. (1997). Surface states of charge carriers and electrical properties of the surface layer of ice. *The Journal of Physical Chemistry B, 101*, 6285–6289.

Petrenko, V. F., & Whitworth, R. W. (1999). *Physics of ice*. Oxford University Press.

Petrenko, V., & Peng, S. (2003). Reduction of ice adhesion to metal by using self-assembling monolayers (SAMs). *Canadian Journal of Physics, 393*(1–2), 387–393.

Petrenko, V., & Sullivan, C. (2003). Methods and systems for removing ice from surfaces. *USA, Patent No. 6653698*

Petrenko, V., & Sullivan, C. (2005). Pulse electrothermal de-icer for power cable. *International, Patent No. 2005083862.*

Petrovic, J. J. (2003). Mechanical properties of ice and snow. *Journal of Materials Science, 38*, 1–6.

Polhman, J. C., & Landers, P. (1982). Present state-of-the-art of transmission line icing. *IEEE Transactions on Power Apparatus System, 101*(8), 2443–2450.

Poots, G. (1996). *Ice and snow accretion on structures*. Research Studies Press, Wiley.

Poots, G. (Ed.). (2000). Ice and snow accretion on structures. *Philosophical Transactions of The Royal Society, London, 358*, 2799–3033.

Poots, G., & Skelton, P. (1995). Thermodynamic models of wet-snow accretion: Axial growth and liquid water content on a fixed conductor. *International Journal of Heat and Fluid Flow, 16*, 43–49.

Prud'Homme, P., et al. (2005a). Hydro-Quebec Trans-Energie line conductor de-icing techniques. In: *Proceedings of 11th International Workshop on Atmospheric Icing of Structures (IWAIS 2005), Montréal, CA.*

Prud'Homme, P., et al. (2005b). Determination of current required to de-ice transmission line conductors. In: *Proceedings of 11th International Workshop on Atmospheric Icing of Structures (IWAIS 2005), Montréal.*

Qi, Y., Chen, S., Sun, H., & Zhang, J. (2019). A multifunctional cool material with contamination-resistant and anti-icing characters based on the application of fluorinated TiO_2. *Solar Energy Materials and Solar Cells, 201.*

Quéré, D. (2002). Rough ideas on wetting. *Physica a: Statistical Mechanics and Its Applications, 313*, 32–46.

Quéré, D. (2005). Non-sticking drops. *Reports on Progress in Physics, 68*, 2495–2532.

Radojcic, M., et al. (2013-09). Comparative testing of different anti-ice coatings for overhead line conductors with special focus on ice accretion, RIV and visual impact. In: *Proceedings of 15 IWAIS, St. John's, NL, Canada.*

Raraty, L., & Tabor, D. (1958). The adhesion and strength properties of ice. *Proceedings of the Royal Society a: Mathematical, Physical and Engineering Sciences, 245*(1241), 184–201.

Reich, A. (1994-01-10). Interface influences upon ice adhesion to airfoil materials. In: *AIAA 32nd Aerospace Sciences Meeting and Exhibit, Reno NV.*

Roberge, M. (2006). *A study of wet snow shedding from an overhead cable* (M.Sc. thesis). Department of Civil Engineering and Applied Mechanics, McGill University, Montreal.

Roberge, M., Farzaneh, M., McClure, G., & Peyrard, C. (2007). Wet snow shedding from an overhead cable part 1: Experimental study. In: *International Workshop on Atmospheric Icing of Structures (IWAIS), Yokohama.*

Roberge, M., McClure, G., Farzaneh, M., & Pellet, L. (2007-10-09). Wet snow shedding from an overhead cable part 2: Evaluating the dynamic response of cables subjected to wet snow shedding. In: *International Workshop on Atmospheric Icing of Structures (IWAIS), Yokohama, Japan*.

Roberge, M., McClure, G., Farzaneh, M., & Pellet, L. (2007-10). Wet snow shedding from an overhead cable part 2: evaluating the dynamic response of cables subjected to wet snow shedding. In: *International Workshop on Atmospheric Icing of Structures (IWAIS), Yokohama*.

Roe, R. (2005). *STS-114 flight readiness review*. NASA Engineering and Safety Center.

Rønneberg, S., et al. (2019). The effect of ice type on ice adhesion. *AIP Advances, 9*(5), 055304.

Ryzhkin, I., & Petrenko, V. (1997). Physical mechanisms responsible for ice adhesion. *The Journal of Physical Chemistry B, 101*(32), 6267–6270.

Safaee, A. (2008). *Nanostructured metal surfaces and their passivation for superhydrophobic and anti-icing applications* (Ph.D. thesis). Université du Québec à Chicoutimi, Canada.

Safaee, A., Sarkar, D., & Farzaneh, M. (2008). Superhydrophobic properties of silver-coated films on copper surface by galvanic exchange reaction. *Applied Surface Science, 254*, 2493–2498.

Saito, H., Takai, K., Takazawa, H., & Yamauchi, G. (1997). Study on snow sticking weight for water repellent coating. *Zairyo/Journal of the Society of Materials Science Japan, 46*(12), 216–219.

Saito, H., Takai, K., & Yamauchi, G. (1997-04). Water- and ice-repellent coatings. *Surface Coatings International, 80*(4), 168–171.

Sakamoto, Y. (2000). Snow accretion on overhead wires. *Philosophical Transactions of the Royal Society, London, 358*, 2941–2970.

Sakamoto, Y., Admirat, P., Lapeyre, J., & Maccagnan, M. (1988). Thermodynamic simulation of wet snow accretion under wind-tunnel conditions. In: *International Workshop on Atmospheric Icing of Structures (IWAIS), Paris* (pp. 180–185)

Saleema, N., Sarkar, D. K., Farzaneh, M., & Sacher, E. (2006-05-07). Superhydrophobic ZnO nanotowers. In: *Proceedings of the 2006 NSTI Nanotechnology Conference, Boston, USA*.

Saotome, H., Yoshoika, M., & Okada, K. (1988). Countermeasures for snow accretion on conductors. In: *Proceedings of the 4th International Workshop on Atmospheric Icing of Structures (IWAIS 1988), Paris*.

Sayward, J. (1979-04). *Seeking low ice adhesion: Special Report 79-11* (1st ed.). US Army Cold Regions Research and Engineering Laboratory.

Seifaddini, M., Saeidi, A., & Farzaneh, M. (2020). Ice stress-strain curve prediction in uniaxial compression loading in the objective of atmospheric icing risk evaluation. *Material Science and Engineering Technology, 51*(5).

Shealy, A. N., Althouse, K. L., & Youtz, R. N. (1952-08). Forty-two years' experience combating sleet accumulations. *Transactions of the American Institute of Electrical Engineers. Part III: Power Apparatus and Systems, 8, 71*(3), 621–628.

Shirtcliffe, N., Thiemann, P., Stratmann, M., & Grundmeier, G. (2001). Chemical structure and morphology of thin, organo-silicon plasma-polymer films as a function of process parameters. In: *Proceedings of the 7th International Conference on Plasma Surface Engineering, Garmisch-Partenkirchen, DE* (pp. 1121–1128).

Shiu, J., Kuo, C., Chen, P., & Mou, C. (2004-01). Fabrication of tunable superhydrophobic surfaces by nanosphere lithography. *Chemistry of Materials, 16*(4), 561–564.

Shore, A. (2000). Capillaroscopy and the measurement of capillary pressure. *British Journal of Clinical Pharmacology, 50*(6), 501–513.

Shulz, M., & Sinapius, M. (2015). Evaluation of different ice adhesion tests for mechanical deicing systems. In: *SAE 2015 International Conference on Icing of Aircraft, Engines, and Structures, Prague*.

Smith, B. W. (2007). *Communication structures* (1st ed.). Thomas Telford Publishing.

Smith, H. B., & Wilder, W. D. (1952-08). Sleet-melting practices-Niagara Mohawk System [includes discussion]. *Transactions of the American Institute of Electrical Engineers. Part III: Power Apparatus and Systems, 71*(3), 631–634.

Smith, J., & Garti, N. (2000). Non-stick polymer-coated articles of manufacture, and process and coatings for the production thereof. *USA, Patent No. 6084020*

Soeno, T., Inokuchi, K., & Shiratori, S. (2004-10). Ultra-water-repellent surface: fabrication of complicated structure of SiO_2 nanoparticles by electrostatic self-assembled films. *Applied Surface Science, 237*(1–4), 543–547.

Sojoudi, H., et al. (2015). Durable and scalable icephobic surfaces: Similarities and distinctions from superhydrophobic surfaces. *Soft Matter, 12*(7), 1938–1963.

Somlo, B., & Gupta, V. (2001). A hydrophobic self-assembled monolayer with improved adhesion to aluminium for deicing application. *Mechanics of Materials, 33*, 471–480.

Soto, S. A. A., & Silva, D. A. D. (2017-09). Maintenance strategies of transmission towers on the Peruvian coast affected by salt pollution. In: *20th International Symposium on High Voltage Engineering, Buenos Aires, Argentina.*

St.Clair, H. P., & Imburgia, C. A. (1954-01). The American gas and electric company 330-kV transmission system. design principles, electrical characteristics, and performance. *Transactions of the American Institute of Electrical Engineers. Part III: Power Apparatus and Systems, 73*(1), 64–69.

Statnett. (2018). *SDOK-39-12, Technical specification—Painting of conductor and string fittings, Revision 4.0.* Statnett SF.

Strobel, M., & Lyons, C. (2011). An essay on contact angle measurements. *Plasma Processes and Polymers, 8*(1), 8–13.

Sullivan, C. R., Petrenko, V. F., McCurdy, J. D., & Kozliouk, V. (2003). Breaking the ice: De-icing power transmission lines with high-frequency, high-voltage excitation. *IEEE Industry Applications Magazine, 9*(5), 49–54.

Susoff, M., Siegmann, K., Pfaffenroth, C., & Hirayama, M. (2013-10). Evaluation of icephobic coatings—Screening of different coatings and influence of roughness. *Applied Surface Science, 282*, 870–879.

Svenska Kraftnät. (1999). *Disturbances in 1999: Ice accretion on insulator strings 400 kV in Skåne and Halland* (in Swedish). Svenska Kraftnät.

Swinny, R. (2011). *Droerivier proteus icing study.* Sunninghill, ZA: ESKOM Internal Research Report RES/RR/11/33123.

T/CEC 184-2018. (2018-11-06). *Test of anti-icing coatings for insulators.* China Electric Power Enterprise Federation.

Takai, K., & Yamauchi, G. (1998). Prevention of snow adhesion to satellite antennas by water-repellent material. *NTT R and D, 47*, 31–36.

Thorsteins, E., & Eliasson, A. J. (1998). *Observation of icing by tension recorders—comparison of conductors.* In: *Proceedings of the 8th International Workshop on Atmospheric Icing of Structures (IWAIS 1998), Reykjavik, IS* (pp. 247–252).

Timoshova, L., Nikiforov, E., & Shkaptsov, V. (2003). *Private comminications with Y. Motlis.* VNIEE.

Tourkine, P., Merrer, M. L., & Quéré, D. (2009). Delayed freezing on water repellent material. *Langmuir, 25*(13), 7214–7216.

USA Department of the Army. (1999). *Ice engineering.* US Army Corps of Engineers.

van Donkelaar, A., et al. (2011, June). Global estimates of ambient fine particulate matter concentrations from satellite-based aerosol optical depth: Development and application. *Environmental Health Perspectives, 118*(6), 847–855.

van Donkelaar, A., et al. (2016, March). Global estimates of fine particulate matter using a combined geophysical—statistical method with information from satellites, models, and monitors. *Environmental Science and Technology, 50*(7), 3762–3772.

van Donkelaar, A., Martin, R. V., Li, C., & Burnett, R. T. (2019). Regional estimates of chemical composition of fine particulate matter using a combined geoscience-statistical method with information from satellites, models, and monitors. *Environmental Science & Technology, 53*, 2595–2611.

Van Oss, C. J., et al. (1992). Surface tension parameters of ice obtained from contact angle data and from positive and negative particle adhesion to advancing freezing fronts. *Journal of Adhesion Science and Technology, 6*, 503.

Varanasi, K. K., et al. (2010). Frost formation and ice adhesion on superhydrophobic surfaces. *Applied Physics Letters, 97*(23), 234102.

Volat, C., Farzaneh, M., & Leblond, A. (2005). De-icing/anti-icing techniques for power lines. Current methods and future direction. In: *Proceedings of 10th International Workshop on Atmospheric Icing of Structures (IWAIS 2005), Montreal.*

VTT Technical Research Centre of Finland Ltd. (2015). *Win power icing atlast—WIceAtlas* [Online]. Available at: https://projectsites.vtt.fi/sites/wiceatlas/www.vtt.fi/sites/wiceatlas.html. Accessed September 23, 2021.

Wagner, P., Furstner, R., Barthlott, W., & Neinhuis, C. (2003-04). Quantitative assessment to the structural basis of water repellence in natural and technical surfaces. *Journal of Experimental Botany, 54*(385), 1295–1303.

Wang, F., Arai, S., & Endo, M. (2004-02). Electrochemical preparation and characterization of nickel/ultra-dispersed PTFE composite films from aqueous solution. *Materials Transactions, 45*(4), 1311–1316.

Wang, F., et al. (2010). Ice accretion on superhydrophobic aluminum surfaces under low-temperature conditions. *Cold Regions Science and Technology, 62*(1), 29–33.

Wang, Z., Cong, Y., & Zhang, B. (2016). Liquid crystal compound anti-ice surface. *Liquid Crystals, 43*(2), 241–248.

Wareing, J. B. (1992-09). *Overhead Lines and climatic environment: A historical and Geographic survey*, Chester, UK: EA Technology Report R2745/T.

Wei, X., et al. (2014-02). Development of anti-icing coatings applied to insulators in China. *IEEE Electrical Insulation Magazine, 30*(1), 42–50.

Wei, X., et al. (2012-12). Study of anti-icing performance of insulator strings bottom coated with semiconductive silicone rubber coating. *IEEE Transactions on Dielectrics and Electrical Insulation, 19*(6), 2063–2072.

Wenzel, R. (1936). Resistance of solid surfaces to wetting by water. *Industrial & Engineering Chemistry, 28*(8), 988–994.

Wenzel, R. N. (1936). Resistance of solid surfaces to wetting by water. *Industrial and Engineering Chemistry, 28*, 988.

Wilen, L. A., Wettlaufer, J. S., Elbaum, M., & Schick, M. (1995). Dispersion-force effects in interfacial melting of ice. *Physical Review B: Condensed Matter, 52*, 12426.

WMO. (2009-12). *COST action 727, measurement and forecasting of atmospheric icing*. World Meteorological Association.

Wolansky, G., & Marmur, A. (1999-10). Apparent contact angles on rough surfaces: The Wenzel equation revisited. *Colloids and Surfaces A: Physicochemical and Engineering Aspects, 10, 156*(1–3), 381–388.

Woodward, I., Schofield, W., Roucoules, V., & Badyal, J. P. S. (2003). Super-hydrophobic surfaces produced by plasma fluorination of polybutadiene films. *Langmuir, 19*(8), 3432–3438.

Wu, X., & Shi, G. (2005-08). Fabrication of a lotus-like micro–nanoscale binary structured surface and wettability modulation from superhydrophilic to superhydrophobic. *Nanotechnology, 16* (10), 2056–2060.

Xiu, Y., et al. (2009). Superhydrophobic surface and method for forming same. *USA, Patent No. 2009011222*

Wang, Y., et al. (2013). Verification of icephobic/anti-icing properties of a superhydrophobic surface. *Applied Material Interfaces, 5*(8), 3370–3381.

Yaji, K., & Homma, H. (2019 01). Insulator flashover caused by salt-contaminated snowstorms—Lessons from the snow damage in Japan. *IEEE Electrical Insulation Magazine, 35*(1), 23–37.

Yamauchi, G., Takai, K., & Saito, H. (2020-07). PTEE based water repellent coating for telecommunication antenna. *IEICE Transactions, E83-C*(7), 1139–1141.

Yan, H., Kurogi, K., Mayama, H., & Tsujii, K. (2005-05). Environmentally stable super water-repellent poly(alkylpyrrole) films. *Angewandte Chemie International Edition, 44*(22), 3453–3456.

Yeong, Y. H., et al. (2015). Atmospheric ice adhesion on water-repellent coatings: Wetting and surface topology effects. *Langmuir, 31*(48), 13107–13116.

Yin, F., Farzaneh, M., & Jiang, X. (2016). Influence of AC electric field on conductor icing. *IEEE Transactions on Dielectics and Electrical Insulation, 23*(4), 2134–2144.

Yu, H., Li, T., & Liu, P. (2019). Influence of ENSO on frequency of wintertime fog days in Eastern China. *Climate Dynamics, 52*, 5099–5113.

Zhang, J., Li, J., & Han, Y. (2004-06). Superhydrophobic PTFE surfaces by extension. *Macromolecular Rapid Communications, 25*(11), 1105–1108.

Zhang, Y., Yu, X., Wu, H., & Wu, J. (2012). Facile fabrication of superhydrophobic nanostructures on aluminum foils with controlled-condensation and delayed-icing effects. *Applied Surface Science, 258*(20), 8253–8257.

Zhang, Z., Qu, L., & Shi, G. (2003-09). Fabrication of highly hydrophobic surfaces of conductive polythiophene. *Journal of Materials Chemistry, 13*(12), 2858–2860.

Zhao, Q., Liu, Y., & Abel, E. W. (2005-02). Surface free energies of electroless Ni–P based composite coatings. *Applied Surface Science, 240*(1–4), 441–451.

Zhu, L., et al. (2013). Ice-phobic coatings based on silicon-oil-infused polydimethylsiloxane. *ACS Applied Materials & Interfaces, 5*(10), 4053–4062.

Zou, M., et al. (2011). Effects of surface roughness and energy on ice adhesion strength. *Applied Surface Science, 257*(8), 3786–3792.

Index

M. Farzaneh and W. A. Chisholm, *Techniques for Protecting Overhead Lines in Winter Conditions*, Compact Studies, https://doi.org/10.1007/978-3-030-87455-1

D

The manufacturer's authorised representative in the EU is Springer Nature Customer Service Centre GmbH, Europaplatz 3, 69115 Heidelberg, Germany. If you have any concerns regarding our products, please contact ProductSafety@springernature.com

Printed and bound by CPI Group (UK) Ltd, Croydon, CR0 4YY
30/09/2026
02225699-0014